20 Years after the Chernobyl Accident: Past, Present and Future

20 YEARS AFTER THE CHERNOBYL ACCIDENT: PAST, PRESENT AND FUTURE

ELENA B. BURLAKOVA
AND
VALERIA I. NAIDICH
EDITORS

Nova Science Publishers, Inc.
New York

For permission to use material from this book please contact us:
Telephone 631-231-7269; Fax 631-231-8175
Web Site: http://www.novapublishers.com

Library of Congress Cataloging-in-Publication Data

20 years after the Chernobyl accident : past, present and future / E. Burlakova.
p. ; cm.
Includes index.
ISBN 1-60021-249-2
1. Radiobiology. 2. Chernobyl Nuclear Accident, Chornobyl', Ukraine, 1986--Environmental aspects. 3. Radioactive pollution--Ukraine. I. Burlakova, E. B. II. Title: Twenty years after the Chernobyl accident.
[DNLM: 1. Chernobyl Nuclear Accident--Ukraine. 2. Radiation Effects--Ukraine. 3. Accidents, Radiation--Ukraine. WN 620 Z999 2006]
QH652.Y33 2006
571.4'5--dc22 2006016754

Published by Nova Science Publishers, Inc. ✦ New York

CONTENTS

PREFACE

In recent 5 years, radiobiology has made a great step forward in development of the new mechanisms forming the basis of various radiation effects, specifically among the mechanisms of low rate irradiation by low doses. The investigations of radiation apoptosis, radiation hormesis, radioadaptive response, and by-stander effect were intensively developed. Of great attention are the processes of premature aging induced by irradiation.

All these questions are discussed in the present collection "20 Years After Chernobyl Accident – Present and Future", which includes the articles by the leading radiobiologists all over the Russian Federation.

The greatest part of the reviews published is devoted to assessments of health of the population irradiated during inhabiting on radionuclide-contaminated territories and people from other regions of the Former USSR, involved in the liquidation on the accident consequences.

Twenty years passed since the Chernobyl Accident have sharply outlined the necessity of summing up the results of investigations performed by Russian scientists and their foreign colleagues, invited for co-operation in the framework of international complex programs. Devoted to the 20th anniversary of the accident, any works by the experts of official organizations such as UN, IAEA and WHO were published.

Many of these experts attempted to turn the public opinion to an idea that the accident consequences appeared not so tragic, as in seemed to be in 1990s. Even suggestions to terminate financial support of the projects on liquidation of the accident consequences were made. The greatest objections of both Russian and foreign scientists (for example, refer to BEIR-7 report by American experts) were induced by conclusions made by these organizations concerning damage to the health of the population dwelling on the radionuclide-contaminated territories. It must be noted that the majority of experts were not aimed at consideration of the accident effect on the health conditions of the population. They were interested only in the changes representing the radiogenic origin according to their expert's point of view, i.e. not the changes induced by the entire accident, but exclusively by irradiation. Therefore, at once, the diseases which might be induced by chronic emotional stress, in particular, changes in the life conditions caused by evacuation of citizens from contaminated territories, etc. were eliminated from the consideration of the medical consequences of the accident.

Obviously, this preliminarily decreased the scope of real damages caused to the population health. Note also that the question if a disease is radiogenic or not is rather disputable. For example, for Japanese scientists, it took half a century after the atomic bomb attack to study and publish the data on non-cancer diseases, which rank infarctions, cerebral thrombosises, digestive apparatus diseases, nervous-psychic dysfunctions, endocrinal diseases, etc. as radiogenic ones. The present collection also includes reviews devoted to non-cancer diseases of thyroid gland in children (Ivanov *et al.*, Zamulaeva *et al.*)chronic diseases and immune insufficiency among liquidators from Moscow and Moscow Oblast, North-West Region and Saint-Petersburg, and Siberia (Oradovskaya). This author characterizes the health state of the liquidators and their disabling, shows the rate of increase of chronic diseases and the immune insufficiency.

It is of great importance that despite the fact that the liquidators are living in different regions of Russia and even in different countries, they possess general features in changes of the health state, of which the increase of the number of the elderly age diseases and premature aging are typical.

Many articles characterize the health of children's population (Zamulaeva *et al.*, Ivanov *et al.*, Baleva *et al.*). Hence, it is shown that non-cancer diseases in children are also of the radiogenic origin.

As mentioned above, the modern stage of radiological medicine development is characterized by a tendency to detect a relation between genetic disorders in the human organism and various diseases. In the above-mentioned articles, as well as in reviews by Sevan'kaev *et al.* and Vorobtsova *et al.*, technical approaches and the results of such investigations are presented. They allow for a belief in possible forecasting of any diseases in remote periods after irradiation basing on cytogenetic characteristics.

At present, of the extreme importance is the knowing of new molecular and regulatory approaches for understanding and explanation of the results, obtained in Chernobyl, which may be hardly explained from positions of the classical radiobiology. Hence, the reviews presented in this collection demonstrate the modern approach to explanation of changes in the health state at the molecular-biological and genetic levels.

This collection also reflects the problems of radioecology related both to agriculture and ecology of aquatic systems (Ryabtsev) and give quite broad description of the regularities of small animal survival (rodents) on contaminated territories (Shishkina *et al.*).

In contrast with multiple official reports, the reviews from the Collection show that Chernobyl had not ended, but vice versa is actively continued, whence the title of the article by Pelevina *et al.* is: 20 years after Chernobyl – is it much or little? The great advantage of this monograph is not only stating of the "Chernobyl" problem, but presenting clear perspectives of our further living and working.

Professor Elena B. Burlakova

ACKNOWLEDGMENT

This Collection was edited by Professor Elena B. Burlakova[1] and Doctor Valeria I. Naidich. The editors are grateful to all of the authors for their submission of their material to be presented in this collection. They are also grateful to Alexei Yu. Borissevitch for translation and complete preparation of the manuscript for publication.

[1] Elena B. Burlakova: Vice Director of N.M. Emanuel Institute of Biochemical Physics; 4 Kosygin Street, 119991 Moscow, Russia; Tel: (7-495) 137-31-27; E-mail: radbio@sky.chph.ras.ru

In: 20 Years after the Chernobyl Accident
Editors: E. Burlakova and V. Naidich, pp. 1-10
ISBN 1-60021-249-2

Chapter 1

THE SIGNIFICANCE OF CYTOGENETIC EXAMINATION FOR ASSESSMENT OF THE CHERNOBYL CATASTROPHE CONSEQUENCES

***V. A. Shevchenko*[1] *and G. P. Snigiryova*[2*]**
[1] N.I. Vavilov Institute of General Genetics,
Russian Academy of Sciences, Moscow, Russia
[2] Federal State Institution "Russian Scientific Center of Roentgeno - Radiology" of
Federal Agency on Public Health Services and Social Development, Moscow, Russia

ABSTRACT

The results of cytogenetic examination of people exposed to irradiation in the Chernobyl accident are presented and the possibility of applying cytogenetic parameters to the estimation of radiation doses perceived and forecasting of the radiation impact consequences is considered. Since 1986 till 2004, a cytogenetic examination of 1,724 participants of the liquidation of the Chernobyl accident consequences (the liquidators) was performed. The radiation dose, calculated in 1986 by the frequency of dicentrics, equaled 0.16 Gy. Since 1992 till 1995, doses perceived by a group of liquidators were determined by the frequency of translocations (FISH-method). For the liquidators who worked in Chernobyl in 1986 exclusively, the average irradiation dose equaled 0.19 Gy, whereas for those who worked regularly in the period of 1986 – 1995 it was 0.39 Gy. It is shown that for the whole period of examination (1986 – 2004) the frequency of dicentrics in blood lymphocytes reliably exceeded the control level. Cytogenetic examination of Bryansk Oblast citizens, performed in 1992 – 1994, detected 5 time elevated level of dicentrics compared with the control. The data obtained are of high significance for forecasting of adverse consequences of irradiation and development of sensitive criteria for early detection of health disorders.

* Main addressee: 86 Profsoyuznaya street, 117997 Moscow, GSP-7, Russia; FGU "RNTsRR Roszdrava", Tel: (7-095) 334-9288; Fax: (7-095) 334-7924; E-mail: Snigiryova@mrids.rssi.ru

Keywords: the Chernobyl accident, liquidators, biodosimetry, chromosome aberrations, dicentrics, translocations, FISH-method, remote consequences of irradiation

Two decades passed since the Chernobyl accident, during which an extensive material on the examination of people exposed to radiation impact was accumulated. Adverse consequences of the accident affected health of many people. First of all, among them the participants of the accident consequences (the liquidators) should be mentioned. They represent the highest irradiated abundant cohort of the suffered population of this country. Citizens of the settlements located in the radioactive fallout zone should also be included in this group.

One of the most urgent tasks of examinations performed is forecasting of the genetic consequences, induced by ionizing radiation impact on the human organism, and estimation of development risks of various pathologies. The advantageous solution of this task is possible in the only case of the existence of certain information about radiation doses perceived. Unfortunately, emergency of accident conditions, which took place at Chernobyl NPP in 1986, make estimation of the true radiation load with the help of physical dosimetry currently present at the moment of accident very difficult. No doubt that biological methods viz. cytogenetic ones, based on the analysis of chromosome aberrations, contribute significantly to solution of this problem. These methods provide a quantitative estimation of the radiation impact on the human organism with respect to his/her individual peculiarities, first of all, personal radiosensitivity, and the state of the organism at the moment of irradiation. The information obtained provides more correct forecasting of possible early and remote consequences of the radiation impact.

The principles of cytogenetic dosimetry method are quite convincingly stipulated in many Russian and foreign investigations, and materials of these investigations formed the basis for developing recommendations by WHO, IAEA and UNSCEAR for practical application of the chromosome analysis of blood lymphocytes to determination of radiation doses.

The radiobiological basis for the application of cytogenetic methods to radiation studies is high sensitivity of blood lymphocytes and the presence of radiation-specific chromosome rearrangements, the dose dependence of which is studied well for the majority of ionizing radiation types.

Radiation may induce various chromosome disorders of the exchange type – dicentrics, polycentrics, centric rings, translocations, and acentric fragments. Of interest for the study of radiation effects, first of all, for the quantitative analysis, is analysis of dicentrics and translocations. The analysis of dicentrics is successfully used in biodosimetric researches for the purpose of determining radiation dose in the early periods (up to a year) after momentary, relatively uniform radiation impact. The advantage of dicentrics analysis is their easy detection on a light microscope without application of any complex processing methods and dyeing of chromosome preparations. The use of this method is rather limited, because being genetically imbalanced, the cells with such aberrations may eliminate during the cell cycle. The uncertainty of dose estimations by the frequency of dicentrics highly increases with the post-radiation period. In this case, the residual frequency of dicentrics in peripheral blood lymphocytes indicates the presence or the absence of any damaging factor of the radiation origin.

For the purposes of biological dosimetry, the analysis of translocations represents more perspective method. This is possible, because translocations, contrary to dicentrics, do not disturb the mitosis proceeding and, therefore, may be freely transferred in the sequence of cell generations, i.e. the frequency of translocations does not change significantly with the post-radiation period.

Spontaneous frequency of dicentrics and translocations is relatively low, giving from 0 to 5 per 10,000 cells for dicentrics, whereas for translocations it is by an order of magnitude higher. By the frequency of chromosome aberrations relatively low doses may be estimated. For dicentrics, the lower sensitivity limit is 0.1 Gy, whereas for translocations it is about 0.25 Gy [1].

Despite the restrictions related with the use of dicentrics analysis for retrospective estimation of the radiation doses, they were widely applied to cytogenetic monitoring in groups of people exposed to irradiation in different emergency situations. The increased level of chromosome aberrations in blood lymphocytes may precede the development of pathological processes in the human organism. Therefore, the information about genome damage is extremely urgent for the risk group formation concerning development of various diseases, including cancer [2, 3].

In this work we present results of cytogenetic examination of people exposed to irradiation by the Chernobyl accident and consider a possibility of cytogenetic parameter application to estimation of radiation doses perceived and forecasting of adverse consequences of the radiation impact.

MATERIALS AND METHODS

Characterization of the Examined Group

Since 1986 till 2004, cytogenetic examination of 1,724 liquidators was performed. In 1986 and 1987, all examined liquidators participated in recovery operations in the 30-kilometer restriction zone. Some of them were also involved in these operations in Chernobyl in the following years until 1995. Total duration of working in the accident zone varied from two weeks up to 13 months. About 60% of liquidators had official data on the radiation dose perceived. The average official dose for the whole examined group was below 0.25 Sv. When working in Chernobyl, examined liquidators have performed various operations. Hence, the dose variation range was 0.01 – 1.36 Sv. This group included experts in dosimetry and deactivation, doctors, scientists, drivers, crane operators, riggers, helicopter pilots, entombment builders, and unskilled workers.

In 1992 – 1994, a cytogenetic examination of 80 citizens of Bryansk Oblast was performed. For the moment of examination, over a half of patients (45 persons) lived in settlements related to the "living zone with the right to compulsory evacuation". More 17 persons lived in the settlements from the "living zone with favorable socioeconomic status".

Before cytogenetic analysis, all examined patients filled in a special form, which included questions about the health state, professional contact with ionizing radiation sources, roentgenological procedures passed, smoking, etc. This information allowed for excluding possible effect of additional factors on the cytogenetic examination results.

The control group included 114 healthy donors aged 21 – 69 (conventional method) and 15 donors aged 39 – 62 (FISH-method), living in Moscow and having no professional contact with ionizing radiation sources.

Cytogenetic examination was performed with the help of conventional method (dicentrics analysis) and FISH-method (translocation analysis).

Lymphocyte Culture and Chromosome Preparation

Lymphocyte cultures and chromosome samples were prepared according to standard protocols [4]. The culture medium included RPMJ-1640 component containing 15% bovine embryonic serum, 2.5% phytohemagglutinin (PHA), 2 mM glutamine with antibiotics and 10 mM 5-bromodeoxyuridine. The cell culture was incubated at 37°C during 48 h. Metaphase chromosome preparations designed for analysis by the conventional method were dyed using the "fluorescence plus Giemsa" technology (FPG-method). This method provides determination the part of cells at the stage of the first mitotic division [5].

Fluorescence in Situ Hybridization (FISH)

Plasmid DNA of chromosome specific Hind III pBS libraries of human chromosomes 1, 4 and 12 [6] was biotinylated as described above [7]. A degenerate α-satellite DNA pancentromeric probe was produced by in vitro amplification using a polymerase chain reaction (PCR) and labeled with digoxigenin [8]. Hybridization, detection of bound biotin-labeled painting probes for the target chromosomes with streptavidin-fluorescein isothyocyanate (FITC) conjugate and the bound digoxigenin-labeled pancentromeric probe with AMCA-labeled (7-amino-4-methylcoumarin-3-acetic acid) antibodies were carried out as described before [8]. Counterstaining was performed with propidium iodide (PI) in an antifade solution [9].

Aberration Scoring

Conventional chromosome analysis was carried out exclusively in complete first division metaphases (M1) identified by uniformly stained sister chromatids. As a rule, 500 – 1,000 metaphases were scored for each person. For scoring of FISH-painted exchanges, each metaphase spread was analyzed with a filter set allowing for a simultaneous observation of FITC and PI fluorescence. The blue AMCA fluorescence of the bound digoxigenin-labeled pancentromeric probe was visualized by using an ultraviolet excitation filter (Zeiss, Oberkochen, Germany). The chromosome morphology was additionally checked with a filter set providing only PI fluorescence. Rearrangements involving detected as dichromatic structures (yellow/red). Dichromatic chromosomes with one blue AMCA centromeric signal were classified either as complete or incomplete symmetrical translocations or insertions. According to the PAINT nomenclature [10], this corresponds to the observation of both t(Ab) and t(Ba); t(Ab) or t(Ba) + ace(b); ins(Aba) or ins(Bab). Deviating from the PAINT system, the respective translocation process was evaluated as a single event. Dichromatic chromosomes with two centromeric signals were classified as dicentrics, dic(AB).

The results obtained were analyzed using standard statistical methods.

RESULTS AND DISCUSSION

Cytogenetic Examination of Liquidators. Biodosometry

When working in Chernobyl, the liquidators were exposed to a short-term, prolonged or fractioned radiation impact of different dose rates. Partial and incomplete information about doses, obtained with the help of physical dosimetry methods, does not allow for an objective estimation of the radiation hazard. Taking into account the fact that the level of chromosome disorders in blood lymphocytes correlates with the radiation dose, more precise estimation of the accident consequences considering both individual and population dosimetry may be obtained by the cytogenetic analysis.

Table 1. The frequency of chromosome aberrations (conventional method) in the group of liquidators, examined in 1986

Groups	Number of patients	Number of cells	(dic ± m)/100 cells	Biological estimation of the dose, Gy (95%)
Liquidators	443	41927	0.33 ± 0.01*	0.16 (0.07 – 0.42)
Control	114	51430	0.02 ± 0.01	

* Reliable difference from the control group: $p < 0.05$; dic are dicentrics

In 1986, cytogenetic examination of 443 liquidators using the conventional method (dicentrics analysis) was performed. Blood samples were taken several days after termination of works in the 30-kilometer restriction zone. Table 1 presents results of this cytogenetic analysis and biological dosimetry for the examined group of liquidators. As follows from the data shown, the frequency of dicentrics is 16.5 higher than the control level. The radiation dose was estimated using the dose-effect calibration curve for dicentrics [11], plotted on the basis of experimental data (acute γ-irradiation of blood with the dose rate 0.103 Gy/min)/ The curve is described by the following linear-quadratic equation:

$$Y = (0.10 \pm 0.03) + (1.5 \pm 0.4)D + (6.3 \pm 0.3)D^2,$$

where Y is the frequency of dicentrics per 100 cells; D is the radiation dose, Gy.

For the examined group of liquidators, the average dose equaled 0.16 Gy. According to other data [12], for a representative group of liquidators from Russia, the average dose equaled 0.12 Gy. For the liquidators been working in 1986 in Chernobyl, the average dose was about 0.16 Gy [13].

Since 1992 till 1995, 52 liquidators participated in the recovery operations in Chernobyl since 1986 were involved in the cytogenetic analysis using the FISH-method. The results of cytogenetic analysis and biological dosimetry for this group are shown in Table 2. The frequency of translocations in the examined group 2.5-fold exceeds the control level. Among the examined group, 17 liquidators regularly worked in Chernobyl in the period of 1986 – 1995, i.e. were exposed to a long-term fractioned radiation impact. The frequency of translocations in this group was 4 times higher than in the control group and 2 times higher than that for a group of liquidators (35 persons), who worked in Chernobyl in 1986 only. The reason is that the liquidators, who came to Chernobyl several times during several years perceived higher total radiation dose compared with these who worked in Chernobyl in 1986 only. The radiation doses were calculated by the frequency of translocations using the calibration dose-effect curve [8]. For the whole examined group of liquidators (52 persons), the average dose equaled 0.27 Gy. For the group regularly worked in Chernobyl in the period of 1986 – 1995, the radiation dose equaled 0.39 Gy, whereas for those who worked in 1986 only – 0.19 Gy. It should be noted that for the liquidators of 1986, the doses estimated by the frequency of dicentrics and translocations correlate well and do not oppose the data obtained by physical dosimetry methods.

The radiation doses of the liquidators, determined with the help of cytogenetic analysis, are equivalent to acute single irradiation doses, because they were calculated by calibration curves, plotted for acute radiation impact with a high dose rate (0.5 Gy/min). Under real conditions, the liquidators were usually exposed to a long-term fractional low rate radiation impact. Therefore in some cases, really perceived doses may be much higher. First of all, this concerns the liquidators, who worked in Chernobyl several times during a long period. This urgent circumstance must be accounted for in the analysis of liquidators' health state.

Table 2. Results of cytogenetic examination (FISH-method) and the average radiation dose in the group of liquidators

Groups of liquidators	Number of patients	Number of cells	Translocations ($F_G \pm m$)/100 cells	Biological estimation of the dose, Gy (95%)
Total	52	44,283	1.20 ± 0.16*	0.27(0.05-0.50)
worked in 1986	35	27,767	0.86 ± 0.13*	0.19(0.05-0.50)
worked in 1986 – 1995	17	15,516	1.81 ± 0.35*	0.39(0.10-0.80)
Control	15	21,953	0.47 ± 0.09	-

* Reliable difference from the control group: $p < 0.05$.

Cytogenetic Monitoring of Liquidators

Figure 1 presents results of long-term analysis of the chromosome aberration frequency, viz. frequency of dicentrics, in blood lymphocytes of the liquidators participated in recovery works in 1986 in Chernobyl. In the first year of monitoring, the average frequency of dicentrics was 0.33 per 100 cells in liquidators that 16.5-fold exceeded the control level. In the following years this parameter decreased gradually compared with the initial year of monitoring (2-fold, approximately). However, during the whole period of monitoring until 2004, the frequency of dicentrics in blood lymphocytes of liquidators remains reliably higher compared with the control: from 0.10 to 0.14 per 100 cells. The data obtained allowed for highly probable conclusion that the cytogenetic effect observed is related to the radiation impact.

The fact of increasing frequency of cytogenetic disorders in the liquidators is confirmed by multiple investigations and is now unambiguous. However, interpretation of the data obtained is rather difficult. It is suggested that genome destabilization represents the main pathogenetic mechanism for development of various somatic diseases. Despite the absence of direct proofs of the influence of elevated chromosome variability of somatic cells on the liquidators' health state, recently, more and more data on the somatic mutation contributing to development of some diseases are accumulated. At present, the investigation results concerning various somatic injuries of health and total morbidity among liquidators is ambiguous still. Many investigators have detected unfavorable tendencies in dynamics of some classes of somatic diseases in the liquidators. An increasing quantity of diseases of cardiovascular and skeletal systems, nervous system and psychological sphere, metabolic disorders and stable immunodeficiency is reported. The dynamic observation for patients demonstrates progressive health deterioration [14]. Gradually, functional disorders transform to organic pathologies, and mild cases to severe ones.

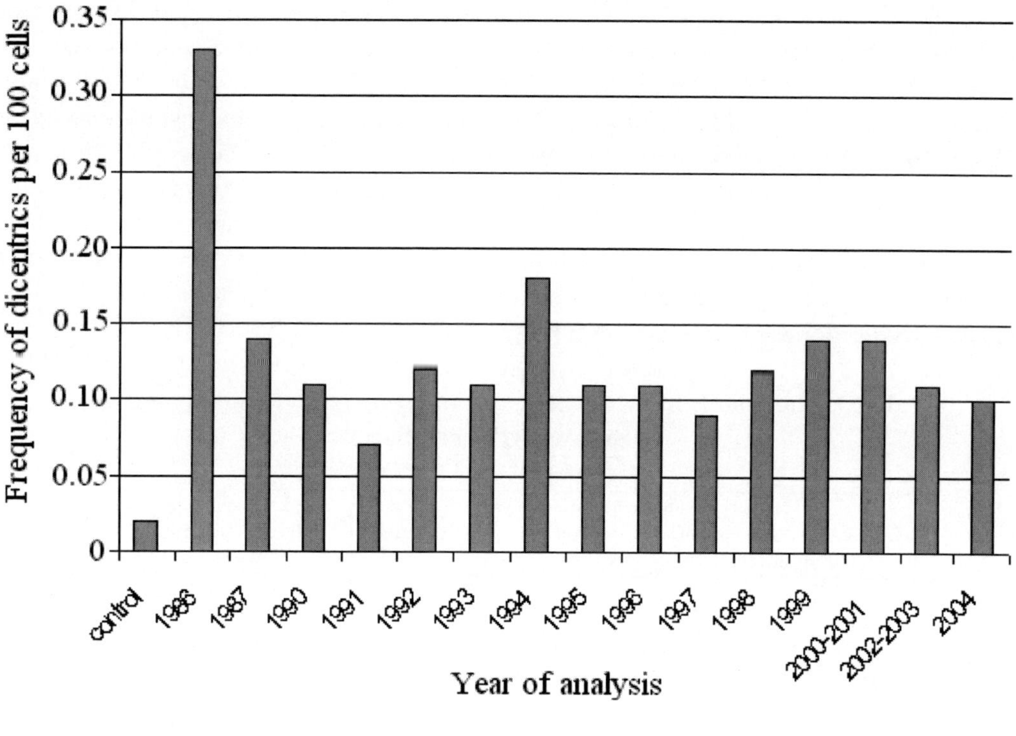

Figure 1. The average frequency of dicentrics in a group of liquidators, examined in different periods after the accident

Cytogenetic monitoring of the liquidators, performed together with the study of their health conditions, will allow for obtaining reliable information about correlation between the frequency of chromosome aberrations and the risk of somatic disease development. At present, there are data in the literature [15, 16] showing a reliable correlation between the frequency of chromosome aberrations and the risk of carcinogenesis. Therefore, it seems very important during the cytogenetic monitoring to separate groups of liquidators with different genetic risks with respect to the frequency of chromosome aberrations in peripheral blood lymphocytes. Patients with high genetic risk require permanent medical monitoring for timely detection and prophylaxis of adverse consequences of radiation.

Cytogenetic Examination of Bryansk Oblast Citizens

In 1992 – 1994, cytogenetic examination of Bryansk Oblast citizens inhabiting in the regions, radionuclide-contaminated by the Chernobyl accident, was performed. It should be noted that due to heterogeneity of contamination, different radionuclide composition of fallout and wide range of biogeochemical characteristics of locality cytogenetic examination of the population for the purpose of estimating the damaging action of radiation became extremely urgent. Under these conditions, data obtained exclusively by physical dosimetry methods may not be used for objective estimation of adverse consequences of irradiation. Table 3 presents examination results of Bryansk Oblast citizens. Since no reliable differences in cytogenetic parameters between groups inhabiting in the zones with different socioeconomic status were determined, all examined citizens were united in a single group. The results of cytogenetic examination obtained were compared with the control data in order to make a conclusion about possible damaging effect of radiation on the human cell genome. In the group of Bryansk Oblast citizens, total frequency of chromosome aberrations equaled 1.43 per 100 cells. This parameter is 2 times higher than in control. Generally, the ionizing radiation induces aberrations of the chromosome type (dicentrics, first of all) in the cells of living organisms. The frequency of dicentrics in the examined group equaled 0.1 per 100 cells, which was 5-fold higher than in the control group. The data obtained correlate well with the results of cytogenetic examination of children and teenagers living on the radionuclide-contaminated territories of Kaluga Oblast [17].

Table 3. The results of cytogenetic examination (conventional method) of Bryansk Oblast citizens

Group	Number of patients	Number of cells	(ab ± m) / 100 cells	(dic± m) / 100 cells	(ace ± m) / 100 cells	(chr ± m) / 100 cells
Bryansk Oblast	80	21,027	1.43 ± 0.08*	0.10 ± 0.02*	0.36 ± 0.04	0.93 ± 0.07
Control	114	51,430	0.66 ± 0.04	0.02 ± 0.01	0.23 ± 0.02	0.41 ± 0.03

* Reliable difference from the control group: $p < 0.05$;

ab – total aberrations;

dic – dicentrics;

R_c – centric rings;

ace – acentric fragments;

chr – aberrations of the chromatid type

The increased level of dicentrics – the markers of the radiation impact, detected 6 – 8 years after the Chernobyl accident in a group of people inhabiting on the contaminated territories, testifies about permanent mutagenic impact of various radiation factors on the human organism.

It should also be noted that the level of chromatid aberrations in the examined group is 2 times higher compared with the control. This type of chromosome aberrations appears as a result of the action of mutagens, mostly in S- and G_2-phases of the cell cycle, which is typical of the overwhelming majority of chemical agents and viruses [18]. The increased level of chromatid disorders observed may be stipulated by chemical pollution of the environment, both as a result of the accident and due to broad application of pesticides and fertilizers in the agricultural production. Moreover, the long-term influence of low-dose radiation from radionuclides, delivered to the human organism from the environment with food, on the genetic apparatus of the cell may not also be excluded.

CONCLUSION

Summarizing the long-term investigations of cytogenetic effects in the liquidators of the Chernobyl accident consequences and people living on the contaminated territories, the following should be mentioned:

- the study of chromosome aberrations in blood lymphocytes is one of the most important and necessary tasks at the examination of people suffered from the accident;
- the results of cytogenetic analysis (conventional and FISH methods) may be successfully applied to indication and quantitative estimation of the radiation impact on the human organism. Doses calculated from the data of cytogenetic examination take into consideration such urgent factors as individual radiosensitivity and the organism condition at the moment of irradiation. Therefore, the information obtained provides more precise forecasting of adverse radiation consequences, including estimation of genetic risks;
- analysis of morbidity in the groups of irradiated persons examined with the help of cytogenetic methods will also provide analysis of a correlation of chromosome aberration frequencies in somatic cells with the risks of development of various pathological processes. This will promote the development of sensitive prognostic indices for early detection of negative changes in the health state of irradiated people.

The authors are grateful to colleagues from cytogenetic laboratory in RNC RR N.N. Novitskaya, E.D. Khazins and A.N. Bogomazova for technical support and help in material designing for the publication.

References

[1] United Nations. Sources and Effects of Ionizing Radiation. UNSCEAR 2000 Report to the General Assembly with Scientific Annexes, United Nations sales publication, No.E.00.IX.4, New York, 2000.

[2] Snigiryova G.P., Lyubchenko P.N., Shevchenko V.A. et al., *Gematol. i Transfuziol.*, 1994, vol. 39(3), pp. 19 – 21. (Rus)

[3] Kholodova N.B., Zubovsky G.A., and Snigiryova G.P., "The state of somatic health and cytogenetic changes in patients in remote periods after low-dose irradiation (the participants of the Chernobyl accident consequence liquidation in 1986 – 1987)", Proc. III Scientific and Applied Regional Conference on the *"Health State of the Liquidators of the Chernobyl Accident Consequences in the Remote Period"*, Moscow, 2004, p. 176. (Rus)

[4] Moorhead P.S., Nowell P.S., Mellma W.J. et al., *Exp. Cell Res.*, 1960, vol. 20, pp. 613 - 616.

[5] Apelt F., Kolin-Gerresheim J., and Bauchinger M., *Mutat. Res.*, 1981, vol. 88, pp. 66 - 72.

[6] Pinkel D., Landegent J., Collins C. et al., Proceedings of the National Academy of Sciences. USA, 1988, vol. 85, pp. 9138 - 9142.

[7] Schmid E., Zitzelsberger H., Brasselmann H., and Bauchinger M., *Int. J. Radiat. Biol.*, 1992, vol. 62, pp. 673 - 678.

[8] Bauchinger M., Schmid E., Zitzelsberger H. et al., *Int. J. Radiat. Biol.*, 1993, vol. 64, pp. 179 - 184.

[9] Johnson G.D. and de C. Nogueira G.M., *Immunological Methods,* 1981, vol. 43, pp. 349 - 350.

[10] Tucker J.D., Morgan W.F., Awa A.A. et al., *Cytogenet. Cell Gen.*, 1995, vol. 68, pp. 211 - 221.

[11] Snigiryova G.P., Bogomazova A.N., Novitskaya N.N. et al., Proc. Intern. Conference "Genetic Consequences of Emergency Radiation Situations", Moscow, RUDN, 2002, pp. 313 - 328.

[12] Vaison A.A., Zhakov I.G., Khizhnikov V.A. et al., *Med. Radiol.*, 1990, pp. 21 – 28. (Rus)

[13] Liberman A.N., Radiation and Reproductive Health, St-Petersburg, 2003, 225 p. (Rus)

[14] Kharchenko V.P., Kholodova N.B., and Snigiryova G.P., *Medical Anthropology,* Minsk, 2004, pp. 213 – 214. (Rus)

[15] Bonassi S.,Znaor A., Norppa H., and Hagmar L., *Cytogenet. Genome Res.*, 2004, vol. 104, pp. 376 - 382.

[16] Rossner P., Boffetts P., Ceppi M. et al., *Environ. Health Perspectives,* 2005, vol. 113(5), pp. 517 - 520.

[17] Sevan'kaev A.V., Mikhailova G.F., Potetnya O.I. et al., *Radiats. Biol. Radioekol.*, 2005, vol. 45(1), pp. 5 – 15. (Rus)

[18] Natarajan T., Boei J., Darroudi F. et al., *Environ. Health Perspectives,* 1996, vol. 104(3), pp. 445 - 448.

In: 20 Years after the Chernobyl Accident
Editors: E. Burlakova and V. Naidich, pp. 11-21
ISBN 1-60021-249-2

Chapter 2

SOMATIC MUTAGENESIS AT T-CELL RECEPTOR LOCUS IN RESIDENTS OF RADINUCLIDE CONTAMINATED AREAS OF THE RUSSIAN FEDERATION AS A RESULT OF THE CHERNOBYL ACCIDENT

I. A. Zamulaeva*[*], *S. G. Smirnova, N. V. Orlova, A. O. Vereshchagina, S. Yu. Chekin, I. A. Smirnova, L. I. Krikunova, V. S. Parshin, V. K. Ivanov and A. S. Saenko
Medical Radiological Research Center, Russian Academy of Medical Sciences, Obninsk, Kaluga Oblast, Russia

ABSTRACT

In 2001 – 2004, the frequency of cells with mutations at T-cell receptor locus (TCR) was studied in 553 residents of radionuclide-contaminated regions of Russia and 154 control persons. The residents of contaminated territories were divided into three groups:

1. irradiated *in utero* by the Chernobyl accident and consequent living at the areas, contaminated in the range of 37 – 555 kBq/m^2 by ^{137}Cs;
2. initial irradiation at the age younger than 14 and living at the areas, contaminated in the range of 20 – 555 kBq/m^2 by ^{137}Cs;
3. initial irradiation at the age older than 18 and living at the most contaminated areas (185 – over 555 kBq/m^2 by ^{137}Cs).

[*] Main addressee: 4 Koroleva Street, 249036 Obninsk, Kaluga Oblast, Russia; MRSC RAMS; Tel: (08439) 74-847; E-mail: zamulaeva@mail.ru

The changes in TCR-mutant cell frequency were most highly expressed, if the radiation impact was initiated at the prenatal development period. In this case, the average frequency of mutant cells was observed 1.5 times higher than in the of the corresponding age control: $4.0 \cdot 10^{-4}$ vs $2.7 \cdot 10^{-4}$ ($p < 0.0001$). The lesser intensive increase of the mutant cell frequency was observed for the group irradiated before the age of 14: 1.3 times higher compared with the control ($3.8 \cdot 10^{-4}$ vs $2.9 \cdot 10^{-4}$, $p = 0.0002$). The persons, who were above 18 at the moment of the radiation impact, were not significantly different by the frequency of mutant cells from the control. The changes in TCR-mutant cell frequency in the persons, irradiated in prenatal and postnatal periods, are different both quantitatively and qualitatively. In the first case, all persons respond to the radiation impact by increasing to a different extent the frequency of TCR-mutant cells. In the second case, only a part of population respond to irradiation by increasing TCR-mutant cell frequency. The part of "responding" persons depends on the age at the moment of the radiation impact and, apparently, the absorbed dose.

For the persons with benign tumors of various localizations and nodules in the thyroid gland, the frequency of mutant cells is higher than for the persons having no such pathology.

Keywords: gene mutations, T-cell receptor locus, ionizing radiation, low doses

After the Chernobyl accident a great part of the territory of the Russian Federation was contaminated by radionuclides, and the residents of these territories were exposed to low dose ionizing radiation, which various biological effects are now unambiguous. At the same time, there are many uncertainties in the estimation of medical consequences of such impact, the most of which are of the malignant tumor type [1]. Since radiation carcinogenesis are mostly induced by changes in the genetic material of somatic cells, it becomes of great interest to study somatic mutagenesis among residents of radionuclide contaminated areas, exposed to low dose radiation as a result of accidents at different nuclear industry enterprises, including Chernobyl NPP.

By now, a considerable data quantity on the frequency of stable and unstable chromosome aberrations in lymphocytes of such persons is accumulated [2 – 4]. Much scanty data are on gene mutations in somatic cells after the radiation impact mentioned. At the same time, the study of the latter seems rather urgent for radiobiology and radiation medicine, because this provides for identification and quantitative assessment of small size change in DNA. Recently, the study of these very mutations has assumed a great importance due to at least several reasons. Firstly, it is found that gene mutations are much more important for solid cancer formation rather than structural mutations. Secondly, the existence of radiation-induced genome instability phenomenon in the descendants of cells exposed to different doses, including low ones [5 – 9], is proved. Also, it is shown that in contrast with the mutations induced by direct radiation impact, the spectrum of mutations, forming in this case, is generally presented by small size DNA changes [5, 10].

In the present work we study the frequency of peripheral blood lymphocytes with mutations at T-cell receptor locus (TCR) in the residents of three Oblasts of the Russian Federation, most heavily radionuclide contaminated as a result of the Chernobyl accident. Previously, the TCR-method was not used for these purposes, though it manifested high sensitivity and information capacity at both estimation of various genotoxic impacts (including radiation) and detection of genetic instability signs, both hereditarily conditioned and radiation induced [11 – 13]. The mutant cell frequency was analyzed in various age groups in the framework of screening examinations performed in 2001 – 2004, i.e. 15 – 19 years after the accident. In some cases, the frequency of mutant cells was analyzed with respect to the presence or the absence of benign tumors of various localizations and thyroid nodules, if these investigations were accompanied by medical examination.

MATERIALS AND METHODS

Examined Persons

The test materials were peripheral blood lymphocytes sampled from 553 residents of three Oblasts of Russia, radionuclide contaminated as a result of the Chernobyl accident, and 154 control persons. Table 1 presents characteristics of examined residents of Kaluga, Bryansk and Orel Oblasts.

Table 1. The examined groups

Residence	No.	Age at the accident moment, years	Examination date, year	^{137}Cs* contamination density, kBq/m^2	Group formation principle
Zhizdrinsk region, Kaluga Oblast	131	*in utero*	2001, 2003	37 - 185	Screening
Klintsy region, Bryansk Oblast	20	*in utero*	2002	185 - 555	Screening
Orel Oblast, including:	**225**	**0 - 14**	**2002**	**20 - 555**	Selection of persons with nodules in thyroid gland and without them during screening
Bolkhovsk region	49	0 - 14		100 - 555	
Mtsensk region	65	0 - 14		40 - 100	
Uritsk region	75	0 - 14		20 - 40	
Kolppnya region	36	0 - 14		20 - 40	
Bryansk Oblast, including:	**177**	**over 18**	**2002 - 2004**	**185 – over 555**	Medicobiological examination of all willing. Selection of persons with benign tumors and without them during the examination
Klintsy region	80	over 18	2002, 2003	185 - 555	
Novozybkov region	97	over 18	2004	over 555	

*[14,16]

The residents of Orel Oblast were subjected to screening ultrasound examination of the thyroid gland, which detected nodules in some persons. For each person with nodules one or two persons without such pathology of the same sex, age and residence were selected. Voluntarily, peripheral blood was sampled from all these persons. Moreover, these persons were questionnaire surveyed for the purpose of determining the dose absorbed by thyroid gland according to the technique, described in detail in ref. [14].

Persons from Bryansk Oblast were suggested ultrasound testing of thyroid gland and examination by mammalogists and gynecologist. This complex medical examination helped in detecting persons with benign tumors in the reproductive system organs, including mammary glands, and healthy persons. Some of them (177 persons) voluntarily agreed for peripheral blood sampling for the purpose of studying parameters of somatic mutagenesis. Thus, 2 groups of Bryansk Oblast residents were formed: one showing benign tumors of various localizations and another without them.

Persons from the control group were not exposed to recorded genotoxic impacts, i.e. were not living on contaminated territories and were not in contact with radiation sources or other adverse factors during professional activities. Three control groups were formed, which were precisely corresponded to the groups of residents on the contaminated territories according to the age distribution.

Determination of Gene Mutations at TCR Locus

The TCR-method is based on the use of monoclonal antibodies to CD3 and CD4-antigenes, labeled by different fluorochromes. It is known that T-lymphocytes possess a complex of T-cell receptor and CD3-antigen expressed on their surface. Since TCR-genes are functionally hemizygous, the products of the only allele present on the surface of lymphocytes. A mutation in the functioning allele leads to the absence of CD3 complex expression on the T-lymphocyte surface. Such mutants are detected with the help of flow cytometry as CD3-negative cells among CD4-positive T-helpers.

The venous blood of examinants was placed to tubes with heparin (Becton Dickinson, USA) and then transported to a laboratory for analysis. Using centrifugation procedure performed in the ficoll-urografin (Paneko Co., Russia, 1.077 density) density gradient, a mononuclear cell fraction was extracted from 3 ml of blood. Hence, blood two-fold diluted by the Hanks solution ((Paneko Co., Russia) were layered over 3 ml ficoll-urografin solution and centrifuged during 30 min at 250*g*. The cells from the mononuclear layer were gathered and washed twice by the Hanks solution and twice by 0.01 M phosphate buffer solution with 0.15 M NaCl (PBS, pH 7.2) by centrifugation at 350*g* during 15 min. For further staining, $(1.0 - 1.5) \cdot 10^6$ cell were taken.

The mononuclear cells were stained according to the technique developed by Kyoizumi *et al.* [15], slightly modified. The cells were added with 10 μl of monoclonal antibodies to CD4 labeled by fluorescein isothiocyanate (FITC), to CD3 and HLA-DR labeled by phycoerythrin (Becton Dickinson Immunocytometry Systems – BDIS, USA) and were incubated in darkness during 30 min. Then the cells were rinsed by centrifugation (300*g*, 5 min) in 1.0 ml PBS containing 1.0% bovine serum albumin (Sigma, USA) and 0.1% NaN_3

(Sigma, USA). The precipitate was added by 0.5 ml PBS; thereafter the suspension of cells was filtered through a nylon filter and fixed in 0.5 ml of 1% formaldehyde solution in PBS.

The samples were analyzed on a flow cytometer FACS Vantage (BDIS, USA) equiped with 488 nm laser (Enterprise 621, Coherent Inc., USA) not later than 24 hours after staining. Cells in the sample were analyzed by four parameters: forward and side light scattering intensity (linear scale), FITC and phycoerythrin fluorescence (logarithmic scale). The fluorescence was measured with the help of bandpass filters: 530/30 nm for FITC and 575/26 nm for phycoerythrin. The laser power equaled 30 mW.

The computer processing was performed with the help of LysysII software (BDIS, USA). A region of undamaged lymphocytes was selected by light scatter analysis and then the fluorescence of cells, FITC and phycoerythrin labeled, was assessed in this region exclusively. The number of cells with CD3+CD4+ immunophenotype typical of T-helper lymphocytes and with the variant CD3-CD4+ immunophenotype, which corresponds to cells with mutations at T-cell receptor locus, was determined. The frequency of mutant cells was calculated as the ratio of the number of CD3-CD4+ cells and the number of CD3+CD4+ cells. In each sample $2 \cdot 10^5$ lymphocytes were analyzed.

The data obtained were statistically processed according to Student's, Mann-Whitney's and Fischer's criteria.

RESULTS AND DISCUSSION

The Frequency of TCR-Mutant Cells at Radiation Impact Initiation Falling on the in Utero Development Period

In 2001 – 2003, the residents of Zhizdrinsk region in Kaluga Oblast exposed to the radiation impact *in utero* immediately after the Chernobyl accident and then in the postnatal period up to the examination were examined. The frequency of TCR-mutant cells was determined in 131 blood samples. The control group was formed of 43 donors of similar age, who lived on the territory of the Russian Federation uncontaminated with radionuclides and had no reliable contact with genotoxic factors. The data on mutant cell frequency in examined groups are shown in Table 2. A statistically significant increase of the frequency of mutant cells in irradiated persons compared with the control group was observed. Analogous results were obtained at the examination of Klintsy region of Bryansk Oblast residents. Therefore, the data on these two regions were united and the results are present in Table 2 and Figure 1. The analysis of united data showed that the statistical value of the differences increased compared with the control group ($p < 0.0001$). Hence, it is found that 32 persons (21%) exposed to radiation in pre- and postnatal periods showed the frequency of TCR-mutant cells exceeding the 95% confidence interval in the current control group e.g. above $5.2 \cdot 10^{-4}$. As we exclude persons with increased frequencies from the analysis, the average frequency (±SE) of mutant cells among the rest persons (N = 119) equals $(3.2 \pm 0.1) \cdot 10^{-4}$, which is statistically different from the control group ($p = 0.02$). Therefore, for this case it may be concluded that the whole cohort of irradiated persons in one degree or another respond to the radiation by increasing frequency of the mutant cells. The average frequency of the mutant cells in males or females were equal (data are not presented).

Table 2. The frequency of TCR-mutant cells in the control group and permanent residents of contaminated regions of Kaluga and Bryansk Oblast exposed to the radiation in prenatal and postnatal periods

Residence	Number of blood samples	The age at the moment of examination, average ±SE, years	Frequency of TCR-mutant cells, $\times10^{-4}$	
			average ±SE	range
Radionuclide uncontaminated territories of Russia	**43**	**16.0 ± 0.8**	**2.7 ± 0.2**	**0.9 - 6.3**
The whole contaminated territories, including:	**151**	**16.1 ± 0.1**	**4.0 ± 0.2** [1]	**0.8 - 14.1**
Zhizdrinsk region of Kaluga Oblast	131	16.0 ± 0.2	4.0 ± 0.2 [2, 3]	0.8 - 14.1
Klintsy region of Bryansk Oblast	20	16.7 ± 0.1	3.6 ± 0.4[2]	1.2 - 6.8

[1] $p < 0.0001$ compared with the control group of persons from radionuclide uncontaminated territories of Russia;

[2] $p < 0.03$ compared with the control group of persons from radionuclide uncontaminated territories of Russia;

[3] $p = 0.40$ compared with the group of residents of Klintsy region of Bryansk Oblast

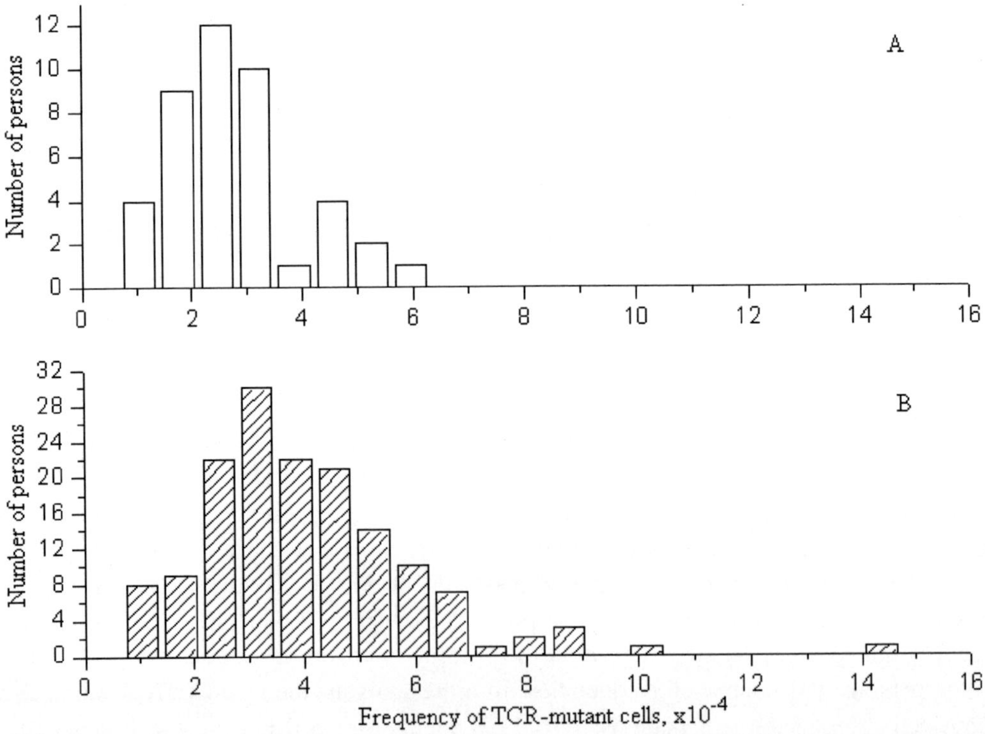

Figure 1. The distribution of TCR-mutant cell frequency in the control group and permanent residents of contaminated territories of Kaluga and Bryansk Oblasts, exposed to the radiation in prenatal and postnatal periods; A – the control group, $N = 43$; B – Irradiation in prenatal and postnatal periods, $N = 151$

The Frequency of TCR-Mutant Cells at the Radiation Impact Initiated at the Child or Teenager Period

In 2002, the gene mutations in residents of some regions in Orel Oblast possessing different levels of contamination by radionuclides were analyzed. All examined persons were born before the Chernobyl accident and were permanent residents of the contaminated territories until the examination. The average frequency of TCR-mutant cells in a randomly selected residents of Orel Oblast was statistically much higher than in the age control group: the corresponding parameters (±SE) equaled $(3.8 \pm 0.2) \cdot 10^{-4}$ and $(2.9 \pm 0.1) \cdot 10^{-4}$ ($p = 0.0002$), respectively. TCR-mutant cell frequency in 16 persons (14%) was above the upper limit of 95% confidence interval in the control (i.e. above $5.5 \cdot 10^{-4}$). Among women and men, the average frequency of mutant cells was equal (the data are not present).

Previously, it has been found with the help of ultrasound investigation that about 10% residents of Orel Oblast possess nodules in the thyroid gland [14]. It is also of interest to assess the features of somatic mutagenesis in the persons with thyroid nodules and without them. Therefore, in the current investigation we have formed two approximately equal by abundance groups from the residents of Orel Oblast with respect to results of ultrasound investigations of the thyroid gland. The distributions by sex and age in these groups were also equal. Separate analysis of TCR-mutant cell frequencies in the persons without thyroid nodules and them was performed (Table 3). The average frequency of TCR-mutant cells in the residents of Orel Oblast with detected nodular pathology was significantly higher than in persons without such pathology. Similar tendency was observed for the comparison of this parameter separately for different regions ($p > 0.05$). 12% of persons without thyroid nodules and 21% of persons with them had increased TCR-mutant cell frequencies (i.e. exceeding the level of $5.5 \cdot 10^{-4}$). Hence, if increased frequencies are excluded from the group without thyroid nodules, mutant cell frequency in the rest persons is identical to the control. It may be concluded that currently, the increasing frequency of mutant cells is related to the presence of a small group of persons showing very high value of this parameter. At the same time, if the analogous operation is implemented on a group of persons possessing thyroid nodules, statistical value of differences is preserved. For example, the average frequency of mutant cells is reduced from $4.4 \cdot 10^{-4}$ to $3.3 \cdot 10^{-4}$, but the latter value exceeds the corresponding parameter in the control ($2.9 \cdot 10^{-4}$, $p = 0.01$).

Thus, for the groups of persons with thyroid nodules and without them, the radiation effects estimated here by the frequency of mutant cells are different. Apparently, this may be related to the difference in irradiation doses perceived. In this connection, it was of interest to analyze individual doses absorbed by the thyroid gland in groups with thyroid nodules and without them. It has been found that the average doses in these groups are similar, equaling (46.4 ± 7.4) and (41.6 ± 7.1) mGy for the residents of Orel Oblast, respectively. Analogous parameters for Bolkhov region are (106.7 ± 24.5) and (111.0 ± 31.1) mGy, respectively. It is the authors' opinion that the data obtained represent an additional proof for higher expressed response to identical radiation impact in persons with thyroid nodules compared with other residents of this region.

The frequency of mutant cells with respect to the contamination level of the residence by ^{137}Cs and the dose absorbed by the thyroid gland was analyzed. A tendency to increase the average frequency of mutant cells among the residents of Bolkhov region, the most highly contaminated by radioactive cesium isotopes, compared with that among the residents of other, lesser contaminated regions of Orel Oblast was observed. Similar tendency was observed for the part of persons with increased frequency of mutant cells (Table 3). As comparing the average frequency of mutant cells among the residents of different regions of Orel Oblast with the average dose absorbed by the thyroid gland, the same tendency to increase the frequency of mutant cells with the dose was observed (Figure 2).

Table 3. The frequency of cells with muteations at T-cell receptor locus in residents of Orel Oblast with respect to the presence or absence of thyroid nodules

Residence	Number of persons	Age at the moment of examination, average ±SE , years	Thyroid nodules	TCR-mutant cell frequencies, ×10^{-4}		The part of persons with increased frequency of mutant cells, %
				average ±SE	range	
Territory of Russia uncontaminated with radionuclides	**86**	**21.4 ± 0.8**	**ND**	**2.9 ± 0.1**	**0.9 - 6.3**	**2.5**
Orel Oblast as a whole, including:	**111**	**21.4 ± 0.3**	+	**4.4 ± 0.3** [1, 2]	**1.4 - 21.5**	**21**
	114	**21.5 ± 0.3**	–	**3.7 ± 0.2** [2]	**1.0 - 13.4**	**12**
Bolkhov region	28	20.4 ± 0.5	+	5.4 ± 0.9 [2, 3, 4]	1.6 - 21.5	25 [5]
	21	20.1 ± 0.6	–	4.0 ± 0.4 [2]	1.4 - 7.2	23 [6]
Kolpnya, Uritsk and Mtsensk regions	83	21.8 ± 0.5	+	4.1 ± 0.2 [2,, 4]	1.9 - 12.7	19
	93	21.8 ± 0.5	–	3.6 ± 0.2 [2]	1.0 - 13.4	10

[1] $p = 0.03$ compared with the residents of the same territory without detected nodular pathology of thyroid gland;

[2] $p < 0.01$ compared with the control from radionuclide uncontaminated territories of Russia;

[3] $p = 0.05$ compared with the residents of lesser contaminated regions (Kolpnya, Uritsk and Mtsensk) and possessing nodules in thyroid gland;

[4] $p = 0.2$ compared with the residents of the same territory without detected nodular pathology of thyroid gland;

[5] $p = 0.17$ compared with the residents of lesser contaminated regions (Kolpnya, Uritsk and Mtsensk) and possessing nodules in thyroid gland;

[6] $p = 0.06$ compared with the residents of lesser contaminated regions (Kolpnya, Uritsk and Mtsensk) and having no nodules in thyroid gland

The Frequency of TCR-Mutant Cells at the Radiation Impact Initiated at the Age Older 18

In 2002 – 2004, the frequencies of TCR-mutant cells in women – the residents of the radionuclide contaminated areas of Bryansk Oblast (Klintsy and Novozybkov regions) were determined. The data on the frequency of TCR-mutant cells were analized with respect to the presence of benign tumors in the reproduction system organs (mostly in mammary glands and uterus -Table 4). A tendency to increase the average frequency of mutant cells in persons with benign tumors compared with the persons having no such pathology was observed: $5.1 \cdot 10^{-4}$ vs $4.3 \cdot 10^{-4}$, respectively ($p = 0.2$). As compared with unexposed control group of the corresponding age, a tendency to increase the frequency of TCR-mutant cells exclusively in the group with benign tumors was observed ($p = 0.2$). The frequency of mutant cells in the residents of Bryansk Oblast having no benign tumors was identical to the control level, even in spite of the fact that this cohort was formed of the residents from the most highly contaminated regions (among the regions studied in the current investigation and all over Russia, as well).

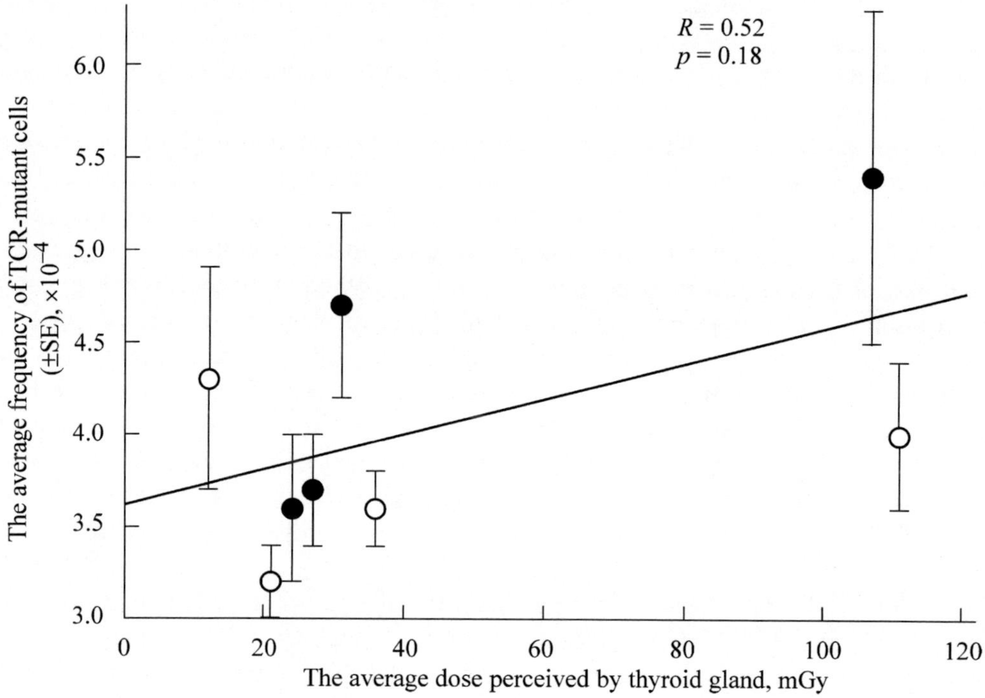

Figure 2. The comparison of the average dose per thyroid gland and the average frequency of TCR-mutant cells in the residents of the monitored regions of Orel Oblast. Dark points mark the groups of persons with thyroid nodules; white points mark the groups of persons having no nodules in the thyroid gland

Table 4. The frequency of TCR-mutant cells in the control persons and residents of Bryansk Oblast with respect to the presence or the absence of benign tumors among them

Examined persons	Number of examinants	Age at the moment of examination, average ±SE, years	TCR-mutant cell frequency, $\times 10^{-4}$	
			average ±SE	range
Residents of radionuclide uncontaminated areas of the Russian Federation (control)	**55**	**40.6 ± 1.2**	**4.3 ± 0.2**	**1.5 - 10.0**
Residents of Bryansk Oblast as a whole, including:	**177**	**40.4 ± 0.5**	**4.9 ± 0.3**[1]	**1.1 - 52.9**
persons with benign tumors	123	40.6 ± 0.6	5.1 ± 0.4[2]	1.1 - 52.9
persons without benign tumors	54	40.0 ± 1.0	4.3 ± 0.3	1.9 - 18.5

[1] $p = 0.3$ compared with the control group;

[2] $p = 0.2$ compared with the control group and the residents of Bryansk Oblast without tumors.

About 11% (20/177) of persons has TCR-mutant cell frequency exceeding the 95% confidence interval in the control group, which is $7.3 \cdot 10^{-4}$. If the persons with increased frequencies are excluded from the analysis, the average frequency of mutant cells in the rest group is identical to that in the control group.

CONCLUSION

Thus, genetic changes in somatic cells estimated in the present article by the frequency of TCR-mutant lymphocytes are the most highly expressed in the residents of contaminated regions, if the radiation impact was initiated at the prenatal period of development. Lesser expressed changes were observed in persons, whose irradiation started in the postnatal period (in the age of 0 – 14 years old). In this case, the frequency of mutant cells was lower in persons having no thyroid nodules than those having such pathology. It may be suggested that a small part of the population of people with nodular pathology of the thyroid gland have been exposed to more intensive (compared with the control group) genotoxic impact, not necessarily of the radiation origin, or these persons demonstrate increased sensitivity to the environment factors as compared with all other population. Finally, the persons been older than 18 at the moment of radiation impact demonstrated statistically equal frequency of mutant cells compared with the control, although a tendency to increase this parameter in persons with benign tumors was determined. We have no data on the frequency of TCR-mutant cells in the case of benign tumors in the control group yet. Therefore, the reason for the change mentioned above is still unclear.

By and large the data obtained indicate that a definite part of the population living in the regions contaminated by radionuclides increases the amount of TCR-mutant cells as a response to low dose radiation. The rest group, which is the greatest in the population, demonstrated no signs of intensified mutagenesis in this locus. The part of "responding" persons depends on the age at the initiation moment of the radiation impact. It is common knowledge that doses absorbed by various organs/tissues of child's and adult's organism are different, even if they are living in the settlement due to the known features of metabolism and the lifestyle, milk drinking, for example. First and foremost this related to thyroid gland, which much more actively accumulated radioactive iodine in children than in adults. It is apparent that the age tendency observed is to some extent explained by different dose load, and to some other extent by the known difference in the radiosensitivity of tissues of growing and already grown-up organism. Therefore, a question arises if the frequency of mutant cells and the part of "responding" persons within the same age range depend on the dose or in the case of the low dose radiation individual features of the organism are crucial. The role of the individual features is confirmed by the fact that at equal radiation doses the mutant cell frequency may be different for persons having thyroid nodules and without them. At the same time, the data on Orel Oblast also allow a suggestion about the presence of a dose dependence of the somatic mutagenesis level. For more particular conclusion further investigations are required.

The researches were performed under financial support of Russian Fund for Fundamental Researches (Grant No. 01-04-096002), Russian Humanitarian Scientific Fund and the Government of Kaluga Oblast (Grant No. 05-06-59607a/c), Federal Agency for Science and

Innovations of the Russian Federation (Grant No. 02.434.11.3003), and the Government of Orel Oblast.

REFERENCES

[1] *United Nations Scientific Committee on the Effects of Atomic Radiation Report to the General Assembly with Scientific Annexes*, "Sources and effects of ionizing radiation", vol. II: Effects, Annex G: "Biological effects at low radiation doses", New York, 2000, pp. 75 - 144.

[2] Shevchenko V.A., Snigireva G.P., Suskov I.I., Akaeva E.A., Elisova T.N., Iofa E.L., Nilova I.N., Kostina L.N., Novitskaia N.N., and Sidorova V.F., *Radiats. Biol. Radioekol.*, 1995, vol. 35(5), pp. 588 – 596. (Rus)

[3] Sevan'kaev A.V., Potetnia O.I., Zhloba A.A., Moiseenko V.V., Potetnia V.I., Ankina M.A., Golub E.V., Epifanova N.I., Zavitaeva T.A., Kozlov V.M. *et al.*, *Radiats. Biol. Radioekol.*, 1995, vol. 35(5), pp. 581 – 588. (Rus)

[4] Sevan'kaev A.V., *Radiats. Biol. Radioekol.*, 2000, vol. 40(5), pp. 589 – 595. (Rus)

[5] Little J.B., *Int. J. Radiat. Biol.*, 1998, vol. 74(6), pp. 663 - 671.

[6] Harms-Ringdahl M., *Mutat. Res.*, 1998, vol. 404, pp. 27 - 33.

[7] Pelevina I.I., Gotlib V.Ya., Kudryashova O.V., Serebryani A.M., and Afanasjev G.G., *Radiats. Biol. Radioekol.*, 1996, vol. 36(4), pp. 546 – 560. (Rus)

[8] Ponnaiya B., Limoli C.L., Corcoran J., Kaplan M.I., Hartmann A., and Morgan W.F., *Int. J. Radiat. Biol.*, 1998, vol. 74(6), pp. 765 - 770.

[9] Pelevina I.I., Aleshchenko A.V., Antoshchina M.M., Gotlib V.Ia., Kudriashova O.V., Semenova L.P., and Serebrianyi A.M., *Radiats. Biol. Radioekol.*, 2003, vol. 43(2), pp. 161 – 166. (Rus)

[10] Nelson S.L., Giver C.R., and Grosovsky A.J., *Carcinogenesis*, 1994, vol. 15(3), pp. 495 - 502.

[11] Kyoizumi S., Umeki S., Akiyama M., Hirai Y., Kusunoki Y., Nakamura N., Endoh K., Konishi J., Sasaki M.S., Mori T., Fujita S., and Cologne J.B., *Mutat. Res.*, 1992, vol. 265, pp. 173 - 180.

[12] Cole J. and Scopek T.R., *Mutat. Res.*, 1994, vol. 304, pp. 33 - 105.

[13] Saenko A. S., Zamulaeva I.A., Smirnova S. G., Orlova N. V., Selivanova E. I., Matveeva N.P., Kaplan M.F., Nugis V.Y., and Tsyb A.F., *Int .J. Radiation Biology*, 1998, vol. 73(6), pp. 613 - 618.

[14] Ivanov V.K., Tzyb A.F., Parshin V.S., Maksutov M.A., Chekin C.Yu., Gorski A.I., Vlasov O.K., Saenko A.S., Sevan'kaev A.V., and Parshkov E.M., *Radiation and Risk. Special Issue*, 2003, 136 p. (Rus)

[15] Kyoizumi S., Akiyama M., Hirai Y., Kusunoki Y., Tanabe K., and Umeki S., *J. Exp. Med.*, 1990, vol. 171, pp. 1981 - 1999.

[16] Ivanov V.K. and Tsyb A.F., Medical Radiological Effects of the Chernobyl Catastrophe on the Population of Russia: Estimation of Radiation Risk, Moscow: Meditsina, 2002. 389 p.

In: 20 Years after the Chernobyl Accident
Editors: E. Burlakova and V. Naidich, pp. 23-38
ISBN 1-60021-249-2

Chapter 3

THE RESULTS OF DYNAMIC CYTOGENETIC STUDY OF CHILDREN AND TEENAGERS LIVING ON THE RADIATION-CONTAMINATED TERRITORIES AFTER CHERNOBYL ACCIDENT

A. V. Sevan'kaev[1], G. F. Mikhailova[1], O. I. Potetnya[1], V. V. Tsepenko[1], I. K. Khvostunov[1], E. V. Golub[1], V. S. Pyatenko[1], O. V. Pozdyishkina[1], N. N. Shepel[1], E. G. Matveenko[1], M. P. Borovikova[2] and N. O. Omaraskhabov[3]

[1]Medical Radiological Research Center RAMS, Obninsk, Russia
[2]Department for Health and Drug Support of Kaluga Administration, Kaluga, Russia
[3]Zhizdrinsk Regional Hospital, Kaluga Oblast, Russia

ABSTRACT

The results of long-term cytogenetic investigations among children and teenagers living on the territories radionuclide contaminated after the Chernobyl accident are presented. The average ^{137}Cs contamination in two regions of inhabitance under comparison equaled 111 and 200 kBq/m^2, and the average whole body absorbed dose received in the period of 1986 – 2001 equaled 6.7 and 11.4 mGy, respectively. In the second region, the averaged thyroid gland doses for all age groups of the population was also found 1.5 time higher than in the first region, at that, for the smallest age (0 – 1) these doses equaled 114.3 and 174.3 mGy, respectively. Cytogenetic observations performed during 17 years after the accident indicated the increased level of chromosomal aberrations in peripheral blood lymphocytes for 30 to 60% of examined patients. For the whole observation period, the average frequency of unstable aberrations (acentric fragments, dicentrics and centric rings) equaled 0.4 per 100 cells (0.22 per 100 cells for the control), and the level of aberrations-markers of radiation impact (dicentrics and centric rings) remained elevated in all periods of examination, varying within 0.04 – 0.19 per 100 cells (0.03 for the control). Parallel investigation of a part of patients using FISH-method for stable aberration frequency (mostly translocations) indicated their level about 3 times exceeding the observed level of dicentrics. Comparatively higher parameters of cytogenetic disorders were observed in the group prenatally irradiated during the accident.

Keywords: the Chernobyl accident, contaminated territories, prenatal irradiation, human lymphocytes, stable and unstable aberrations, chromatid aberrations, multiaberrant cells, carcinogenic risk

Eighteen years have already passed since the Chernobyl accident – the most large-scale nuclear catastrophe in the Mankind history in times of peace. The consequence of this catastrophe was radioactive contamination of large territories by long-living radioisotopes. These territories are the inhabitance for highly abundant populations of people, who, therefore, to one extent or another are exposed to continuous chronic radiation impact. There an unexampled situation appeared, when dozen and even hundred thousand children (we are speaking exclusively about the population of Russia) were not only exposed to acute radiation at the moment of the Chernobyl accident, including incorporation of iodine radionuclides, but are still living exposed to continuous impact of low radiation doses. It is commonly known that child's organism is the most sensitive to radiation impact, especially, in the antenatal period of development. Therefore, children form the critical group of the population, mostly injured by the Chernobyl accident.

The carcinogenic effect of ionizing radiation was numerously demonstrated in various epidemiological investigations. According to modern ideas, carcinogenesis is based on mutation changes in somatic cells of the organism [1 – 3]. The reliable correlation between cancer formation risk and the frequency of chromosomal aberrations has been demonstrated on a quite abundant group (about 3,000 people) [4, 5]. Therefore, persons with high level of structural damages of chromosomes may be related to a group with increased carcinogenic risk [1]. Translocations are the most frequent chromosomal anomalies typical of hemoblastoses and some solid tumors [6]. Among survivals from Hiroshima and Nagasaki a good correlation in the dose dependence between translocations in bone marrow cells and leukemias, as well as solid tumors, was shown [7].

As a consequence, the level of chromosomal aberrations observed may represent a forecasting test for early detection of persons with elevated carcinogenic risk, i.e. already prior to development of a malignant degeneration of a cell. It is the authors' opinion that this is the main aim, the necessity and the value of performance of a long-term cytogenetic monitoring of the population living on radiation-contaminated territories, under conditions of chronic low-dose radiation.

It should be noted that immediately after the Chernobyl accident, Governmental Long-Term Program on Liquidation of the Accident Consequences was established, in the framework of which regular cytogenetic examination of the accident liquidators and the population living on the contaminated territories was envisaged. Unfortunately, collapse of the USSR and consequent hard economic development of the Russian Federation stopped targeted financial support of these investigations. Nevertheless, the systematic cytogenetic examination of children and teenagers living in radiation contaminated regions of Kaluga and Bryansk Oblasts, which we initiated in 1989, is still continued. This paper presents results of dynamic cytogenetic study for the mentioned cohort living on contaminated territories of two regions in Kaluga Oblast.

Object and Methods of Investigation

The groups under study included children and teenagers, permanent residents of Zhizdrinsk and Ul'yanovsk districts of Kaluga Oblast. Totally, in Zhizdrinsk and Ul'yanovsk districts 333 and 407 children and teenagers were examined, respectively. This gave 740 patients, 149 among which were exposed to prenatal radiation during the accident, distributed by the fetal age at the moment of the accident (26 April, 1986) as follows: 0 – 8 weeks – 36 persons (24.2%); 9 – 28 weeks – 76 persons (51%); 29 – 40 weeks – 37 persons (24.8%). The rest 591 persons were grouped as follows: 540 patients born before the accident and aged 0 – 18 at the moment of the accident, and 51 patients born after the accident, later than 07 March, 1987 (see Table 2). For the observation period of 1989 – 2003, in Ul'yanovsk district repeated cytogenetic examinations were performed 5 times, and in Zhizdrinsk district – 7 times. Hence, a part of children and teenagers were repeatedly examined from 2 to 6 times. The control group included 87 children and teenagers (examined in 1991 and 1993) living on non-contaminated territories.

Blood (1 – 3 ml) was sampled from ulnary veins of examined patients to sterile test tubes with heparin solution (200 un/blood ml) and transported to a laboratory, where lymphocyte culture were processed. Then chromosome substance was prepared and analyzed at standard staining (by Gimza) of chromosomes and by fluorescence *in situ* hybridization method (FISH-method). The composition of the cultural medium was the following: whole blood – 0.8 ml; MEM medium – 6.16 ml; inactivated bovine serum – 1.6 ml; L-glutamine – 0.08 ml; an antibiotic solution – 0.08 ml; phytohemagglutinin (PHA) – 0.15 ml. The flasks were hold at 37°C in a thermostat for incubation during 48 h. To block mitosis at the metaphase stage 2 h before the incubation end, the flasks were added with 0.2 μg demecolcine solution per 1 ml medium.

After cultivation, the cell culture was poured to centrifugal test tubes and processed at 1,000 rot/min rate during 15 min in order to precipitate cells. Supernatant fluid was removed by a water-suction pump; the culture was added by a hypotonic solution (0.75 mol/l KCl), preliminarily heated up to 37°C, and resuspended the deposit in it. Then tubes with the cell culture were exposed to a water bath (37°C) during 10 – 12 min. After hypotensing it was centrifuged one more time under the same conditions with subsequent removal of supernatant fluid.

For the purpose of cell fixing, the deposit was resuspended in 1 – 1.5 ml of fresh fixing agent (a mixture of ethyl alcohol and glacial acetic acid in proportion 3:1) in a shaker and the volume was added up to 10 ml.

The preparations were made immediately after the fixation end. Suspension of the cells was applied by drops to glass slides, wetted with distilled water and preliminarily cooled. After drying the preparations were hydrolyzed in 5N hydrochloric acid during 10 min.

Chromosome preparations were dyed by the Gimza method.

Cytogenetic analysis were performed on a light microscope under immersion with 90×10 amplification.

All types of chromosomal aberrations detectable without karyotyping were analyzed. Among aberrations of the *chromosome type* are acentric fragments (paired and point), centric rings and dicentrics. These are the so-called aberrations of unstable type. According to modern ideas, among them dicentrics and centric rings represent aberrations-markers of the radiation impact, i.e. detection of this type of aberrations in peripheral blood lymphocytes indicates exposure of the organism to mutagens of the radiation origin.

The standard method of metaphase analysis, i.e. without karyotyping, allows for registering a definite part of *stable aberrations*, too. These are the so-called anomalous monocentrics, which in the most cases are the results of a reciprocal translocation. They were also taken into account in this study. According to the latest data [8], this method is quite suitable for registering up to 20% of the initial translocations.

Among aberrations of the *chromatid type* acentric fragments (singular deletions and isodeletions) and exchange aberrations (as a whole) were considered.

The study performed by FISH-method used a mixture of biotin-modified (via PCR) DNA-probes for chromosomes 2 and 3. Prior to hybridization preparations were treated by RNAase (100 μg/ml in 2×SSC) during 1 h at 37°C. For the purpose of DNA denaturation, the preparation under the cover-glass was added with a solution consisting of 70% deionized formamide, 2×SSC and 10mol/l phosphate buffer (pH 7.0). The preparation was placed to a hot bench at 80°C for 4 min with subsequent passing through a set of ethyl alcohol dilutions: 70%, 90%, 100%. Then the preparations were dried. Chromosomal DNA probes were mixed with hybridization buffer and denatured during 5 min at 70°C, then placed first to an ice-cold water for 5 min and after that to a water bath at 37°C for 45 min. For chromosomes 2 and 3, a mixture of DNA probes was applied to a preparation under the cover-glass and for hybridization placed to a thermostat at 37°C for the night. After hybridization, the preparations were rinsed in 50% formamide, then in 0.1×SSC and 0.05% Tween 20 (SIGMA, Germany). The next stage was immunochemical staining; biotin-modified DNA probes were detected with the help of FITS-Avidin D (Vector Laboratories, Germany) and biotin-modified Anti-Avidin D (Vector Laboratories, Germany). FITS-Avidin D (5 mg/ml, 3×30 min) and Anti-Avidin D (5 mg/ml, 2×30 min) were successively applied on the preparation, which was then incubated at room temperature and rinsed in 0.05% Tween 20 (3×5 min). Thereafter, the preparations were passed through a set of ethyl alcohol dilutions (70%, 90%, 100%) and dried. Then PI-Vectashield was applied.

Metaphases were analyzed using a fluorescent microscope of Zeiss Company with dual filter (AHF), specific for FITC and PI.

Stable aberrations of chromosomes were registered according to PAINT (Protocol of Aberration Identification and Nomenclature Terminology) classification system [9]. At the standard chromosome staining (by G-band), from 300 to 500 cells for each examined patient were analyzed, and by FISH-method – from 350 to 1,500 cells. After estimation of the translocation frequency, formed by 2 and 3 pairs of painted chromosomes composing 14.9% of genome, the frequency of translocations per genome was estimated by the formula, suggested by Lucas J.N. *et al.* [10].

The data were statistically processed by commonly used parametric and nonparametric criteria, realized in ORIGIN and MS EXCEL software.

RESULTS AND DISCUSSION

One of the basic and mandatory factors of the radiation epidemiological investigations is knowledge of the radiation conditions and radiation doses perceived by the population in the areas, where such investigations are performed, without which any scientifically justified estimation of the radiation hazard and forecasting of late medical consequences of the radiation impact for human populations under study are impossible. Just by 2000 using the specified data, total accumulated doses of the global irradiation and thyroid gland doses for the population living on the contaminated territories after the Chernobyl accident were reconstructed, which then were approved by Ministry of Health of the Russian Federation and adopted by All-Russia Scientific Commission on Radiation Protection (ARSCRP) [11]. For the population of Kaluga Oblast, the absorbed doses were reconstructed in the Russian State Medical-Dosimetric Register (RSMDR, Obninsk).

The average ^{137}Cs contamination in the studied areas, as well as total accumulated whole body doses and doses per thyroid gland for the population of these areas are presented in Table 1. It is obvious that by the average ^{137}Cs contamination of Zhizdrinsk and Ul'yanovsk districts were twice different, approximately (111 and 200 kBq/m^2, respectively), though due to irregular fallout of ^{137}Cs the contamination density in settlements were significantly different from the average values by the whole zone. In accordance with the density of radioactive contamination and average whole body doses accumulated by the population of these districts in the period of 1986 – 2001, were also different by two times, approximately – 6.7 and 11.4 mGy, respectively.

Experts from RSMDR have also reconstructed specified average age-dependent and individual (if radiometry for 1986 was obtained) thyroid gland doses for the population of Kaluga Oblast. As follows from the calculations performed, in 1986, averaged thyroid gland doses for the population of all age groups of Ul'yanovsk district were also found 1.5 times higher than for Zhizdrinsk district. Hence, maximal thyroid gland doses received by residents of both districts were caused by predominant delivery of radioactive iodine with milk, calculated for the smallest age of 0 – 1 (174.0 and 114.3 mGy by the districts mentioned, respectively). Further on, with increasing age these doses were decreasing and became minimal at the age of 18, giving 29.2 and 19.1 mGy for Ul'yanovsk and Zhizdrinsk districts, respectively.

Table 1. ^{137}Cs contamination and district-averaged thyroid gland doses for the population of Kaluga Oblast [11]

District	^{137}Cs contamination, kBq/m^2	Whole body absorbed dose at 2001 y, mGy	Thyroid gland dose, mGy					
			Age, years					
			0–1	1–3	3–8	8–13	13–18	≥18
Zhizdrinsk	111	6.7	114.3	86.8	53.2	30.4	23.9	19.1
Ul'yanovsk	200	11.4	174.0	136.6	82.2	47.3	36.1	29.2

Table 2 presents data on the cohort of patients examined in two contaminated districts of Kaluga Oblast. The most abundant group was presented by children and teenagers. Totally, cytogenetic examination was applied to 740 persons-residents of Ul'yanovsk (333 persons) and Zhizdrinsk (407 persons) districts. In 1989 – 1998, in the first district examinations were performed 5 times; in 1990 – 2003, in the second district examinations were performed 7 times. Hence, some part of children were repeatedly examined from 2 to 6 times (refer to Table 2).

Table 2. The cohort of examined children and teenagers from Kaluga Oblast

District of residence	Total number of examinants (men and women)	Year of birth (min-max)	Total blood samples in the period of 1989-2003	Distribution of persons on number of examinations					
				1	2	3	4	5	6
Zhizdrinsk	407 (191 - 216)	1966 - 1989	717	196	140	48	19	3	1
Ul'yanovsk	333 (161 - 172)	1968 - 1989	543	181	102	42	8	-	-
Total	740 (352 - 388)	1966 - 1989	1260	377	242	90	27	3	1

The main results of dynamic cytogenetic observation for the cohort are presented in Table 3. Hence, the first examination of Ul'yanovsk district residents was performed 2.7 years after the Chernobyl accident and the latest – 12.5 years after, for Zhizdrinsk district – 4.4 and 17 years after the accident, respectively.

In the period of 1989 – 1998, the 5-time consecutive cytogenetic examination in Ul'yanovsk district generally involved 543 persons with respect to repeated examination of the same patients (refer to Table 2). Hence, almost 150,000 metaphases were analyzed, on average, 300 metaphases from each patient. In percents, the number of patients with increased concentration of cells with chromosomal aberrations (compared with the control) varied from 28.9 to 53.8%. The frequency of unstable chromosomal aberrations (acentric fragments, dicentrics and centric rings) in them varied from 0.33 to 0.46 per 100 cells giving, on average, (0.39 ± 0.02) for the observation period, which is reliably higher than the control level (0.22 ± 0.03). It should be noted that the average level of radiation aberration markers (dicentrics and centric rings) has reliably exceeded the control level in all observation periods varying in the ranges of 0.04 – 0.09 and 0.01 – 0.08 per 100 cells in relation to 0.02 and 0.01 per 100 cells in the control, respectively.

Totally, the 7-time cytogenetic examinations, performed on Zhizdrinsk district in the period of 1990 – 2003, invoked 717 patients with respect to repeated examination of the same patients. Hence, over 250,000 metaphases were analyzed, on average, about 350 metaphases per each patient. In percents, the number of patients with increased concentration of cells with chromosomal aberrations (compared with the control) varied from 27.8 to 61.6%, i.e. in the range similar to the examined residents from Ul'yanovsk district. According to the observation periods, the frequency of unstable chromosomal aberrations varied in similar ranges, from 0.28 to 0.59 per 100 cells, giving, on average, (0.40 ± 0.01), which reliably exceeds the control level. Similar to the residents of Ul'yanovsk district, in the overwhelming majority of observations, the average frequency of aberrations-markers of the radiation impact (dicentrics and centric rings) in this cohort reliably exceeded the control level, varying in the ranges of 0.03 – 0.11 and 0.01 – 0.08 per 100 cells, respectively.

Table 3. The average frequency of chromosomal aberrations per 100 cells in children and teenagers – the residents of contaminated districts of Kaluga Oblast

Year of examination	Time after the accident, years	Number of examined patients	Average age*, years	Number of cells	Aberrant cells**, %	Percentage of patients with aberrant cells above control ***
Zhizdrinsk district						
1990	4.4	73	7.2	14600	0.58±0.04	61.6
1991	5.4	96	8.4	18226	0.47±0.04	48.0
1994	8.4	109	11.0	31870	0.27±0.02	27.8
1995	9.4	133	12.9	39867	0.44±0.02	36.8
1998	12.2	88	13.9	44000	0.40±0.02	55.6
2001	15.4	128	15.1	63491	0.37±0.02	49.2
2003	17.0	90	16.7	45000	0.40±0.02	52.2
All years		717		257054	0.40±0.01	
Ul'yanovsk district						
1989	2.7	17	13.1	4920	0.30±0.07	23.5
1991	5.0	146	9.1	28459	0.46±0.03	53.4
1992	6.0	192	13.3	36781	0.37±0.03	47.3
1994	7.8	83	11.8	24400	0.34±0.03	28.9
1998	12.5	105	15.1	52380	0.42±0.02	53.8
All years		543		147140	0.40±0.01	
Control		87	12.4	26878	0.23±0.03	

Year of examination	Chromosome type					Chromatid type	MAC/1,000 cells
	Acentrics	Centric rings	Dicentrics	Total****	Anomalous monocentrics		
Zhizdrinsk district							
1990	0.45±0.06	0.08±0.02	0.07±0.02	0.59±0.06	0	0.62±0.07	0
1991	0.30±0.04	0.04±0.01	0.07±0.02	0.41±0.05	0.07±0.02	0.56±0.06	0.21±0.11
1994	0.17±0.02	0.01±0.01	0.10±0.02	0.28±0.03	0.01±0.01	0.58±0.04	0
1995	0.33±0.03	0.05±0.01	0.07±0.01	0.45±0.03	0.02±0.01	0.45±0.03	0.03±0.03
1998	0.32±0.03	0.01±0.01	0.03±0.01	0.36±0.03	0.05±0.01	0.52±0.03	0
2001	0.26±0.02	0.03±0.01	0.11±0.01	0.40±0.03	0.03±0.01	0.86±0.04	0.54±0.09
2003	0.30±0.03	0.02±0.01	0.08±0.01	0.39±0.03	0.02±0.01	0.73±0.04	0.04±0.03
Total years	0.29±0.01	0.03±0.01	0.08±0.01	0.40±0.01	0.03±0.01	0.64±0.02	0.12±0.06
Ul'yanovsk district							
1989	0.24±0.07	0.08±0.04	0	0.33±0.08	0	1.22±0.16	0
1991	0.36±0.04	0.05±0.01	0.05±0.01	0.46±0.04	0.02±0.01	0.45±0.04	0.14±0.07
1992	0.32±0.03	0.01±0.01	0.05±0.01	0.38±0.03	0.02±0.01	0.59±0.04	0.19±0.07
1994	0.29±0.03	0.02±0.01	0.04±0.01	0.35±0.04	0.004±0.004	1.16±0.07	0.08±0.06
1998	0.25±0.02	0.04±0.01	0.09±0.01	0.38±0.03	0.07±0.01	0.77±0.04	0.05±0.03
All years	0.29±0.01	0.03±0.01	0.06±0.01	0.39±0.02	0.04±0.01	0.73±0.02	0.09±0.04
Control	0.19±0.03	0.01±0.01	0.02±0.01	0.22±0.03	0.01±0.01	0.41±0.04	0.04±0.04

* - at the moment of observation;

** - with aberration of the chromosome type;

*** - with 95% probability;

**** - with no consideration of anomalous monocentrics.

Among aberrations of the chromosome type, the so-called anomalous monocentrics, according to the formation mechanism mostly representing symmetrical reciprocal translocations, because they cause no obstacles for mitotic fission of cells and are freely converted to daughter cells, were also taken into account. According to the data obtained [8], the standard (excluding karyotyping) method of metaphase analysis of chromosome aberrations allows detecting up to 20% of translocations. Table 3 shows that in all observation periods the frequency of observable translocations in the residents of both districts exceeded the control level giving, on average, 0.03 – 0.04 per 100 cells.

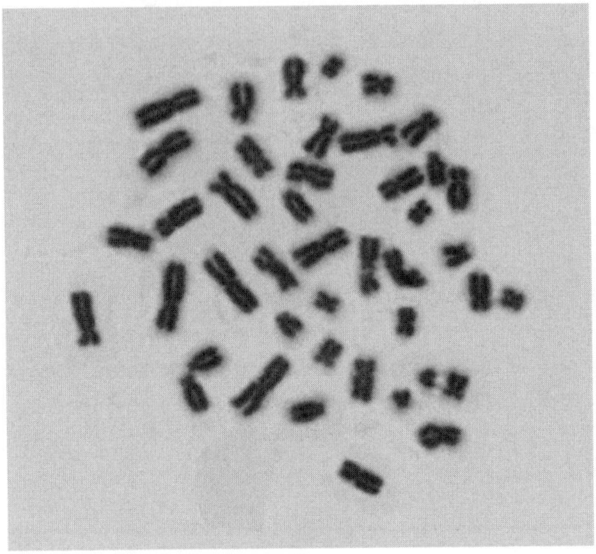

A

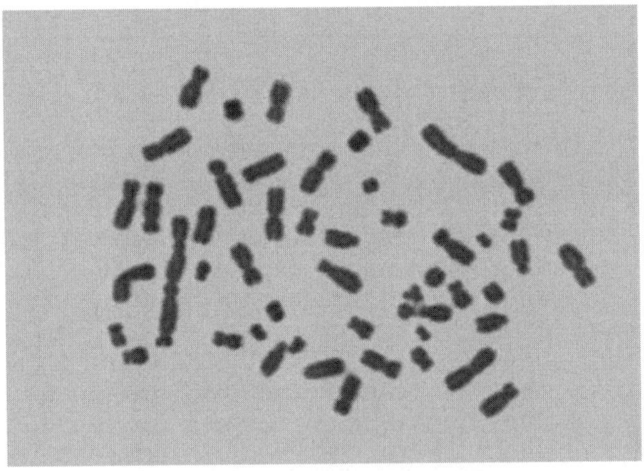

B

Figure 1. Microphotographs of normal (A) and multiaberrant (B) cells.

Beside the chromosomal aberrations, we also considered chromatid aberrations. They are usually divided into 2 main classes: fragments (deletions) and exchanges. For the control cultures of human lymphocytes and, as testified by results of the current investigation, the residents of contaminated territories, among observed chromatid deletions, the overwhelming majority was presented by singular fragments and extremely rarely – by paired (isodeletions), and among chromatid exchanges – by symmetrical and asymmetrical exchanges. According to the majority of investigations and the authors' results [12], the culture of lymphocytes in control persons shows the ratio between chromosome type aberrations (with no respect to translocations) and chromatid type equals 1:1.5 – 2, on average. Hence, as shown before [12], the frequency of chromatid aberrations increases with the age and, therefore, the mentioned ratio may shift more towards chromatid aberrations.

Table 3 shows that for the residents of both districts the level of chromatid aberrations was increased in all periods of observation giving, on average, 0.73 and 0.64 per 100 cells for Ul'yanovsk and Zhizdrinsk districts, respectively (0.41 in the control). Both in control and residents of Ul'yanovsk districts, the ratio between chromosome and chromatid aberrations was practically equal (1:2 and 1:1.9), whereas for the residents of Zhizdinsk district it was somewhat lower (1:1.5).

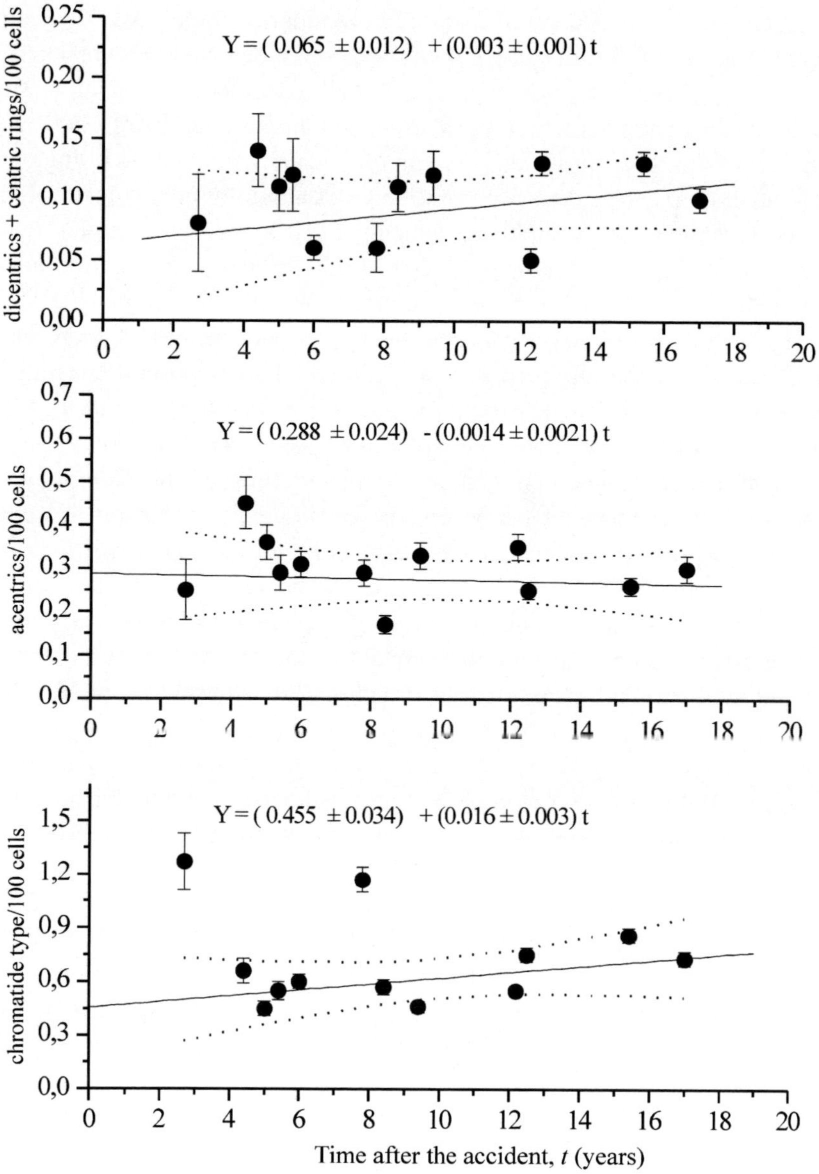

Figure 2. Statistic analysis of dynamics of various aberration types in the cohort of examined patients, two districts combined.

The increased level of chromosomal aberrations in human lymphocytes is usually related to the impact of radiation mutagens on the organism, whereas chromatid aberrations are related to the influence of chemical agents, including medical preparations, as well as to viral infections (hepatitis, erythema centrifugum, flu, etc.). As shown in Table 3, in some years of cytogenetic observations, the examined patients possessed comparatively higher level of chromatid aberrations. Apparently, this was induced by any viral infections: adenoviral, flu epidemics and some prophylactic medical measures, immunizations, for example.

Finally, the last cytogenetic index, which was used in our investigation, is the occurrence of the so-called multiaberrant cells (MAC) or "rogue cells" in blood lymphocytes. Cytogenetically, the MAC phenomenon is demonstrated in the fact that among traditional quantity of metaphases (200 – 300 cells) from an individual a single, less frequently 2 cells containing a great amount of chromosomal aberrations were detected (Figure 1). There were cases, similar to the present one, when practically all genome of the cell is structurally damaged and involved in complex structural rearrangements. Hence, the rest cells either have no chromosomal aberrations, or these cells are singular, usually containing a single aberration.

For the first time, MAC were detected about 35 years ago in lymphocytes of Indians from Latin America possessing 1/200 metaphase frequency [13]. Since the following two years of observation demonstrated a considerable reduction of this frequency (to 1/5,000 metaphases), it was suggested that the reason for MAC occurrence in this cohort had been a tropical infection [14]. However, in the next years the analogous occurrences of MAC in different places all over the world, including Europe, were reported. This important question is not the topic of the present review, because it requires special, more comprehensive analysis and discussion. We would like just to note that MAC have been detected already in the initial cytogenetic tests of various cohorts of patients – the victims of the Chernobyl accident. Therefore, first of all, the occurrence of these cells was related to radiation consequences of the accident [15], most probably, to incorporated effect of "hot" particles [16].

Our tests performed on the cohort of the Chernobyl accident victims also demonstrated the occurrence of MAC in some patients in various groups of population: persons evacuated from the accident zone, residents from contaminated territories, liquidators of the accident. However, we did not succeed unambiguous relating the occurrence of MAC in blood lymphocytes to radiation factors of the accident, including possible effect of incorporated "hot" particles [17]. According to opinions of some authors [18, 19], purely biological etiology of these specific structural damages of chromosomes in human cells may not also be excluded, for example, as a result of direct viral infection impact or activation of latent retroviruses (transposones), contained in the cell genome. Apparently, for the benefit of this hypothesis the results of Zhizdrinsk district residents' examination performed in 2001 may be an indirect confirmation, when the highest average frequency of chromatid aberrations (0.86 per 100 cells) among all years of observation was observed. These aberrations may also be induced by a viral infection, and in the case mentioned were accompanied by the highest MAC frequency (0.54 per 1,000 cells) (Table 3). This is just a suggestion still. By now, there is no a general opinion on the MAC etiology and their biological role on the organism.

Despite the fact that the ^{137}Cs contamination and average dose received by the population in the districts under consideration are twice different (Table 1), neither the spectrum of chromosome aberrations, nor their total frequencies indicated any reliable differences (Table 3). Therefore, to obtain clearer dynamics of cytogenetic disorders for the residents of the contaminated territories and for obtaining broader statistics results from both districts were combined, and dynamics observed was presented in Figure 2. Clearly, the average frequency of dicentrics and centric rings (aberrations – markers of a radiation impact) shows a weak tendency to increase with time, which is much clearer for chromatid aberrations. Hence, the level of acentric aberrations remained constant in all periods of observation.

The elevated level of the aberrations – the markers of radiation impact observed in human populations, the residents of the territories contaminated by the Chernobyl accident, showing no tendency to decrease during 17 years after it, indicates continuous radiation mutagenic environmental impact on these populations. Table 3 shows that ^{137}Cs contamination (200 and 111 kBq/m^2) on the territories of Ul'yanovsk and Zhizdrinsk districts and the average doses–received by the population of these districts in 2001 (11.4 and 6.7 mGy) are not so high. Nevertheless, the average frequencies of dicentrics and cetric rings observed (totally, 0.09 and 0.11 per 100 cells, respectively) in lymphocytes of the examine patients are by 3 and 3.7 times high than the control level, respectively (Table 3).

On the one hand, the observed tendency to increase for the frequency of chromatid aberrations in the examined populations, untypical of radiation mutagenesis, may testify about increasing global anthropogenic, predominantly chemical contamination of the environment. On the other hand, they may also partly be related to the radiation impact, because it is known that low doses of radiation induce genome instability in descendants of irradiated cells, which is also manifested in the form of aberrations, mostly as simple rearrangements – occurrence of cells with micronuclei, acentric fragments of the chromosome and chromatid type, and sister chromatid exchanges [20].

Presently, we have a common opinion that the prenatal period of the organism development is the most sensitive to the radiation impact. It includes a sequence of important consecutive periods of the human organism formation. Therefore, any adverse impact on the embryo and developing fetus, of course, including the radiation impact, may then cause either its death or congenital malformations, formation of tumors in the childhood, etc. Hence, sensitivity to such impacts and their expressivity will be generally defined by the prenatal development stage. The results of prenatal irradiation of survivals in atomic bombings of Hiroshima and Nagasaki indicate the period of 8 – 15 weeks of pregnancy as the most risky for disordering the cerebrum development and formation of subsequent mental deficiency [21, 22]. Though no clear dose dependence of mental deficiency was observed in this case, nevertheless, its occurrence was detected at doses about 1 Gy or higher.

It is known that structural disturbances of chromosomes represent the earliest manifestations of the cell response to the radiation impact and can be accurately calculated. Therefore, of special interest is the comparative estimation of cytogenetic disturbances in people exposed as a result of Chernobyl accident in prenatal and postnatal periods. During the observation period (1989 – 2003), the combined group from both districts included 149 patients exposed to prenatal radiation impact at the moment of the accident and still living on the contaminated territories. They were–separated into three groups with respect to the fetal age at the moment of the accident: 0 – 8, 9 – 28 and 29 – 40 weeks (Table 4).

Table 4. The cohort of prenatally irradiated children living on radiation-contaminated territories of Kaluga Oblast

Fetal age, weeks	Born in the period	Number of examined patients (men and women)	Blood samples taken in the period of 1989–2003	Patient distribution-on the number of examinations			
				1	2	3	4
0 – 8	13 December,1986 – 10 March, 1987	36 (13–23)	61	15	18	2	1
9 – 28	26 July, 1986 – 12 December, 1986	76 (31–43)	139	24	42	9	1
29 – 40	26 April, 1986 – 25 July, 1986	37 (17–20)	98	4	10	18	5
Total period	26 April, 1986 – 10 March, 1987	149 (61–86)	298	43	70	29	7

Table 5 presents the results of comparative analysis of the main cytogenetic indices, determined for the accident consequence effects in prenatal and postnatal periods of development. It is obvious that, in general, differences in the cytogenetic effects between the examined two groups were observed in the total frequency of dicentrics and centric rings only, giving 0.13 and 0.09 per 100 cells, as well as chromatid aberrations – 0.76 and 0.64 per 100 cells, respectively. Comparatively higher expressive cytogenetic effect of the accident consequences was observed for the first and the second periods of prenatal development (0 – 28 weeks), whereas at the third period (29 – 40 weeks) it was reliably lower. The level of aberrations observed in prenatal and postnatal irradiated groups demonstrated reliable excess over the control level (Table 3).

Table 5. The average frequency of chromosomal aberrations per 100 cells in children and teenagers, prenatally and postnatally irradiated in the Chernobyl accident

Parameters		Period of development of irradiated patients, weeks				
		0 – 8	9 – 28	29 – 40	prenatal	postnatal
Number of patients		36	76	37	149	591
Number of blood samples		61	139	98	298	962
Number of cells		24,353	62,340	43,771	129,964	273,230
Number of aberrant cells*, %		0.46 ± 0.03	0.43 ± 0.02	0.35 ± 0.02	0.41 ± 0.01	0.39 ± 0.01
Chromosome type	Acentrics	0.29 ± 0.03	0.30 + 0.02	0.26 ± 0.02	0.28 ± 0.01	0.30 ± 0.01
	Dicentrics and centric rings	0.14 ± 0.03	0.14 ± 0.02	0.10 ± 0.02	0.13 ± 0.01	0.09 ± 0.01
	Total**	0.45 ± 0.04	0.44 ± 0.03	0.36 ± 0.03	0.41 ± 0.02	0.38 ± 0.01
	Anomalous monocentrics	0.03 ± 0.01	0.03 ± 0.01	0.03 ± 0.01	0.03 ± 0.01	0.03 ± 0.01
Chromatid type	Deletions and isodeletions	0.57 ± 0.05	0.64 ± 0.03	0.74 ± 0.04	0.66 ± 0.02	0.56 ± 0.01
	Exchanges	0.12 ± 0.02	0.10 ± 0.01	0.06 ± 0.01	0.09 ± 0.01	0.08 ± 0.01
	Total	0.69 ± 0.05	0.74 ± 0.03	0.80 ± 0.04	0.76 ± 0.02	0.64 ± 0.02

* - with aberrations of the chromosome type;

** - with no consideration of anomalous monocentrics

The level of cytogenetic damages at prenatal irradiation of 42 children due to the Chernobyl accident, exceeding the control in the late period, was also found in ref. [23]. In addition, the authors applied a differential method of chromosome G-staining, which showed significantly higher yield of stable aberrations (translocations, para- and pericentric inversions, terminal and interstitial deletions, and insertions) compared with the yield of unstable aberrations.

In this study, we have also applied the FISH-method for chromosomes 2 and 3, which form 14.9% of the cell genome, to the analysis of stable chromosomal aberrations in a group of the patients (21 persons). Unfortunately, none of prenatally irradiated during the accident was included in this group. Table 6 shows that 66.7% of cases e.g. 2/3 part of the examined patients (14 of 21 persons) demonstrated any stable aberration in lymphocytes. The average frequency of translocations in this group reliably exceeded the control level: 0.37 and 0.11 per genome/100 cells, respectively. Since these patients were also examined at routine staining of unstable aberration chromosomes, it would be of interest to compare frequencies of dicentrics and translocations, observed for the same patient. Theoretically, the expected ratio of the frequencies at the radiation impact should be 1:1.

Table 6. The frequency of stable chromosomal aberrations in children and teenagers living on the contaminated territories of Ul'yanovsk district (examined in 1998)

Exami-nant No.	Number of cells	Number of aberrant cells	Number of aberrations				Translocation frequency per genome/ 100 cells	Frequency of dicentrics per 100 cells*
			Trans-locations complete	Trans-locatio ns incomp lete	Deleted chromo-somes	Insertions		
1	1,500	0	0	0	0	0	0	0
2	1,500	0	0	0	0	0	0	0
3	1,500	3	1	1	1	0	0.59±0.24	0
4	700	1	0	1	0	0	0.63±0.36	0.02±0.02
5	600	1	0	1	0	0	0.74±0.42	0
6	1,500	1	0	0	1	0	0	0
7	1,245	0	0	0	0	0	0	0
8	1,000	2	0	2	0	0	0.89±0.36	0
9	353	1	0	0	1	0	0	0
10	616	5	1	3	1	0	2.89±0.71	0.02±0.02
11	1,500	1	0	0	1	0	0	0
12	1,500	0	0	0	0	0	0	0.40±0.28
13	1,500	1	0	1	0	0	0.30±0.17	0
14	1,500	2	1	1	0	0	0.59±0.17	0
15	1,500	2	0	0	1	1	0	0
16	566	1	0	0	1	0	0	0
17	1,076	0	0	0	0	0	0	0
18	570	0	0	0	0	0	0	0
19	464	0	0	0	0	0	0	0
20	684	1	0	1	0	1	0.65±0.37	0
21	1,500	3	2	0	1	0	0.59±0.24	0
Averaged in group							0.37±0.04	0.02±0.01
Control							0.11±0.05	0.02±0.01

* - the results were obtained for the standard routine chromosome staining.

Table 6 shows that only 14.3% of cases (3 of 21 patients) possessed dicentrics, whereas translocations were observed in 42.9% (9 of 21 patients), i.e. 3 time more frequent. This can be explained in the only way. Since renovation of the lymphoid cell population of peripheral blood proceeds with the death of both normal and aberrant lymphocytes; hence, elimination of cells with stable and unstable aberrations is equiprobable. Then the observed increased level of translocations is related to their delivery to peripheral blood from irradiated stem cells. These results testify about the perspective of the study of stable chromosomal aberrations, first of all, in the Chernobyl cohort in late periods. These aberrations are preserved in the organism for a long time and represent the potential source of malignant degeneration of cells.

As mentioned above, the examined cohort included three groups of patients: born before the accident (before 26 April, 1986), prenatally irradiated (in the period of 26 April, 1986 – 10 March, 1987) and born after the accident (after 10 March, 1987). Thus, the latter group really consisted of persons born almost a year after the accident and, consequently, escaped from additional radiation by radioactive iodine. Therefore, it becomes possible to estimate the relative contribution of additional radiation by radioactive iodine, which might effect on the persons from the first two groups, into the observed cytogenetic effect.

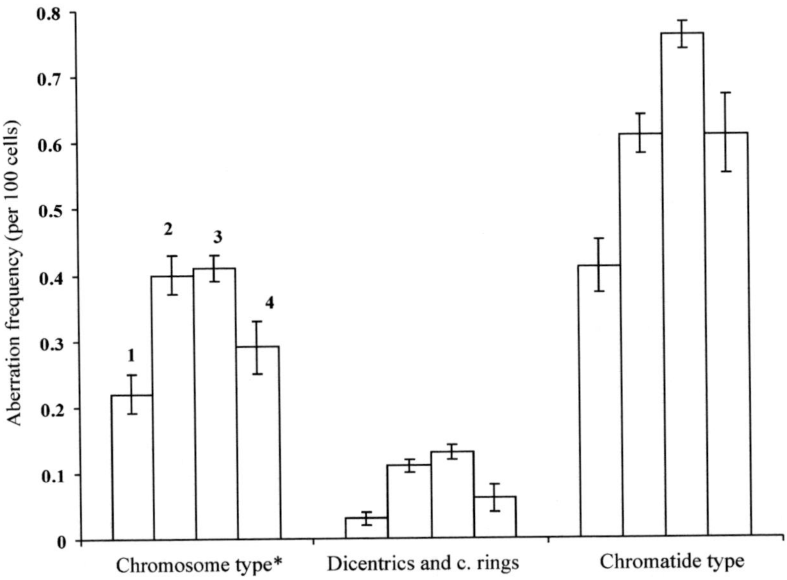

Figure 3. The frequency of chromosome aberrations in groups of children with different radiation scenarios: * – without consideration of anomalous monocentrics; 1 – control group; 2 – children born before the accident (aged 0 – 1); 3 – prenatally irradiated children; 4 – children born 1 year after the accident (aged 0 – 1)

Figure 3 presents the results for the main cytogenetic indices in children born before the accident (aged 0 – 1 by the accident moment), prenatally irradiated and born after the accident, i.e. after 10 March, 1987 (also aged 0 – 1), as well as the control group. It is obvious that both total frequency of chromosome aberrations and total frequency of dicentrics and centric rings in the groups of children born before the accident and prenatally irradiated ones reliably exceed these parameters for children born almost a year after the accident. This allows a conclusion that higher level of cytogenetic damages, observed in the first two groups of children, is related to an additional effect of radioactive iodine, whereas comparatively lower frequency of aberrations in the third group of children is related exclusively to the radiation impact, induced by living on the contaminated territories. The chromatid type of aberrations was found reliably increased only for prenatally irradiated persons compared with the mentioned two groups.

To conclude the discussion, we would like to note the following. As determined by epidemiological investigations, in the post-accident period the population of Belarus, Ukraine and Russia living on the contaminated territories indicates steady increase of general morbidity. The increase of thyroid gland diseases is the most high. Data from the literature and the results of authors' investigations indicate the increased level of cytogenetic disturbances in somatic cells of these populations of people. There is no doubt that the main reason for development of both processes is related to consequences of the Chernobyl accident and, first of all, to chronic impact of low doses of radiation during living on the contaminated territories. Though no direct proofs of influence of increased mutability of somatic cells on people's health have been found yet, there are more and more data on a definite contribution of somatic mutations in development of some diseases, such as diabetes, heart diseases, atherosclerosis, heaves, cancer diseases, etc.

As already mentioned, we now have quite clear proofs for the reliable correlation between the frequency of chromosomal aberrations and the risk of cancer formation. Stable aberrations, translocations, in particular, long-living in the organism, are specifically potentially dangerous as possible inductors of the cell malignant degeneration. The above results of our investigation indicate their significant prevalence in the observed spectrum of aberrations in persons living on the contaminated territories. That is why, implementation of regular tests of the cytogenetic disturbance level in somatic cells of human populations living on the contaminated territories are of so high importance. These investigations may be a prognostic test for early detection of patients with increased risk of cancer formation.

ACKNOWLEDGMENTS

This investigation was partly supported by RFBR grant No. 01-04-96002.

REFERENCES

[1] Sorsa M., Wilbourn J., and Vainio H., *Mechanisms of Carcinogenesis in Risk Identification*, Lyon, IARS, 1992, pp. 543 - 554.

[2] Turusov V.S., *Carcinogenesis*, Moscow, Nauchnyi mir, 2000, pp. 251 – 259. (in Russian)

[3] Wild C.P., Law G.R., and Roman E., *Mutat. Res.*, 2002, vol. 499(1), pp. 3 - 12.

[4] Hagmar L., Brogger A., Hansteen J.L. *et al.*, *Cancer Res.*, 1994, vol. 54, pp. 2919 - 2922.

[5] Hagmar L., Bonassi S., Stromberg U. *et al.*, *Cancer Res.*, 1998, vol. 58, pp. 4117 - 4121.

[6] Fleishman E.V., *Carcinogenesis*, Moscow, Nauchnyi mir, 2000, pp. 342 – 360. (in Russian)
[7] Bond V.P., *Stem Cells*, 1995, vol. 13(Suppl. 1), pp. 21 - 29.
[8] Nakano M., Kodama Y., Ohtaki K. *et al.*, *Int. J. Radiat. Biol.*, 2001, vol. 77(9), pp. 971 - 977.
[9] Lucas J.N., Awa A.A., Straume T. *et al.*, *Int. J. Radiat. Biol.*, 1992, vol. 62, pp. 53 - 63.
[10] Tucker J.D., Morgan W.F., Awa A.A. *et al.*, *Mutat. Res.*, 1995, vol. 347, pp. 21 - 24.
[11] Vlasov O.K., God'ko A.M., Shishkanov N.G. *et al.*, *The Chernobyl Hangover*, 2001, Iss. 3, Kaluga, Oblizdat, pp. 30 – 38. (in Russian)
[12] Sevan'kaev A.V., Kozlov V.M., Guzeev G.G. *et al.*, *Genetika*, 1974, vol. 10(6), pp. 114 – 120. (Rus)
[13] Bloom A.D., Neel J.V., Choi K.W. *et al.*, *Proc. Nat. Acad. Sci. USA*, 1970, vol. 66(3), pp. 920 - 927.
[14] Bloom A.D., Neel J.V., Tsuchimoto T. *et al.*, *Cytogen. Cell Genet.*, 1973, vol. 12, pp. 175 - 186.
[15] Pilinskaya M.A., Shemetun A.M., Dybskyi S.S. *et al.*, *Tsitologia i Genetika*, 1994, vol. 28(1), pp. 27 – 32. (Rus)
[16] Domracheva E.V., Rivkind N.B., Shklovsky-Kordi I.E. *et al.*, *Problemy Gematologii i Perelivaniya Krovi*, 1997, No. 2, p. 12. (in Russian)
[17] Sevankaev A.V., Tsyb A.F., Lloyd D.C. *et al.*, *Int. J. Radiat. Biol.*, 1993, vol. 63(3), pp. 361 - 367.
[18] Neel J.V., Awa A.A., Kodama Y. *et al.*, *Proc. Nat. Acad. Sci. USA*, 1992, vol. 89(15), pp. 6973 - 6977.
[19] Lazutka J.R., *Mutat. Res.*, 1996, vol. 350(2), pp. 315 - 329.
[20] Pelevina I.I., Aleshchenko A.V., Antoshchina M.M. *et al.*, *Radiats. Biol. Radioekol.*, 2003, vol. 43(2), pp. 161 – 166. (in Russian)
[21] Pochin E.E., *Radiat. Prot. Bull.*, 1988, No. 95, pp. 10 - 12.
[22] Mole R.H., *Int. J. Radiat. Biol.*, 1990, vol. 57(4), pp. 647 - 663.
[23] Stepanova E.I., Misharina Zh.A., and Vdovenko V.Yu., *Radiats. Biol. Radioekol.*, 2002, vol. 42(6), pp. 700 – 703. (in Russian)

In: 20 Years after the Chernobyl Accident
Editors: E. Burlakova and V. Naidich, pp. 39-54

ISBN 1-60021-249-2

Chapter 4

THE PROBLEM OF INDUCED GENOME INSTABILITY AS THE BASIS FOR ELEVATED MORBIDITY AMONG CHILDREN EXPOSED BY LOW-RATE RADIATION AT SMALL DOSES

I. I. Suskov[1]*, N. S. Kuz'mina[1], L. S. Baleva[2] and A. E. Sipyagina[2]

[1] N.I. Vavilov Institute of General Genetics, Russian Academy of Sciences, Moscow, Russia

[2] Federal Children's Scientific-Practical Center of Radiation Safety, Moscow, Russia

INTRODUCTION

Main results of the complex study of genome instability in children inhabiting on the territories, contaminated by radionuclides due to the Cernobyl accident (Novozybkov district of Bryansk region, 16 – 18 Ci/km^2 by ^{137}Cs), and exposed to low-rate irradiation at different stages of ontogenetic development: postnatally irradiated in 1986 (born before the accident), prenatally irradiated during the accident in 1986, born after the accident in 1987 – 1992 and 1994 – 2000 from irradiated parents. In all studied groups of irradiated children elevated frequencies of any radiation-induced chromosome aberrations were observed, as well as reduced activity of genome DNA unscheduled synthesis in lymphocytes, and peculiarities in individual heterozygosity of genes encoding structural and enzymatic blood proteins. Children living on the radionuclide-contaminated territories demonstrated increased radiosensitivity of lymphocyte genomes to testing irradiation *in vitro* and the peculiarities of chromosome aberration and sister chromatid exchange frequency dynamics in three generations of cells. The data obtained suggest systemic character of dysgenomic effect, the

* Main addressee: N.I. Vavilov IGG, RAS, 119991 Moscow, Russia; Tel: (7-095) 135–7172; Fax: (7–095) 132–8962; E-mail: suskov@vigg.ru.

reality of genome instability induction in child's growing organism exposed by low-rate radiation at small doses, the expression which is defined by individual genotypic characteristics of the organism. Biological significance of the phenomenon of postradiation genome instability, its relation to the health state and pathogenetic role in development of somatic pathology are postulated.

Already 20 years the immense cohorts of the population of Russia, Ukraine and Belarus are living on the territories, contaminated by radionuclides from the Chernobyl accident. Only the Russian Federation has 14 regions (Belgorod, Bryansk, Voronezh, Kaluga, Kursk, Leningrad, Lipetsk, Orel, Ryazan', Tambov, Tula, Penza, Mordovia Republic), 114 districts of which have increased radiation background, and the population of which exceeds one and half million adults and children. It should also be taken into account that hundred thousands people are living in Urals (Chelyabinsk region) and Siberia (Tomsk region, Altai Krai), on the territories contaminated by radionuclides and chemical reagents [1, 2].

Recently, beside the increased frequency of chromosome aberrations [3 – 6, etc.], an increase of total somatic morbidity, nervous-psychic deadaptation, a tendency for diseases to become chronic, increasing frequency of immunodeficiency state, anematization, etc. development were observed, which is mostly typical of children, whose organisms are higher sensitive to radiation [7 – 11]. It has been calculated that possible genetic consequences of the Chernobyl catastrophe (hereditary diseases, birth defect, spontaneous abortions, stillbirths) will not exceed 1% of spontaneous level [12], but increased levels of chromosome aberrations in the human organism may induce the above-listed states and precede the development of chronic fatigue, secondary immunodeficiency, premature aging, reproductive dysfunctions and neoplastic aberration development [8, 13, 14].

The classical radiobiology and radiation genetics methods have fixed two phenomena as follows:

1. nonthreshold character of radiation doses – any minimal dose, as hitting a cell, causes DNA molecule breaks in it and, consequently, gene mutations and chromosome aberrations;
2. biological intensification of induced gene/chromosome mutations as a consequence of reduplication in proliferating cells of the organism [15, 16].

In recent decade, significant data indicating the remote effects of radiation – induced genome instability appeared as various disorders of the genetic apparatus in descendants of multiply divided cells, which had been exposed to moderate and high dose radiation both *in vitro* and *in vivo* – were published [17 – 20]. Analysis of the world literature data indicates noncorrespondence of induced genome instability regularities to the classical conception of the radiation genetics (target theory and hit principle) [21].

In this connection the question of estimating the reality of genome instability induction in the child organism exposed to low-intensity radiation at small doses appears extremely urgent. This review presents the main results of the study of this new genetic phenomenon in children exposed to low-rate irradiation at various stages of ontogenetic development and discusses possible biological significance of post-radiation genome instability and its relation to the health state.

Materials and Methods

A group of 90 children inhabiting on radioactive contaminated territories (Novozybkov district of Bryansk region, 16 – 18 Ci/km^2 by ^{137}Cs) and exposed to low-rate irradiation at different stages of ontogenetic development: postnatally irradiated in 1986 (born before the accident, 32 persons), prenatally irradiated during the accident in 1986 (15 persons), born after the accident in 1987 – 1992 (19 persons) and in 1994 – 2000 (24 persons) from irradiated parents. A sub-group with high dose load on thyroid gland (200 rem or higher) was separated from the group of children born before the accident. The periods of children's living on radionuclide-contaminated territories varied from 1.5 to 16 years. A group of 15 children lived on radiation contaminated territories: prenatally irradiated during the accident in 1986 (5 persons), born after the accident in 1987 – 1991 (5 persons) and in 1994 – 1998 (5 persons) from parents irradiated in the age of adolescence, were examined for genome instability expression in 3 consecutive cell generations with the use of testing irradiation of blood lymphocytes *in vitro* (^{137}Cs, 476 cGy/min dose rate). The control group was formed of children from clean regions (16 persons). The group for examination was formed by the patients of Federal Children's Scientific-Practical Center of Radiation Safety possessing various disorders of the health state: central nervous system, digestive system, respiratory system, urinary excretion system, endocrine system, etc. Control group was formed by children with similar pathology.

Cultivation of peripheral blood lymphocytes, preparation of slides of metaphase chromosomes and analysis of chromosome aberrations were performed according to common technique pursuant to WHO and IAEA recommendations [22, 23]. Identification of the cell generation number and SCD frequency calculation were performed by sister chromatid differential painting technique with the use of 5-bromodeoxyuridine (5-BDU) [24]. The repair activity of genome DNA in lymphocytes was estimated by the level of unscheduled synthesis using ^{3}H-thymidine. Pheno/genotypes of 6 monomorphous (Alb, Hbα. Hbβ, EsA, EsB, PGM_2) and 7 polymorphous (Hp, Tf, Gc, EsD, PGM_1, AcP, 6-PGD) blood proteins/enzymes, encoded by 13 loci genes, were determined by electrophoregrams in histochemically painted polyacrylamide and starch gels [25].

Results and Discussion

Dysgenomic Effects in Children's Lymphocytes of the Examined Group

The results of cytogenetic examination indicate increased average levels of any radiation-induced chromosome aberrations in all above-listed groups of children living on the radiation-contaminated territories. The average frequencies of aberrant genomes in these children are 1.5 – 2.8 times higher than in the control group (Table 1) [26]. Irradiated children demonstrated high interindividual variability of cytogenetic disorders. Clearly, the highest frequencies of chromosome aberrations and aberrant genomes were detected in children with highly dose loaded thyroid gland that correlates with the results of works of other investigators [4].

Table 1. Frequencies and types of structural chromosome aberrations in genomes of lymphocytes of children from examined groups

Examined children			Number of analyzed metaphases	Aberrant metaphases (genomes), %	Chromatid type aberrations, frequency per 100 cell genomes	Chromosome type aberrations, frequency per 100 cell genomes	
						Paired fragments+ centromeric breaks+ deletions	Dicentrics + rings + reciprocal translocations + inversions
Children living at radionuclide-contaminated territories	Born before the accident	With recorded dose for thyroid gland (13 pers.)	3672	3.21±0.37 *** (1.33 ÷ 6.0)	1.87 ±0.30 (0.33 ÷ 3.67)	0.87±0.20*** (0 ÷ 2.22)	0.57±0.13 *** (0 ÷1.33)
		With unrecorded dose for thyroid gland (19 pers.)	5440	1.73±0.24 * (0.33 ÷ 4.0)	1.14 ±0.21 (0.33 ÷ 3.57)	0.40±0.09** (0 ÷ 1)	0.21±0.06* (0 ÷ 0.67)
	Born after the accident	Prenatally irradiated in 1986 (15 pers.)	4385	2.21±0.22** (0.5 ÷ 3.6)	1.57±0.19 * (0.5 ÷ 3.33)	0.39±0.09** (0 ÷ 1.2)	0.25±0.07* (0 ÷ 0.83)
		Born in 1987 - 1992 (19 pers.)	5281	2.44±0.33 *** (1.0 ÷ 5.91)	1.55±0.31 (0 ÷ 5.5)	0.64±0.11*** (0 ÷ 2.0)	0.34±0.14* (0 ÷ 2.5)
		Born in 1994 - 2000 (24 pers.)	6271	1.66±0.21 (0.5 ÷ 4.0)	1.20±0.19 (0 ÷ 2.8)	0.33±0.07 ** (0 ÷ 1.5)	0.18±0.10 (0 ÷ 2.0)
Children born from fathers – liquidators (43 pers.)		10251	2.61±0.26 *** (0.33 ÷ 7.5)	1.86±0.21* (0 ÷ 6.5)	0.50±0.08 *** (0 ÷ 1.62)	0.27±0.07** (0 ÷ 1.62)	Children born from fathers – liquidators (43 pers.)
Control (16 pers.)		4160	1.13±0.14 (0.33÷ 2.0)	1.08±0.14 (0.33÷2.0)	0.05±0.04 (0÷0.5)	0	Control (16 pers.)

No significant difference in the frequencies of aberrant genomes between groups of children living on the radiation-contaminated territories was observed, though calculated average doses of external and internal technogenic irradiation by long-living radionuclides, accumulated by these children, were found different (Table 1, Figure 1). Analogous results were also obtained at comparative analysis of the frequencies of other cytogenetic parameters (Table 1). The data obtained may testify about either individual radiosensitivity or radiobiological features of long-term low-rate action of radiation at small doses on the child's organism. Apparently, beside direct mutagenic effects of radiation, a definite part of examined children, born after the accident from irradiated parents, demonstrates a transgeneration phenomenon of somatic cell genome instability. Moreover, the parents of children born in 1994 – 2000 were usually exposed to radiation in the teen (adolescent) age, which is considered as one of the most sensitive periods of ontogenesis [27]. A reliable correlation between aberrant genome frequency in children born in 1997 – 2000 and their mothers was observed (r = 0.41, p = 0.028). Reality of the radiation-induced genome instability phenomenon in the child's organism is confirmed by the data testifying about elevated frequencies of dysgenomic effects in children born from irradiated fathers – the liquidators of the Chernobyl accident, now living in clean regions (Table 1) [28], which correlates with the results of other works [29, 30].

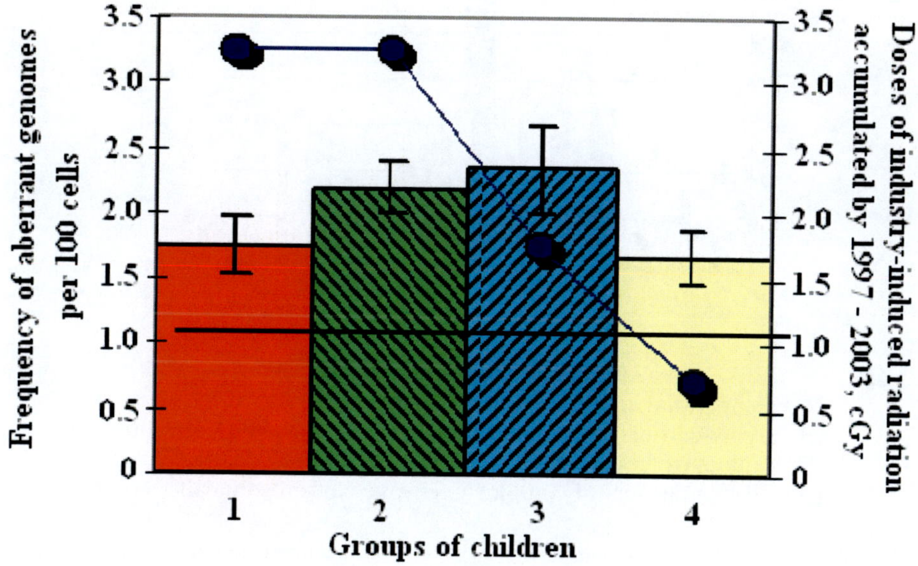

—•— – the frequency of aberrant genomes in children of the control group

Figure 1. Dose loads (○) and aberrant genome frequencies (□) in children living on the radionuclide-contaminated territories: 1 – children born in the period of 1982 – the beginning of 1986; 2 – children born in 1986 (prenatal irradiation); 3 – children born in 1987 – 1992; 4 – children born in 1994 – 2002

The increased levels of chromosome aberrations, detected in lymphocytes of peripheral blood of children living on the radionuclide-contaminated territories, may be explained by three reasons:

- as a result of a radiation quantum/an ionizing particle (or radical) hitting genome (nucleus) of a lymphocyte circulating in peripheral blood at G_0 stage;
- as a result of "multiplication" of a definite part of aberrant cells and their delivery to peripheral blood from irradiated precursors;
- as a result of chromosome/genome instability induction in descendants of multiply divided irradiated cells of growing organism – the initial cell-zygote and consequent stem cells.

Peculiarity of the situation is that the examined cohorts of children living under the conditions of long-term action of low dose radiation and exposed to it at various stages of ontogenetic development demonstrate any combinations of all the above-listed types of chromosomal disorders, typical of both direct mutagenic effect of radiation and after-radiation induction of genome instability [21].

The majority of children–carriers of the radiation-induced chromosome aberrations have a reduced repair activity of genome DNA either after γ- and UV-irradiation; however, disorders induced by γ-irradiation are predominant [26].

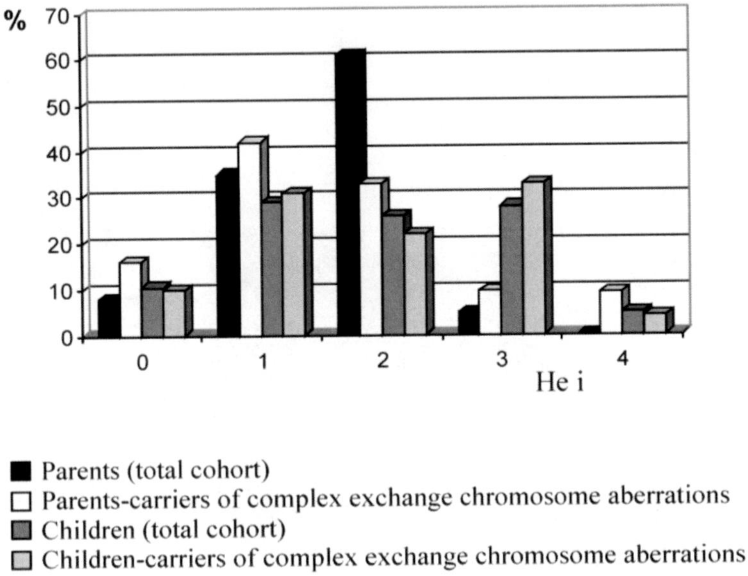

Figure 2. Distribution of children and their parents living on radionuclide-contaminated territories of Bryansk region by individual heterozygosity : 0 – homozygotes; 1 – monoheterozygotes; 2 – diheterozygotes; 3 – triheterozygotes; 4 – tetraheterozygotes.

Apparently, high inter-individual variability of dysgenomic effect frequencies is stipulated by genotypic features of the organism. Though radiation-induced mutant alleles encoding the studied monomorphous (**Alb, Hbα. Hbβ, EsA, EsB, PGM_2**) and polymorphous (Hp, Tf, Gc, EsD, PGM_1, AcP, 6-PGD) blood proteins/enzymes were not determined, peculiarities in distribution by the levels of individual heterozygosity of children and their parents were determined (Figure 2). If the group of parents – carriers of complex exchange chromosome aberrations demonstrates abruptly decreased the part of diheterozygous genotypes compared with that in the general group, for children the part of genotypes with the average individual heterozygosity is reduced in both cohorts. Moreover, for children the part of polyheterozygous genotypes significantly exceeds this parameter observed for the parents. On the one hand, the data obtained suggest the significance of genotypic background of an organism in the induction of chromosome aberrations; on the other hand, they indicate that the level of gene combinatorics effects on the biological resistance of an organism, because at the moment of the examination all children had disorders in the health state. The data obtained agree with the concept of an optimal level of population heterozygosity [31].

Table 2. Frequencies of exchange aberration of the chromosomal type in the first mitosis metaphases of intact and irradiated in vitro lymphocytes of children from examined groups

Examined children	Dose, cGy	Exchange chromosome aberrations (frequency per 100 cell)
Control (5 persons)	0	0
	10	0.33 ± 0.09
	100	8.35 ± 0.39
Born in 1986 (prenatally irradiated) (5 persons)	0	0.32 ± 0.09
	10	0.83 ± 0.14
	100	15.36 ± 1.39
Born in 1987 – 1991 (5 persons)	0	0.50 ± 0.17
	10	1.16 ± 0.25
	100	17.19 ± 2.07
Born in 1994 – 1998 (5 persons)	0	0.16 ± 0.16
	10	0.82 ± 0.33
	100	14.18 ± 2.46

Genome Instability Expression in Vitro in Lymphocytes of Examined Children

In children living on radionuclide-contaminated territories either prenatally irradiated or born after the accident from irradiated parents, the frequency of exchange chromosome aberrations after testing irradiation *in vitro* by 10 and 100 cGy are reliably higher than in the control group (Table 2). This indicates an increased radiosensitivity of somatic cell genomes of children exposed to a long-term, low-rate and low-dose radiation *in vivo* and is one of the characteristics of genome instability. High inter-individual variability of the lymphocyte sensitivity to testing irradiation *in vitro*, found higher for the children born after the accident from irradiated parents (Figure 3), was determined. A positive correlation between the level of exchange chromosome aberrations in intact lymphocytes (irradiated *in vivo*) and their sensitivity to irradiation *in vitro* ($r = 0.71$ and $r = 0.54$ for irradiation *in vitro at* 10 and 100 cGy doses, respectively; $p < 0.01$). Other authors have also determined the increased sensitivity of chromosomes to testing irradiation *in vitro* for children born from the parents exposed to radiation impact (the Chernobyl accident liquidators and parents exposed to therapeutic impact of ionizing radiation), as well as children evacuated from the zones with adverse environmental conditions induced by the Chernobyl accident [29]. The examinations of children and adults exposed to chronic action of radiation at small doses (induced by the Chernobyl accident and radionuclide contamination of Techa River), performed by other investigators, indicated decreasing ability to adaptive response; hence, increased radiosensitivity after exposure to adaptive irradiation was observed more frequently than in the control group [19, 32, 33].

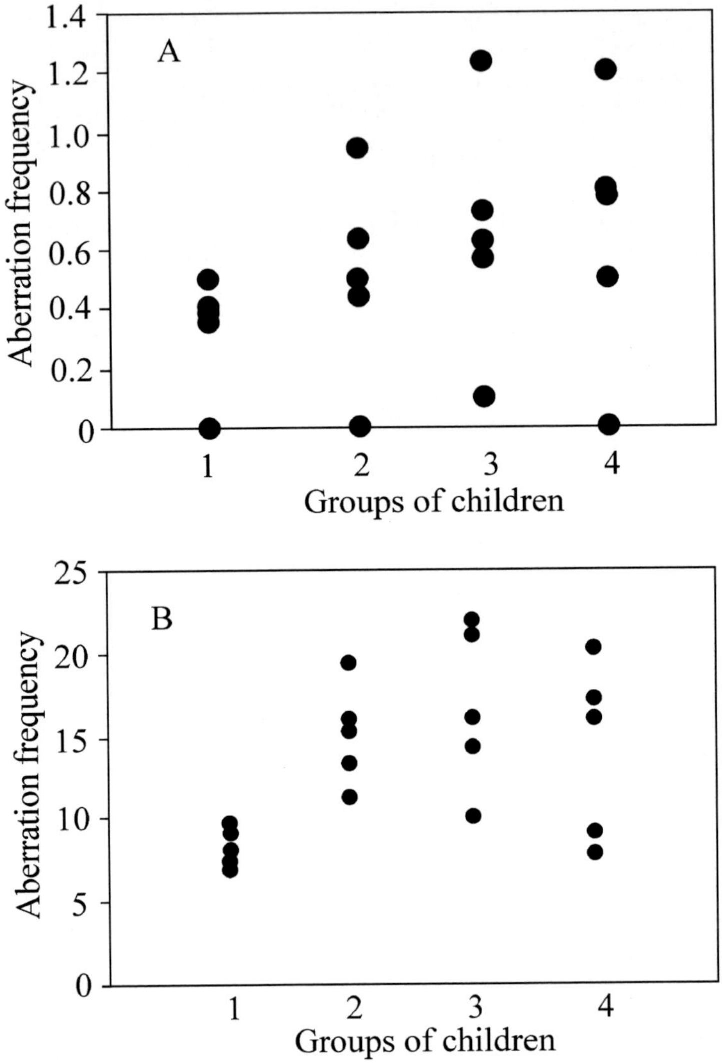

Figure 3. Individual frequencies of exchange aberrations of the chromosomal type induced by testing irradiation of lymphocytes *in vitro* by 10 cGy (A) and 100 cGy (B) doses in examined children: 1 – control group; 2 – prenatally irradiated children; 3 – irradiated children born in 1987 – 1991; 4 – irradiated children born in 1994 – 1998.

Compared with the control group, the advanced elevation of paired fragments + centromeric breaks frequencies in 3 consecutive mitoses of lymphocytes, intact and irradiated *in vitro* by 10 cGy dose, of children exposed to low rate irradiation also indicates the induced after-radiation chromosome instability, which accompanies reduplication of chromosome aberrations and their preservation on subsequent cell generations (Figure 4) [34]. For example, the rate of elevation frequency of these cytogenetic disorders in intact lymphocytes, estimated by the linear regression coefficients, equaled 0.30, 0.66 and 0.56 per 100 genomes for 1 cell generation in children prenatally irradiated in 1986, children born in 1987 – 1991 and in 1994 – 1998, respectively. This appeared 2 – 4 times higher compared with the group from clean regions (0.16 per 100 genomes). For lymphocytes irradiated *in vitro* by 10 cGy dose, this parameter equaled 2.50, 1.74 and 1.16 per 100 genomes for 1 cell generation in children prenatally irradiated in 1986, children born in 1987 – 1991 and in 1994 – 1998, respectively. This appeared 4 – 9 times higher compared with the group from clean regions (0.28 per 100 genomes). Apparently, a sharp elevation of frequencies of these dysgenomic effects in the third mitosis of lymphocytes irradiated *in vitro* by 10 cGy dose [34] may be explained by existence of superlong living local potential DNA changes, already described in 1960s, able to pass "shared" through several DNA syntheses before they are realized in chromosome aberrations [35]. In this case, higher expression of genome instability was provided by an additional testing irradiation of cells *in vitro* that resulted in potentiation of primary genome damages, induced by a long-term action of low-dose low-rate irradiation.

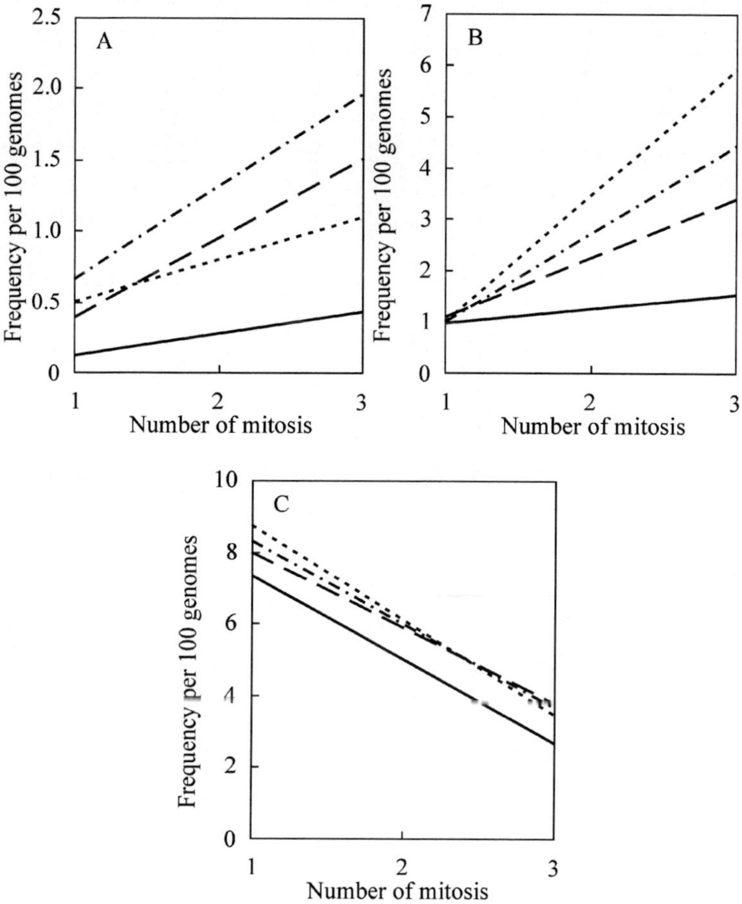

Figure 4. Regression curves of paired fragments + centromeric breaks frequency dynamics in lymphocyte genomes of children from examined groups; A – intact lymphocytes; B – lymphocytes irradiated *in vitro* by 10 cGy dose; C – lymphocytes irradiated *in vitro* by 100 cGy dose; —- control group; —·—·—- children born in 1987 – 1991; - - - - - children irradiated prenatally in 1986; ———- children born in 1994 – 1998.

The different directions of time-effect curves for the frequencies of these dysgenomic effects and aberrant genomes after lymphocyte irradiation *in vitro* by 10 and 100 cGy doses confirms the differential significance of the radiation damage of cells by low and medium doses. Contrary to superlow and low doses, medium and high doses of radiation induce significant genomic and cytoplasmic damages that reduce viability of the cells and, consequently, their elimination in the sequence of cellular cycles. However, either low (10 cGy) or medium (100 cGy) doses induced elimination in consecutive cellular cycles of complex unstable chromosome aberrations (dicentrics + rings) with preservation of complex stable aberrations (inversions + translocations) at the same level [34].

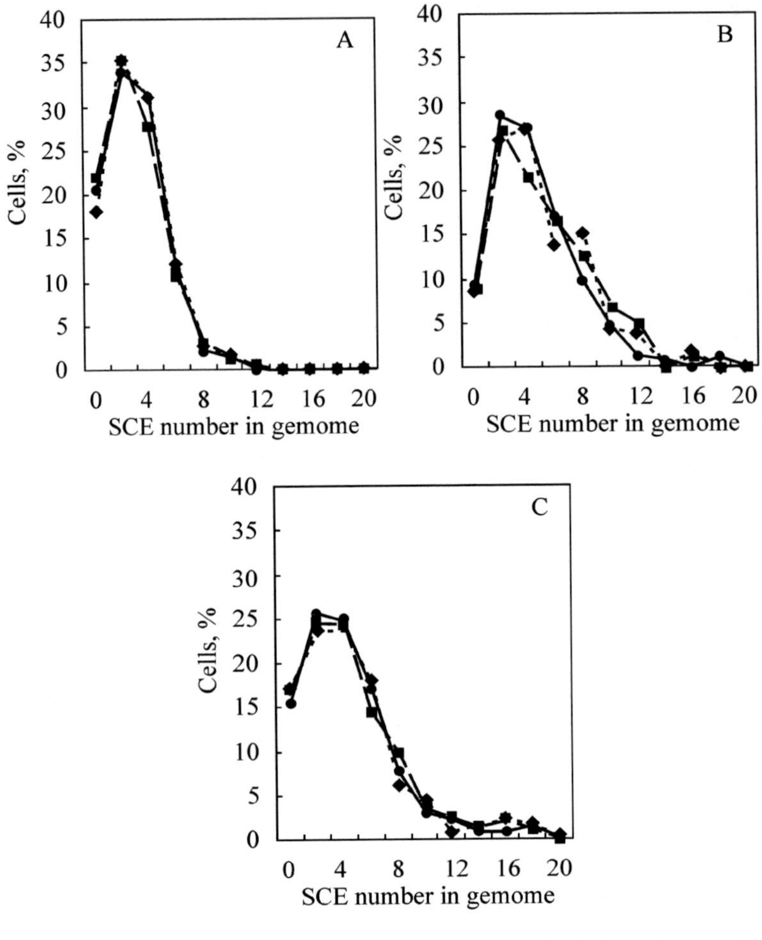

Figure 5. Distribution of intact and irradiated *in vitro* the third mitosis lymphocytes by the number of reciprocal SCE in children of the examined groups; A – control group; B – children prenatally irradiated in 1986; C – children born in 1987 – 1991; —●— - intact lymphocytes; —◆— - lymphocytes irradiated *in vitro* by 10 cGy dose; ——■—— - lymphocytes irradiated *in vitro* by 100 cGy dose.

In the examined children, the expression of genome instability was also studied using the sister chromatid exchange (SCE) test. For the children prenatally irradiated in 1986 during the accident and the children born after the accident in 1987 – 1991, the SCE frequency in intact lymphocytes of the second mitosis (the first and the second cellular cycle exchanges) equaled (6.49 ± 0.17) and (6.42 ± 0.19) per genome, respectively, which closely approached the control value ((6.23 ± 0.19) per genome). These data correlate well with the results obtained by other investigators [5]. No significant differences in the average frequencies and distributions by SCE cells in genomes of the second mitosis lymphocytes, both intact and irradiated *in vitro* by 10 and 100 cGy dose for either children from the control group or the children living on the radionuclide-contaminated territories ($p > 0.05$) were determined. These results correlate well with the results, previously obtained at the cell irradiation at G_0 stage [36].

However, for the children living on radionuclide-contaminated territories, the average reciprocal SCE frequencies determined in metaphases of the 3rd mitosis of lymphocytes, both intact and irradiated *in vitro* (4.3 – 4.9 per genome), reliably exceed these of the control group (2.9 – 3.0 per genome), which is due to higher number of cells possessing SCE equal 8 or higher. Hence, no significant differences in the average frequencies and distributions by SCE cells in the 3rd mitosis lymphocytes both intact and irradiated *in vitro* by 10 and 100 cGy doses either in the children of the control group or the children living on radionuclide-contaminated territories ($p > 0.05$) (Figure 5). The results obtained testify about DNA – molecular genome instability induction in the descendants of divided cells that may be a result of realization of long-living genomic damages in lymphopoietic cells – precursors of adult lymphocytes, induced by chronic low-dose radiation impact. The occurrence of genomes with high SCE amount in generations 6 – 16 of cells exposed to 0.5 Gy γ-irradiation was determined [37], which was not observed in non-irradiated populations.

CONCLUSION

Thus, the data obtained testify about systemic character of dysgenomic effects, reality of genome instability induction in somatic cells of children exposed to low-rate radiation at small doses, the expression of which is determined by individual genotypic peculiarities of the organism. Cytogenetic efficiency of the low rate action of the industrial radiation on the child's organism per specific dose (1 cGy) is much higher than after short-term irradiation of lymphocytes *in vitro* by medium/high doses.

First of all, this relates to the levels of cytogenetic disorders in children born after the accident from irradiated parents and in children exposed to prenatal irradiation during the accident in 1986 (Table 3) [38]. With respect to the above-said, this fact may be explained as follows. Under long-term impact of radiation at small doses, the spontaneous damages in matrix DNA derivatives occurring in the norm during molecular-genetic processes (replications, transcriptions, translations, segregations of daughter structures) in proliferating cells are accompanied by the radiation-induced damages. Unrepaired damages are either realized in DNA/chromosome breaks or gene/chromosome mutations, or remain as potential damages, a part of which represents long-living ones. The latter may be the initial material for inducing genome instability. It is typical that under the action of superlow/low doses of industrial ionizing radiation the ability to multiple cell divisions in proliferating systems is the same, which causes cellular amplification of both primarily induced and remoted dysgenomic effects [38]. It should be taken into account that free radicals of peroxide oxidation of lipids, accumulated in the cell as a result of water radiolysis, damage the genetic apparatus and membrane structures of the cell, and may also realize the "by-stander" effect [39]. The normal state demonstrates a balance between concentrations of reactive oxygen species and activity of the antioxidant system. However, in the case of chronic low dose radiation this balance is disturbed. It is noted that the children living on radionuclide-contaminated territories demonstrate the state of deep peroxide stress [40]. A negative correlation between the level of chromosome damages and concentration of antioxidants was also determined [41]. Thus, the polygenomic imbalance, dysfunctions of cells, tissues and organs appear in the organism that affects differentiation processes reduces biological stability of the organism and increase risks of multifactorial diseases, which is observed in children living on the territories with increased background radiation [7 – 10].

Table 3. The frequency of aberrant genomes and chromatid breaks in children living in Novozybkov region of Bryansk Oblast (16 – 18 Ci/km^2 by ^{137}Cs)

Examined children	Industry-induced doses, accumulated by 1997 - 2002, cGy	Average number of aberrant genomes per 100 cells/1 cGy	Average number of chromatid breaks per 100 genomes/1 cGy
Born before the accident in 1982 - 1986 (19 persons)	3.0 - 3.5	0.17±0.12	0.45±0.12
prenatally irradiated in 1986 (15 persons)	3.0 - 3.5	0.30±0.09	0.68±0.15
born after the accident in 1987 - 1992 (19 persons)	1.5 - 2.0	0.75±0.25	1.76±0.53
born after the accident in 1994-2000 (24 persons)	0.4 - 0.6	1.07±0.38	2.90±1.77
Children living on clean territories	10 cGy (irradiation *in vitro* by ^{137}Cs)	0.14±0.03	0.30±0.07

Recently, a question about relation between the level of genome destabilization in the organism cells and development of any radiation-induced pathology becomes urgent. Our investigations have not detected any reliable differences in the frequencies of simple and exchange chromosome aberrations – the radiation impact markers with respect to the morbidity structure.

For the children with the nervous system diseases (vegetovascular dystonia, asthenoneurotic dysfunctions), including severe disturbances of central nervous system (encephalopathies, hypertension-hydrocephalic syndrome, epilepsy), only frequency of chromatid fragments exceeded the corresponding index in the group of children with diseases of the infectious-allergic origin ($p < 0.02$) (Figure 6). The highest sensitivity to testing irradiation *in vitro* and expressivity of dysgenomic effects in the 3rd mitosis lymphocytes were observed in several children with central nervous system diseases (2 persons), chronic gastritis (1 person), chronic pneumonia (1 person), and inborn anomalies of extremities (1 person).

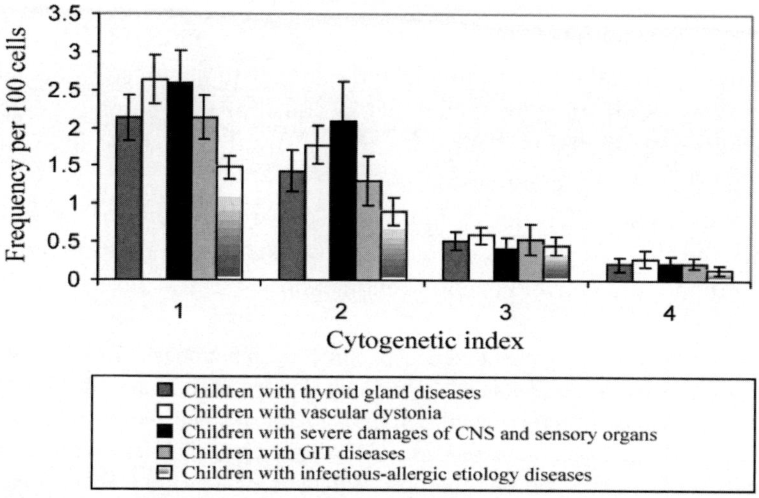

Figure 6. The frequency of cytogenetic disorders depending on the morbidity structure among examined children living on the radionuclide-contaminated territories; 1 - aberrant cells; 2 – single chromatid fragments; 3 – paired fragments + centromeric breaks + deletions; 4 – dicentrics + rings + rec. translocations + inversions.

The analysis of the immune system – one of the most important special systems providing biological stability and adaptation potential of the organism – has shown that the majority of children with expressed dysgenomic effects demonstrated combined abnormalities involving cellular and humoral immunity, and the system of phagocytes [42]. All possible types of the immune response were determined: autoimmune, immunodeficiency, and active immune response. The changes on the immunogram increased with the expressivity of clinical symptoms, which depend on duration of chronic low dose radiation impact and the individual peculiarities of adaptation-compensatory reactions. For example, the existing identical dose loads on the thyroid gland caused by ^{131}I do not lead to 100% development of autoimmune thyroiditis. For this group, the autoimmune type of the immunogram was observed only for 50% of patients. The rest children demonstrated other clinical implications such as chronic diseases of gastrointestinal tract, nephropahies and allergic diseases, which indicated a nonspecific character of the radiation mutagenesis in formation organ/polyorgan insufficiency in irradiated children. At present, the combination revealed cytogenetic and immunological deviations are analyzed depending on the morbidity structure, and their possible clinical-pathological meaning is under discussion.

Though the data obtained require further investigations, it is apparent that radiation is not the mutagenic factor inducing specific pathology. The exclusion is several tumor diseases, leukemias, in particular [13]. Despite the active elimination of dysgenomic cells (apoptosis), which is testified about by high level of CD95+ cells with FAS APO I marker – the apoptosis predictors in irradiated children (according to V.C. Suskova), as a result of permanent occurrence of primary damages under chronic radiation impact and their cellular "reproduction", the organism persists drastically broad spectrum of cell with various dysgenomic effects: mutations and casual suppression/expression of genes, structural mutations of chromosomes (deletions, inversions, translocations) and mono-/trisomies that cause transition to hereditary adverse recessive mutations from heterozygous to hemi-

/homozygous state. As a consequence, the penetration/expressivity of cellular geno-/phenotypes are changed [38]. Apparently, the above-mentioned intensifies phenotypic realization of multifactorial diseases, to which the organism demonstrates hereditary aptitude, as well as to conversion of some functional dysfunctions to chronic pathologies. Hence, development of any pathology of any organ is also stipulated by individual features of radionuclide distribution in the organism.

The above results and their discussion, which prove the reality of the genome instability phenomenon in the child's organism exposed by low-rate radiation at small doses, suggest the following definition of this novel genetic phenomenon:

> Genome instability is a phenomenon of structure-functional variability of the genetic material, which occurs in descendants of multiply divided cells, exposed to radiation impact (or any other genotoxic agents), manifested by DNA breaks and reconformations, chromatin recompacting, chromosome aberrations, SCE, aneu/polyploidies, casual suppression/expression of genes, gene and chromosome mutations, which induces the genome balance disturbance and is accompanied by cellular dysfunction development, neoplastic transformations, induction of apoptosis and cell death [21].

Thus, the Chernobyl accident has elucidated many tasks for modern radiation genetics and radiobiology of man, in particular, the peculiarities of child and adult organism responses to low dose radiation impact, real estimation of genetic risk not only as gametal (gene/chromosome) mutations, but also remote radiation effects, such as induction of genome instability in somatic cells of the organisms of children permanently living on the radionuclide-contaminated territories and the children of special monitoring category – the descendants of parents, irradiated at different stages of ontogenesis. The solution of these tasks requires complex population-radiobiological studies. Further family-wise study of the induction regularities of the genome instability and genotypic peculiarities of the organism, their casual effects on clinical-immunological indices and estimation of individual heritability coefficients for multifactorial diseases is required for detection of children possessing increased radiation risk, development of adequate diagnostic criteria for various premorbid states and selection of adequate medicoprophylactic and treating measures.

This work was financially supported by FCNTP "Preferred directions in science and technology development".

References

[1] Il'inskikh N.N., Bulatov V.I., Adam A.M. *et al.*, *Radiation Ecogenetics of Russia*, Novosibirsk, Krokus, 1998, 290 p. (Rus)

[2] Rikhvanov L.P., *General and Regional Questions of Radioecology*, Tomsk, Tomsk Polytechnical University, 1997, 384 p. (Rus)

[3] Pilinskaya M.A., Shemetun A.M., Dyibsky S.S. *et al.*, *Vestnik RAMN*, 2001, No. 10, pp. 80 – 84. (Rus)

[4] Sevan'kaev A.V., Potetnya O.I., Zhloba A.A. *et al.*, *Radiats. Biol. Radioekol.*, 1995, vol. 35(5), pp. 581 – 588. (Rus)

[5] Eliseeva I.M., Iofa E.L., Stoyan E.F., and Shevchenko V.A., *Radiats. Biol. Radioekol.*, 1994, vol. 34(2), pp. 163 – 171. (Rus)
[6] Stepanova E.I., Misharina J.A., and Vdovenko V.U., *Proc. Intern. Conf. "Genetic Consequences of Emergency Radiation Situations"*, Moscow: PH of RPF, 2002, pp. 185 - 187.
[7] Baleva L.S., Sipyagina A.E., Lavrent'eva E.B. *et al.*, *Dispensary Monitoring of Various Cohorts of Children's Population Exposed to the Radiation Factor Impact, Manual for Doctors*, Approved by Scientific Council MH RF, Minzdrav, Moscow, 2003, 14 p. (Rus)
[8] Sipyagina A.E., Baleva L.S., Suskov I.I. *et al.*, *Rossiyskii Vestnik Perinatologii i Pediatrii*, 2004, vol. 49(5), pp. 43 – 47. (Rus)
[9] Baleva L.S., Lavrent'eva E.B., Sokha L.G. *et al.*, "The reasons for disability among children exposed to radiation impact and the principles of their medical and social rehabilitation", *Manual for Doctors*, Approved by Scientific Council MH RF, Minzdrav, Moscow, 2003, 13 p. (Rus)
[10] Baleva L.S., Sipyagina A.E., Terletskaya R.N. *et al.*, "The health state of the children – the descendants of the first and the second generations of irradiated parents, principles of medical observationand rehabilitation", *Manual for Doctors*, Approved by Scientific Council MH RF, Minzdrav, Moscow, 2004, 12 p. (Rus)
[11] "Malignant neoplasms in children living on the territories, radiation-contaminated by the Chernobyl accident", *Statistical Materials Collection*, Ed. L.S. Baleva, MH RF, Moscow, 2004, 47 p. (Rus)
[12] Bochkov N.P., *Thes. Rep. Third Congress on Radiation Investigations*, 1997, vol. II, p 95. (Rus)
[13] Domracheva E.V., Aseeva E.A., Udovichenko A.I. *et al.*, *Radiats. Biol. Radioekol.*, 2002, vol. 42(6), pp. 76 - 79. (Rus)
[14] Il'inskikh N.N., Yurkin A.Yu., Il'inskikh I.N., and Il'inskikh E.N., *Thes. Rep. Intern. Conf. "Problems of Radiation Genetics at the Edge of the Centuries"*, Moscow, RUDN, 2000, p. 279. (Rus)
[15] Li D.E., *The Radiation Effect on Living Cells*, Moscow, Atomizdat, 1963, 287 p. (Rus)
[16] Timofeev-Resovsky N.V., Ivanov V.I., and Korogodin V.I., *The Target Theory in Radiobiology*, Moscow, Atomizdat, 1968, pp. 3 – 228. (Rus)
[17] Harms-Ringdahl M., *Mutat. Res.*, 1998, vol. 404, pp. 27 - 33.
[18] Mazurik V.K. and Mikhailov V.F., *Radiats. Biol. Radioekol.*, 2001, vol. 41(3), pp. 272 – 289. (Rus)
[19] Pelevina I.I., Aleshchenko A.V., Antoshchina M.M. *et al.*, *Radiats. Biol. Radioekol.*, 2003, vol. 43(2), pp. 161 – 166. (Rus)
[20] Vorobtsova I.E., *Radiats. Biol. Radioekol.*, 2003, vol. 43(6), pp. 639 – 643. (Rus)
[21] Suskov I.I. and Kuz'mina N.S., *Radiats. Biol. Radioekol.*, 2001, vol. 41(5), pp. 606 – 614. (Rus)
[22] *WHO. Methods for the Analysis of Human Chromosome Aberrations*, Ed. K.E. Buckton and H.J. Evans, WHO, Geneva, 1973.
[23] *Biological Dosimetry: Chromosomal Aberration Analysis for Dose Assessment. Technical Reports*, Ser. 260, Vienna: Int. Atomic Energy, 1986, 69 p.
[24] Pyatkin E.K. and Nugis V.Yu., *Radiobiologia*, 1980, vol. XX(6), pp. 871 – 876. (Rus)

[25] Harris H. and Hopkinson D.A., *Handbook of Enzyme Electrophoresis in Human Genetics*, Amsterdam, NH Publ. Co., 1978.

[26] Kuz'mina N.S., Suskov I.I., Sadovnikov V.P., and Sorochinskyi M.V., *Ekologicheskie Systemy i Pribory*, 2003, No. 7, pp. 41 – 46. (Rus)

[27] Sipyagina A.E., *Children's Health and Radiation: Urgent Problems and Solutions*, Ed. L.S. Baleva and A.D. Tsaregorodtsev, Moscow, Media Sfera, 2001, pp. 30 – 34. (Rus)

[28] Suskov I.I., Kuz'mina N.S., Shevchenko A.V. *et al., Proc. Russian Scientific Conference on "Medical Aspects Of Radiation and Chemical Safety"*, St.-Petersburg, Izd. Voennomeditsinskaya Akademiya, 2001, pp. 247 – 248. (Rus)

[29] Vorob'eva M.V., "The study of chromosome radiosensitivity in children from irradiated parents", *Candidate Dissertation Thesis*, St-Petersburg, TsNIRRI MZRF, 1995, 16 p. (Rus)

[30] Stepanova E.I., Vanyurikhina E.A., Kondrasheva V.G., and Shcheplyagina L.A., *Pediatria*, 1996, No. 1, pp. 63 – 67. (Rus)

[31] Altukhov Yu.P., *Dynamics of the Population Gene Pool under Anthropogenic Impacts*, Moscow, Nauka, 2004, 619. (Rus)

[32] Akleev A.V., Aleshchenko A.V., Antoshchina M.M. *et al., Radiats. Biol. Rdioekol.*, 2002, vol. 42(6), pp. 696 – 699. (Rus)

[33] Zasukhina G.D., Vasil'eva I.M., and Semyachkina A.N., *Thes. Rep. Intern. Conference "Problems of Radiation Genetics at the Edge of the Centuries"*, Moscow, RUDN, 2000, p. 110. (Rus)

[34] Kuz'mina N.S. and Suskov I.I., *Radiats. Biol. Rdioekol.*, 2002, vol. 42(6), pp. 735 – 739. (Rus)

[35] Dubinin N.P., *Potential Changes in DNA and Mutations*, Moscow, Nauka, 1978, 246 p. (Rus)

[36] Goh K.O., *Radiol.*, 1980, vol. 15, pp. 332 - 334.

[37] Pelevina I.I., Gotlib V.Ya., Kudryashova O.V. *eta al., Tsitologia*, 1998, vol. 43(5), pp. 67 – 477. (Rus)

[38] Suskov I.I. and Kuz'mina N.S., *Radiats. Biol. Rdioekol.*, 2003, vol. 43(2), pp. 150 – 152. (Rus)

[39] Burlakova E.B., Mikhailov V.F., and Mazurik V.K., *Radiats. Biol. Rdioekol.*, 2001, vol. 41(5), pp. 489 – 499. (Rus)

[40] Durnov L.A., Baikova V.N., and Mayakova S.A., *The Health of Children and Radiation: Urgent Problems and Solutions*, Ed. L.S. Baleva and A.D. Tsaregorodtsev, Moscow, Media Sfera, 2001, pp. 65 – 68. (Rus)

[41] Neifakh E.A., Alimbekova A.I., and Suskov I.I., *Radiats. Biol. Rdioekol.*, 2002, vol. 42(6), pp. 615 – 623. (Rus)

[42] Susukov I.I., Kuz'mina N.S., Suskova V.S. *et al., Proc. III Intern. Symp. "Cronic Radiation Impact: Medicobiological Effects"*, Chelyabinsk, 2005 (in press). (Rus)

In: 20 Years after the Chernobyl Accident
Editors: E. Burlakova and V. Naidich, pp. 55-74

ISBN 1-60021-249-2

Chapter 5

COMPLEX CYTOGENETIC CHARACTERISTICS PEOPLE SUFFERED FROM THE CHERNOBYL ACCIDENT

I.E. Vorobtsova* and A.V. Semenov

Central Research Institute of Roentgenology and Radiology, St. Petersburg, Russia

ABSTRACT

Cytogenetic investigations the Chernobyl clean-up workers, their children, persons evacuated from radionuclide contaminated areas (adults and children), as well as the veterans of "extra risks divisions", exposed to radiation in various radiation accidents, nuclear tests, etc. were performed. Compared with the norm, the exposed groups had increased level of unstable and stable (FISH-method) chromosomal aberrations, micronuclei in lymphocytes and erythrocytes, HPRT-mutations. Functional genome validity was reduced, which was displayed by its instability, increased radiosensitivity *in vitro* and elevated accumulation of translocations with the age. A positive correlation between the level of chromosomal aberrations, detected by the routine method, and the frequency of cardiovascular and gastrointestinal diseases in the groups under study was observed. Basing on the analysis of more than 40,000 lymphocytes, performed in eight laboratories, *in vitro* calibration dose-response curve for dicentrics was generated in the dose range of 0 – 100 cGy. It was shown that whole-body irradiation of cancer patients (daily at the single dose of 11.5 cGy up to total dose of γ-radiation equal to 57.5 cGy) yielded the lower frequency of dicentrics per dose unit than irradiation of blood of the same patients *in vitro* in the similar dose range. No difference between *in vivo* and *in vitro* dose-responses for translocations was observed.

Keywords: humans, ionizing radiation, lymphocytes, chromosomal aberrations, dose-response *in vitro* and *in vivo*, genome instability, age, morbidity

---

* The main addressee: 70/4 Leningradskaya street, Pesochnyi, 197758 St.-Petersburg, Russia; CRIRR; Tel: (812) 596-87-79; Fax: (812) 596-67-05; E-mail: radgen@IV9171.spb.edu

In the XXth century, the application of ionizing radiation for military, industrial and medical purposes increased the radiation impact on some cohorts of people. Various emergency situations have much higher increased the background radiation of the planet and the amount of irradiated people. This required forecasting of possible late medicobiological and genetic effects for the human population. The consequences of relatively high dose ionizing radiation impact on humans are studied quite well. However, the majority of radiation impacts from the industrial sources are characterized by low doses and low dose rates. In this case, it is difficult to estimate the risk of negative effects for people's health using direct epidemiological observations. The studies of disorders, arised in cells as a response to such a radiation, becomes highly urgent. They allow to understand the initial stages of the processes inducing development of cancer and noncancer forms of late radiation pathology, as well as the predisposition to these states in the progeny of irradiated parents.

At low radiation impacts a considerable role in development of a somatic pathology plays the individual genetic structure, which, finally, defines the individual risk of disease development or decrease of the life quality. Definition of the initial stages of radiation damage and genetic factors promoting or preventing their progress in the pathology of tissues and organs will provide a more justified formation of the risk group in human populations as well as the design of the required protection measures.

A priori it is obvious that DNA damages must affect the cell functions and its viability, as well as the fitness of tissues and, finally, the health state. Being very long and not 100% probable, the transformation of various radiation-induced DNA damages to a distinguishable pathology at the organism level is studied not well enough. Various types of radiation-induced DNA damages may cause different cellular effects: death, genetic instability, the change of proliferative potencies, adaptive response of radioresistance, etc. A complex and monitoring estimation of the state of the cell genome in people exposed to low-dose radiation impact seems to be an efficient approach to the estimation of potential danger of such an impact for irradiated persons themselves and for their descendants.

Since 1986, the Laboratory on Radiation Genetics of Central Research Institute of Roentgenology and Radiology (CRIRR) in Saint-Petersburg studies the cytogenetic effects in persons, suffered from the Chernobyl accident, and their significance in development of the late somatic pathology. The patients examined were sampled from the following cohorts: liquidators of the accident consequences (except endured an acute radiation disease); children of the liquidators, born after their parents took part in the clean-up works; persons, evacuated from radionuclide contaminated regions to Saint-Petersburg. In addition the examination was applied to the people called to "veterans of special risk division", who were exposed to radiation during nuclear tests, nuclear waste burial, radiation accidents, taken place on nuclear plants and on atomic submarines. Simultaneously, people who had no professional or any casual contact with ionizing radiation and their children were studied, as well as children of cancer patients undergone radio- and chemotherapy (the positive control) before conception. The groups were formed on the basis of Saint-Petersburg Division of Professional Pathology – TMD No.20, Leningrad Regional Clinical Hospital and Consulting Center in Saint-Petersburg Pediatric Medical Academy.

The investigations were performed in the following main directions:

1. Characteristics of the morphological state of genome by the frequency of unstable chromosomal aberrations (CA), micronuclei (MN), stable CA, and the frequency of GPA and HPRT-mutations in blood cells;
2. Characteristics of the functional state of genome by the chromosomal sensitivity of lymphocytes to challenge irradiation (*in vitro*) and by the age dynamics of stable CA (*in vivo*);
3. A correlation between the degree of genome damages and the risk of somatic pathology development;
4. Applied investigations, aimed to verification of person absorbed doses: construction of dose-effect curve for dicentrics in lymphocytes irradiated *in vitro* and comparison of the dose-responses for CA in human lymphocytes irradiated *in vitro* and *in vivo*.

Partly the investigations were carried out in the framework of the international collaborative projects "Radiation genotoxicity from the Chernobyl accident" (Lawrence Livermore National Laboratory, USA) and "Biological dosimetry for persons irradiated by the Chernobyl accident" (CES/CIS), as well as national programs "Complex ecological program for investigations of the Chernobyl accident consequences" and "Children of Chernobyl".

Materials and Methods

The examined groups: AC – adult control (20 – 85 years old); CC – children control (2 – 18 years old); LQ – liquidators (24 – 66 years old); LC – liquidators' children (0.5 – 7 years old); EA – evacuated adults (26 – 45 years old); EC – evacuated children (2 – 19 years old); VT – veterans of special risk divisions (48 – 76 years old); PC – children of cancer patients, born after exposure to radio- and chemotherapy (3 – 13 years old). A personal form was filled in for every examinant. Beside the particulars, it included data on the kind, time and place of work, officially estimated absorbed dose (LQ, VT), place and time of inhabitance on the contaminated territories (EA, EC), time of parents' participation in clean-up works after Chernobyl accident (LC), diseases, X-ray diagnostic procedures, drug therapy (for all groups), harmful habits – smoking, alcohol consumption (LQ, VT, EA, AC), terms and character of radiation- and chemotherapy (PC).

In studies targeted at the generation of the dose-effect curves for the unstable (dicentrics) and stable (translocations) chromosome exchanges in lymphocytes the blood of three healthy donors was used. For comparison of dose responses of chromosome aberrations in lymphocytes irradiated *in vivo* and *in vitro*, blood from 5 cancer patients, was sampled prior to the radiation treatment and after every procedure of whole-body irradiation. Detail description of healthy donors and patients are presented elsewhere [1–4].

Lymphocytes Culture and Preparation of Chromosome Slide

Cultivation of lymphocytes, harvesting procedure and slide preparations were performed in accordance with the standard protocol described earlier [1-4]. For the cell cycle control lymphocytes were cultivated in the presence of BrdU (9.23 μg/ml).

Preparation of Micronucleus Slides

To prepare micronucleus lymphocyte slides, blood was cultivated during 72 h, and 24 h prior the harvesting the culture was added by cytochalasine B (6 μg/ml). Hypotony was performed in a cooled solution of 0.55% KCl during 1 min 15 s at 4°C. After centrifugation (1000 rot/min) the suspension was three times fixed in methanol/glacial acetic acid mixture and dropped to cooled wet slides, dried in the air and stained by 3% Giemsa dye.

Micronuclei in erythrocytes were analyzed on blood smears, dried in the air, fixed in 96% ethanol and stained by 2% Giemsa dye.

Fluorescent Hybridization in Situ

The procedure of fluorescent hybridization *in situ* used (the FISH-method) is comprehensively described [4, 7]. As a rule the cocktail from the target chromosomes 1, 4 and 12 was used. Some slides were hybridized with directly Spectrum Orange labeled DNA probes to chromosomes 1, 2 and 4 or with directly FITC (chromosomes 1, 2 and 3) and Texas Red (chromosomes 4, 5 and 6) labeled DNA probes. These slides were prepared and analyzed in Lawrence Livermore National Laboratory (LLNL) in the framework of the collaborative Russian-American Project [8]. In the applied studies labeled DNA probes to chromosomes 1, 4 and 8 and pancentromeric probes were used, the immunochemical detection of which were carried out in accordance with the recommended protocol (Cambio, UK).

Microscopic Analysis of Slides

On the chromosome slides prepared for routine cytogenetic analysis, dicentric and ring chromosomes, double and single fragments in the first division cycle metaphases containing (46 ± 1) chromosomes were scored. Micronuclei in binucleated lymphocytes were identified according to common criteria [9] at examination of 1,000 – 2,000 binucleated cells. Micronuclei in erythrocytes were determined as small, round, thickly stained structures on blood smears. Three hundred or more fields of vision, covered with a frame with about 100 cells were overlooked, i.e. about 30,000 erythrocytes were analyzed. The slides were analyzed under light microscopes with ×900 – 1,000 amplification. The slides for stable chromosomal aberrations were analyzed under fluorescent microscope Axioplan (Carl Zeiss) with ×1,000 amplification, using a triple band pass filter, for simultaneous observation of green (FITC), red (Texas-Red) and yellow (FITC+Texas-Red) fluorescence. The first cycle metaphases containing six painted chromosomes were only analyzed; total number of chromosomes in the cell was not determined. Beside reciprocal (complete) and terminal (incomplete) translocations, insertions, dicentrics and acentrics, as well as complex CA were registrated. The genome frequency of events was determined by dividing the number of aberrations observed on the cell equivalent. The cell equivalent was calculated by the formula:

$N \times 2.05 f_P(1 - f_P)$ [10],

where N is the number of cells scored; f_P is the part of genome painted.

Mutation Tests

The methods of registration of GPA-mutations in erythrocytes and HPRT-mutations in lymphocytes, used as the parameters of genetoxicity of the Chernobyl accident in LLNL at the examination of the liquidators and control donors, are discussed in details elsewhere [8].

Blood Irradiation in Vitro

For the estimation of genome stability, the donor blood was exposed to γ-radiation from ^{137}Cs *in vitro* at the dose 150 cGy (with the dose rate was 30 cGy/min). For generation of dose–effect calibration curves, blood of healthy donors were exposed to radiation at doses of 10, 25, 50, 75 and 100 cGy of γ-radiation from ^{60}Co (with the dose rate was 50 cGy/min) [1] or of 10, 25, 30, 40, 50, 75, 100 and 150 cGy of γ-radiation from ^{137}Cs (unit the dose rate was 37 cGy/min) [2].

To compare dose-responses of CA yield at *in vivo* and *in vitro* irradiation of lymphocytes, blood of cancer patients, sampled before the patient treatment, was irradiated at the doses of 10, 20, 30, 40, 50 cGy or 8, 16, 24, 32, 40 cGy of γ-radiation from ^{60}Co (with the dose rate 2 cGy/min).

Whole-Body Irradiation of Patients

Prior to local radiation therapy the patients were exposed to whole-body γ-radiation at a single dose of 11.5 cGy from ^{60}Co unit up to total dose of 57.5 cGy (at the dose rate between 0.95 – 1.23 cGy/min). Time period between the irradiation procedures varied from 1 to 3 days. The technique of whole-body irradiation was developed in CRIRR and is applied as a pretreatment in therapy of some forms of cancer [11, 12]. Prior to whole-body irradiation and after each irradiation procedure, blood from patients was sampled and chromosome slides were prepared in accordance with the standart protocol. The conditions of irradiation of patients and irradiation of their blood *in vitro* are presented in more detail in [3, 4].

Statistical Processing of the Results

The significance of differences was estimated using of Student's t-test, Fisher's exact test and χ^2 test. The regression analysis of the data was performed using computer software Poisson Iteratively Reweighed Least Squares and Statistica for Windows 6.0.

RESULTS AND DISCUSSION

Characteristics of Cytogenetic Status of people survived the Chernobyl Accident and Other Emergency Radiation Exposures

Data on the frequency of unstable CA in different groups of examined adults and children are shown in Table 1. All exposed groups are characterized by significantly higher level of various CA types as compared to the corresponding controls. The frequency of dicentrics was increased in the groups of persons, which were exposed to direct irradiation, except the EA group, which is the smallest among all and, therefore, the CA level in it may not be objectively estimated. The highest frequency of dicentrics and rings was observed in veterans of "special risk division". Apparently, the radiation doses absorbed by these people were high enough, because even several decades after exposure they still possess elevated frequency of unstable chromosomal exchanges. In all exposed group, the part of people with exchange chromosomal aberrations was higher than in the control group (23.3% vs. 12.9%, $p<0.05$, χ^2 test). The results obtained coincide with the data of other authors, who have also determined the increased level of dicentrics and rings in the liquidators of the Chernobyl accident consequences and other irradiated cohorts in the late period after the radiation impact [13–18].

Table 1. The number of chromosomal aberrations (per 100 cells), detected by routine method in persons of different groups

Group	Children			
	CC	EC	LC	PC
Number of people	52	64	60	15
Number of cells	9826	11853	12495	2672
Aberrant cells	1.67±0.13	2.51±0.14**	2.30±0.13**	2.66±0.31**
Total aberrations	1.71±0.13	2.57±0.15**	2.41±0.14**	2.73±0.32**
Dicentrics and rings	0.06±0.02	0.14±0.03**	0.07±0.02	0.11±0.06
Double fragments	0.67±0.08	0.79±0.08	0.86±0.08	0.97±0.19
Single fragments	0.92±0.10	1.56±0.11**	1.42±0.11**	1.65±0.25**

Group	Adults			
	AC	EA	LQ	VT
Number of people	93	19	174	22
Number of cells	17438	3730	32605	5477
Aberrant cells	2.03±0.11	2.71±0.27*	2.59±0.09**	3.54±0.25**
Total aberrations	2.14±0.11	2.73±0.27*	2.70±0.09**	3.76±0.26**
Dicentrics and rings	0.11±0.02	0.13±0.06	0.17±0.02*	0.55±0.10**
Double fragments	0.94±0.07	1.07±0.17	1.30±0.06**	1.42±0.16**
Single fragments	0.97±0.07	1.42±0.19*	1.12±0.06	1.55±0.17**

Differences from the control: * $p < 0.05$; ** $p < 0.01$ (Student's t-test)
CC – control children; EC – evacuated children; LC – children of liquidators; PC – children of patients; AC – adult control; EA – evacuated adults; LQ – liquidators; VT – veterans of special risk division

People exposed to radiation at the prezygotic stage of development (LC and PC) are characterized by increased level of single fragments compared with the corresponding control (CC). The same is observed for other exposed groups, though the difference with the control is not always significant. The increased frequency of single fragments in different Chernobyl cohorts is demonstrated in studies [19, 20]. Since this very type of CA is typical for spontaneous chromosomal mutagenesis, one may suggest that it proceeds more intensive than in norm both in directly irradiated persons and in children of irradiated parents due to genome instability, induced by the radiation impact. This is also confirmed by the fact that total frequency of excess double and single fragments in persons from the groups LQ, VT, EA and EC possessing dicentrics and rings, was significantly higher than in persons from the same groups, which such CA were not detected (1.4 ± 0.09 vs. 1.0 ± 0.05; $p < 0.05$, Student's t-test).

Table 2. The number of chromosomal aberrations (per 100 cells) for the liquidators of 1986 – 1987, examined in different years

Group	Number of people	Age, years	Number of cells	Number of CA	Dicentrics + centric rings	Double fragments	Single fragments
1	88	21-58	14350	3.5±0.19[a]	0.4±0.07[a]	1.9±0.14[a]	1.3±0.12[a]
2	73	23-48	11463	2.2±0.14[b]	0.2±0.04[b]	1.0±0.09[b]	1.0±0.09[b]
3	37	20-45	6132	1.4±0.15	0.1±0.04	0.7±0.11	0.7±0.11

1 – liquidators of 1986 – 1987, examined in 1987;
2 – liquidators of 1986 – 1987, examined in 1990 – 1992;
3 – control donors;
[a] – differences from group 2: $p < 0.05$ (Student's t-test);
[b] – differences from group 3: $p < 0.05$ (Student's t-test)

For retrospective dosimetry, it is of importance to imagine the temporary dynamics of CA frequency. Table 2 presents such results for the liquidators, worked in Chernobyl in 1986 – 1987 and examined after different periods of time: group 1-in 1987 and group 2-in 1990 – 1992 [21]. Group 3 – the control group consisted of people cytogenetically examined in the laboratory in the same years. As shown in Table 2, both groups of liquidators possess increased frequency of all CA types compared with the control. At the same time, it is obvious that the number of CA decreases through time. In other studies, higher frequency of dicentrics 4 – 6 – 10 years after the accident was also observed in liquidators as compared with the control [17, 22]. The frequencies of fragments in people with exchanges are 5 and 3 times higher in groups 1 and 2, respectively, compared with group 3. At the same time, the frequency of the same types of aberrations in the liquidators having no exchange rearrangements are much lower than in the liquidators with such rearrangements, are practically equal for groups 1 and 2 and for control.

According to the data of Russian State Dosimetric Register (RSDR) the average radiation doses absorbed by liquidators in 1986 and 1987 equaled 0.17 and 0.13 Gy, respectively [23]. This corresponds to the average data of physical dosimetry, calculated from the personal form information: 0.17 and 0.18 Gy for the liquidators worked in 1986 and 1987, respectively. An attempt was made for retrospective estimation of doses, absorbed by liquidators, using the results of cytogenetic analysis. To determine collective dose, the method of CA distribution reduction by lymphocytes to the Poisson distribution was used [24]. For the purpose of calculating the collective dose, both groups of liquidators were combined, because they were identical by the period of work in Chernobyl. For the acute period, the restored CA frequency equaled 4.78% and corresponded to the radiation dose, calculated by the *in vitro* calibration curve constructed in [25], of 0.20±0.05 Gy. Thus the values obtained in three estimations of the collective dose were practically identical.

Table 3. The number of micronuclei (per 1,000 cells) in people from different groups

Group	Lymphocytes		Erythrocytes
	Spontaneous level	Induced *in vitro* (1.5 Gy)	Spontaneous level
CC	8.11 ± 0.16 (19)	150 ± 10 (18)	0.01 ± 0.002 (24)
EC	9.04 ± 0.12* (43)	205 ± 14 (18)	0.07 ± 0.006* (22)
AC	11.12 ± 0.22 (12)	–	0.04 ± 0.003 (30)
EA	22.11 ± 0.30* (11)	–	0.08 ± 0.006* (25)

- Differences from the control: $p < 0.01$ (Student's t- test); in brackets – the number of examined persons.
- Explanations of abbreviations see in footnotes for Table 1.

Table 3 presents data on the frequency of micronuclei in lymphocytes and erythrocytes in the groups of persons, examined for these cytogenetic parameters. Both lymphocytes *in vitro* and erythrocytes *in vivo* [26] demonstrated increased MN level in adults and children evacuated from radionuclide contaminated zones as compared with the corresponding age control.

Table 4 presents the results published earlier [27-29] on the frequency of stable and unstable chromosome exchanges detected by FISH-method in various groups. In all exposed groups the translocation frequency significantly exceeds the corresponding control values. For the groups LQ and VT, a reliably higher frequency of dicentrics is also observed. It has been shown [8] that the radiations impact due to Chernobyl accident affects the translocation frequency much more significant than smoking and ageing. It should be noted that though in many studies performed by FISH-method the increased level of stable CA in different exposed groups was found [8, 30, 31], some investigators have not observed this phenomenon [32, 33].

Table 4. The number of exchange chromosomal aberrations (per 100 cells), detected by FISH-method in people from different groups

Group		Average age, years	Number of			
			people	Cell equivalents	Translocations	Dicentrics
Control	Adults	40.1 ± 4.2	134	93773	0.78 ± 0.03	0.14 ± 0.01
	Children	11.3 ± 7.5	19	6851	0.16 ± 0.05	0.06 ± 0.03
Evacuated	Adults	38.3 ± 11.2	19	8537	1.38 ± 0.13*	0.15 ± 0.04
	Children	14.6 ± 9.1	15	5286	0.62 ± 0.11*	0.13 ± 0.05
Liquidators		42.1 ± 3.0	275	179199	1.08 ± 0.02*	0.18 ± 0.01*
Veterans		61.3 ± 8.5	33	17393	2.41 ± 0.12*	0.25 ± 0.04*

* Difference with the control: $p < 0.05$ (Student's t-test)

Table 5. The number of chromosomal aberrations (per 100 cells) induced in vitro in people from different groups

Group	Children				Adults		
	CC	EC	LC	PC	AC	EA	LQ
Number of people	32	45	52	13	45	12	114
Number of cells	4219	5356	5727	1661	5237	1350	14370
Aberrant cells	34.5±0.9	37.9±0.7**	41.7±0.7**	35.0±1.2*	34.5±0.7	39.9±1.3**	29.4±0.4**
Total CA	40.3±0.9	50.8±0.7**	55.9±0.7**	44.6±1.2*	46.0±0.7	51.0±1.4**	37.6±0.4**
Dicentrics and rings	21.5±0.8	27.6±0.6**	31.2±0.6**	24.6±1.1*	21.8±0.6	25.5±1.2**	17.9±0.3**
Double fragments	17.1±0.7	16.7±0.5	17.9±0.5	16.7±0.9	18.6±0.5	19.9±1.1	15.3±0.3**
Single fragments	1.70±0.25	2.68±0.22**	2.51±0.21*	3.07±0.42*	2.54±0.22	2.59±0.43	2.16±0.12

Differences from the control: * – $p < 0.05$; ** – $p < 0.01$ (Student's t-test).

Explanations of abbreviations see in footnotes for Table 1.

The results of mutation tests were ambiguous: the frequency of mutations in HPRT-locus [8] and T-cell receptor locus [34], and the level of mutant T-helpers [19] were found to be increased compared with the control for persons exposed to long-term radiation impact, whereas the GPA-mutation frequency in erythrocytes either did not change [8, 34] or also increased [35].

During two decades passed since the Chernobyl accident a lot of data on cytogenetic and molecular-biological estimation of the genome state in various cohorts of victims were published. They provide evidence that the overwhelming majority of these people, as well as others exposed to uncontrolled low-dose radiation (Semipalatinsk, Novaya Zemlya, South Urals, etc.) possessed various chromosome and gene mutations with the frequencies exceeded the control level. At present, one of the most important tasks is review of all existing data and their analysis. It is urgent to estimate representativness of groups studied, control adequacy, contributions from confounding factors into the effects observed, informative values of different mutation tests, their abilities and restrictions for biological reconstruction of radiation doses and for selection the groups of risk in respect of adverse biological (medical) effects. It is the authors' opinion that the important direction of future investigations is monitoring of the same persons in order to follow time dynamics of the radiation induced genetic disorders, in parallel to incidence of various diseases.

Functional State of the Genome in People Suffered from Radiation Accidents

The increased frequency of single fragments (that is of CA typical for spontaneous mutagenesis) in people exposed to a low-dose radiation (Table 1), gives reason to suggest the instability, of their genome. For direct estimation of lymphocyte chromosome stability, the blood from people of different exposed groups was irradiated *in vitro* at the dose of 150 cGy, and the frequency of induced CA was estimated (Table 5). Both children from LC and from PC groups (prezygotic irradiation) as well as persons from EC and EA groups (postnatal irradiation) demonstrate the increased sensitivity of chromosomes to challenge irradiation *in vitro*: levels of all CA types in these groups was found to be higher than in corresponding controls. It is worth to note that the distributions of people from groups on the number of induced CA were shifted as a whole towards the higher values of CA. It means that uniform response of previously irradiated cells to mutagenic insults take place. Analogous data were obtained when MN frequency (Table 3) was taken as the parameter of lymphocyte radiosensitivity. The inverse situation was observed for the liquidators' group. Since due to their professional duties the majority of liquidators had already contacted the ionizing radiation sources before the Chernobyl accident, one may suggest an occurrence of an original "adaptive response" in their cells, which protect them against the induction of genome instability by the radiation impact from the Chernobyl accident. It is of interest that in contrast to the other groups, spontaneous frequency of single fragments (Table 1) in the liquidators was not elevated.

One more approach to the estimation of the genome functional state was the comparison of the age dynamics of stable CA in exposed and control groups [27, 28]. It is known that the number of genome, chromosome and gene mutations in animal and human cells increased with the age. This testifies on the inferiority of DNA repair mechanism and/or increasing genome instability with the age.

Stable chromosome exchanges (translocations) are most appropriate for the study of the age dynamics of structural chromosome disorders. Being nonlethal for the cells, they are not eliminated through cell divisions and may accumulate with the age. If the low-dose radiation induces a decrease of functional ability of the cell, then translocations in the exposed groups will be accumulated faster with the age. As shown by the histogram in Figure 1, the irradiated persons of the same age with the control ones are characterized by higher frequencies of stable chromosome exchanges in blood lymphocytes.

For both groups, the individual yields of translocations fitted well the quadratic age-response model ($p<0.05$):

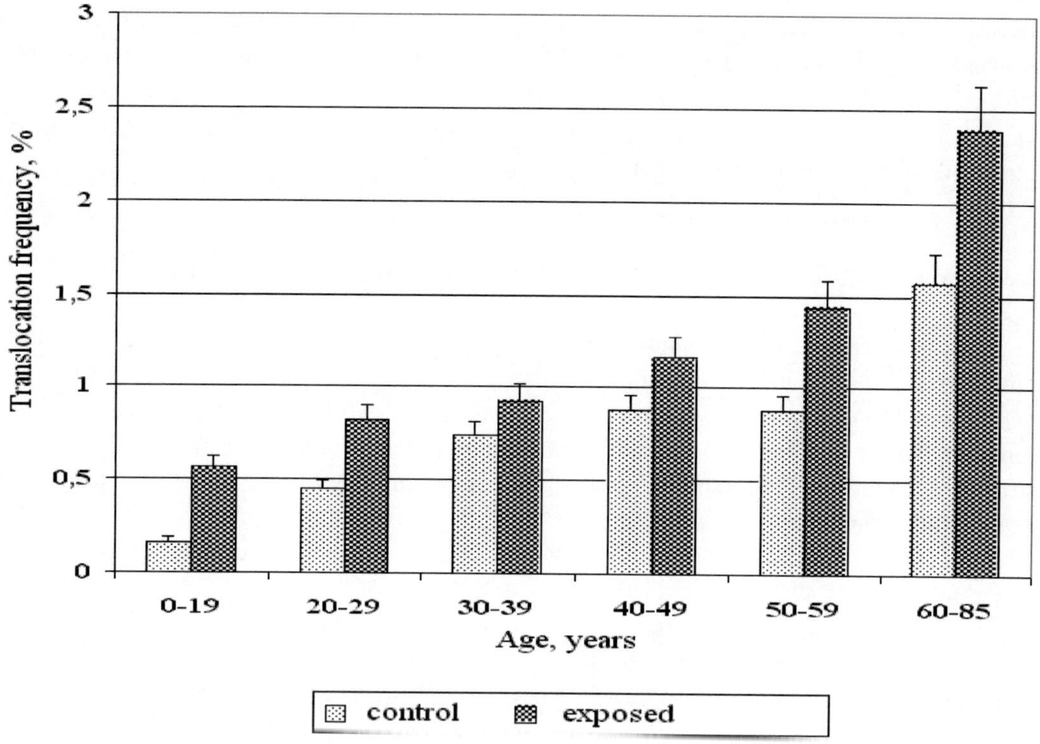

Figure 1. The age dynamics of the frequency of translocations, detected by FISH-method in exposed and control groups.

The resulting equations were: $Y = (0.21 \pm 0.08) + (0.32 \pm 0.05)X^2$, for the control donors, and $Y = (0.42 \pm 0.11) + (0.51 \pm 0.06)X^2$, for the exposed group people, where Y is the genomic translocation frequency (per 100 cells); X is the age (years). The comparison of β-coefficients from these equations shows that exposed persons aged faster as estimated by the frequency of translocations than in norm e.g. their "biological age" appeared to be greater than the calendar one. In this connection, it is of interest to note that on the base of translocations level the age of patient with the syndrome of hereditary premature aging (Werner syndrome) was estimated to be 55 – 65 years, whereas he was only 26 years old [36].

The investigations of the functional state of genome in people exposed to low-dose radiation are much scanty than those studied the mutation endpoints. At the same time, in the recent 15 years many works performed on the cells *in vitro* have shown that the low-dose radiation may induce various functional shifts, genetic instability, in particular [37]. The studies, performed on the children of cancer patients, undergone to radio- and chemotherapy before conception [38], was chronologically the first one demonstrated the genome instability of the persons irradiated at the prezygotic stage of development. Prior to this work was the similar investigation, carried out on the experimental animals [39, 40]. At present, analogous results were obtained by others [19, 41, 42]. A correlation between genome instability and the carcinogenesis risk, demonstrated in animal models [see ref 40], allows to consider the genetic instability as a prognostic marker of carcinogenesis in exposed persons. It is obvious that further investigations in this direction are extremely urgent now. A set of works by I.I. Pelevina *et al.* [41] also contribute to the problem of functional characteristics of irradiated genome. Authors studied the ability of the cells exposed to low dose ionizing radiation *in vitro* and *in vivo* to adaptive response (AO) – the way by with cells protect themselves against extreme impacts. The change of ability to develop adaptive response was demonstrated for the lymphocytes of children living in the radionuclide contaminated areas and of the liquidators of the Chernobyl accident consequences.

The Correlation between the Number of Genome Damages and Morbidity

Some evidence on the functional importance of genome damages may provide a comparison of the morbidity parameters among persons with low and high levels of mutation events in cells. As it was shown, the persons with high level of CA are characterized by an increased risk of cancer compared with the persons possessing low CA frequency [43]. The estimation method was based on the division of the examined group to quartiles in accordance with the individual CA frequencies.

In the current work, the same approach was used to estimate the correlation between the individual level of cytogenetic disorders and the risk of occurrence of various somatic pathologies (diseases). For this aim all examined persons (exposed and control), were ranged in the ascending order of each parameter values. Then these raws were divided into 4 approximately equal parts, the so-called quartiles. Further on, mean value of the parameter was determined for each quartile. Persons with the lowest parameter values (25% of the selection) formed the first quartile, and persons with the highest parameter values (25% of the selection) formed the fourth quartile (Table 6). Totally, 329 patients of 891 persons examined possessed the highest values of parameters estimated. Some people were examined by 2, 3, 4 or 5 parameters and, thus results of the cytogenetic analysis, performed by different methods and their combinations, indicated that totally 191 persons were characterized by the highest levels of genetic disorders. The correlation between morbidity and the level of genetic damages was analyzed for a cohort of adults, examined on the spontaneous frequency of CA, detected by the routine method. The morbidity of persons with low (L) CA level (the first quartile) and the persons with high (H) CA level (the fourth quartile) was estimated by three most valuable systemic pathologies: diseases of gastrointestinal track (GIT), cardiovascular pathology (CVP) and acute respiratory viral infections (ARVI). In the L group, the part of persons with GIT and CVP diseases equaled 23%, and in the group H – 45.2% ($p > 0.01$, χ^2 test). No differences in ARVI frequency for groups L and H was observed. The analogous analysis performed among the atomic industry workers also found a positive correlation between CA frequency and the risk rate of various diseases (G.P. Snigireva, personal communication).

The study of correlation between morbidity and CA level in the children's group is in progress now as well as the study of a correlation between health parameters and other cytogenetic endpoints studied. A selection of the most informative endpoints for forecasting the possible risk of somatic pathologies in future seems to be very important.

Table 6. Cytogenetic characteristics of persons, attributed to different quartiles in accordance with the individual values of estimated parameters

	Quartile No.		1	2	3	4
Spontaneous frequency of	Chromosomal aberrations, %	Number of persons	116	118	117	104
		M ± SE	0.64±0.04	1.76±0.02	2.89±0.03	5.24±0.14
	Translocations, %	Number of persons	119	121	119	119
		M ± SE	0.19±0.01	0.59±0.01	1.01±0.01	2.23±0.09
	Micronuclei, ‰	Number of persons	20	21	22	21
		M ± SE	2.85±0.33	6.01±0.18	9.60±0.27	20.57±1.54
Frequency of induced *in vitro*	Chromosomal aberra-tions, %	Number of persons	67	71	61	65
		M ± SE	25.7±0.5	33.7±0.2	38.8±0.2	48.2±0.7
	Micronuclei, ‰	Number of persons	20	21	19	20
		M ± SE	82.8±5.6	119.1±1.8	160.3±3.2	224.7±7.5

It should be noted that in parallel to cytogenetic and molecular-biological results, a great amount of independent data on medical consequences of various radiation accidents was accumulated. Hence, the classification and analysis of these data is necessary to perform in order to attribute the cellular effects to effects on the higher levels of organization (tissues and the whole organism). Analysis of the correlation between CA level and the risk of cancer and noncancer diseases seems to be quite perspective approach to estimation of the role of radiation-induced genome damages in the late somatic pathology. Organization of the working groups with the representation of physicians providing verified diagnoses and biologists performing cytogenetic and molecular-biological investigations is likely to be the necessary condition for implementations of such works.

Investigations Directed to the Assessment of Accumulated Doses

Traditionally, in epidemiology the risks of development of late (mostly stochastic) radiation consequences are calculated from individual (or collective) absorbed doses. In the absence of reliable physical dosimetry the radiation doses are estimated with the help of dose-dependent biological parameters, unstable chromosome exchanges (dicentrics) in lymphocytes, in particular. The level of dicentrics observed after irradiation or calculated by the residual frequency for the early postexposure period is compared to the dose-effect curve, obtained for these aberrations at blood irradiation *in vitro*. In some competent laboratories such curves were constructed [44]. Usually, they are also used by other investigators who are reconstructing the absorbed doses. It is also obvious that for the accurancy of biodosimetry self calibration dose-effect curve should be generated by each laboratory. It is also quite important to estimate the interlaboratory variability of parameters of these curves. Both above-mentioned tasks were being resolved in the framework of the International Project ECP-6 (CES/CIS) "Biological dosimetry for persons irradiated by the Chernobyl accident".

The dose–response for dicentric, detected by the routine method in the dose range of 10 – 150 cGy of γ-irradiation from ^{137}Cs, was studied on lymphocytes of healthy laboratory donor (a woman 23 years old). The data were fitted satisfactorily to the linear-quadratic function:

$$Y = (0.07 \pm 0.33)\cdot10^{-4} + (4.52 \pm 1.81)\cdot10^{-2}D + (6.54 \pm 1.77)\cdot10^{-2}D^2,$$

where Y is the frequency of dicentrics per cell; D is the dose in Gy [2].

Blood sampled from two other healthy donors was irradiated in the laboratory headed by Prof. A.V. Sevan'kaev (MSRC, Obninsk, Russia), metaphase slides were distributed between the project partners. Each of them analyzed of about 1,000 cells from each donor for 0, 10, 25 and 50 cGy doses and about 500 cells for 75 and 100 cGy doses. All data were processed statistically in NRPB (Chilton, UK) [45, 46].

Table 7. Coefficients of linear-quadratic dose–effect models for dicentrics (routine analysis)

Laboratory No	C ± SE×10^{-4}	α ± SE×10^{-2}	β ± SE×10^{-2}
1	8.61±6.43	3.08±1.11	6.27±1.60
2	9.56±13.75	1.17±2.03	6.60±3.17
3	5.76±6.98	1.32±1.24	9.22±1.98
4	4.91±5.00	5.77±1.31	2.07±1.90
5	6.91±7.26	-0.45±0.88	9.58±1.43
6	5.85±5.78	4.56±1.22	5.45±1.79
7	5.06±5.69	3.92±1.26	8.91±1.98
8	5.57±7.56	1.33±1.30	7.45±2.04
Total	4.21±1.68	2.71±0.36	6.95±0.56

Table 7 presents results for every laboratory and total results for all partners [45]. Clearly, the parameters of linear-quadratic models describing the dose-effect dependencies, obtained in different laboratories, are significantly different. This is specific particularly for the linear coefficient in the equation, which is used for the reconstruction of doses at protracted low-dose irradiation of people. Nevertheless, when the results of all partners were combined, the coefficients in the dose-response curve equations coincided well with the data available from literature [44]. This gives grounds for using the dose-response curve for dicentrics composed by analysis of more then 40,000 cells in total as the calibration one for biodosimetric estimations at least by these eight laboratories.

The introduction of FISH-method into cytogenetic practice allowed for registration of stable chromosomal aberrations and using the frequency of these CA for verification of doses both in the early and the late periods after the radiation impact. The results of FISH analysis of lymphocytes of the same donors performed by four laboratories are presented in Table 8. The interlaboratory differences in linear-quadratic equation coefficients for translocations appeared to be much higher than for dicentrics (Table 7). Apparently, this is caused by several circumstances: the use of different mixtures of chromosome-specific DNA-probes by different laboratories, insufficient quantity of analyzed cells, different number of donors examined in each laboratory. At the same time, good coincidence was observed for the dose-response curve parameters calculated by the total data for both types of exchange CA: dicentrics (Giemsa) and translocations (Tables 7 and 8). For bicolor dicentrics (the FISH-analysis), according to the authors' data (laboratory No. 7), the coefficients of linear-quadratic dose-response model equaled: $C = (1.10 \pm 1.05)\cdot 10^{-4}$, $\alpha = (3.82 \pm 1.73)\cdot 10^{-2}$, $\beta = (8.93 \pm 2.59)\cdot 10^{-4}$. It is close enough to coefficients shown in Table 7 for dicentrics detected by routine scoring.

Table 8. Coefficients of linear-quadratic dose–effect models for translocations (FISH-method)

Laboratory No	$C \pm SE \times 10^{-3}$	$\alpha \pm SE \times 10^{-2}$	$\beta \pm SE \times 10^{-2}$
1	44.92±23.77	-17.26±15.53	33.42±11.28
3	2.68±1.91	2.08±1.96	2.33±2.47
6	5.55±3.80	1.76±4.21	11.98±5.27
7	5.67±3.88	6.41±4.00	6.59±5.29
Total	5.60±1.95	3.65±1.96	7.73±2.50

The next direction of our biodosimetric investigations was comparison of the dose-response for dicentrics and translocations after *in vitro* and *in vivo* irradiation of donor blood. The application of *in vitro* dose-response calibration curves for reconstruction of persons absorbed doses suggests identity of the cytogenetic response of lymphocytes to irradiation *in vivo* and *in vitro*. There are just several works, in which correctness of this assumption was checked on the blood from cancer patients, undergone whole-body irradiation at comparatively low doses [47 – 49]. In these works, the *in vivo* dose-responses for dicentrics were determined on the base of results obtained on different patients. Each of them received different single doses of acute exposures, similar to those applied for irradiation of lymphocytes *in vitro*. Obviously the dose estimation in the case of protracted radiation exposure, which took place after Chernobyl accident, would be more correct if calibration curves were constructed for the similar regimen of irradiation of lymphocytes *in vitro*. However, this is impossible due to inability of sampled lymphocytes to survive for a long time.

For the cases of accidental protracted radiation exposure, the reconstruction of the individual doses could be performed in the first approximation by the dose-response curve, constructed on the base of analysis of chromosome exchanges in cancer patients, exposed to whole-body fractionated irradiation at low doses before local radiation therapy. We have generated such curves for dicentrics and translocations according to the data of FPG and FISH techniques for five patients exposed to daily whole-body irradiation at the single dose of 11.5 cGy up to total dose of 57.5 cGy [4]. The dose-response curves for both types of exchange aberrations were also obtained on the patient's blood of sampled before whole-body treatment and irradiated *in vitro* at the same dose ranges.

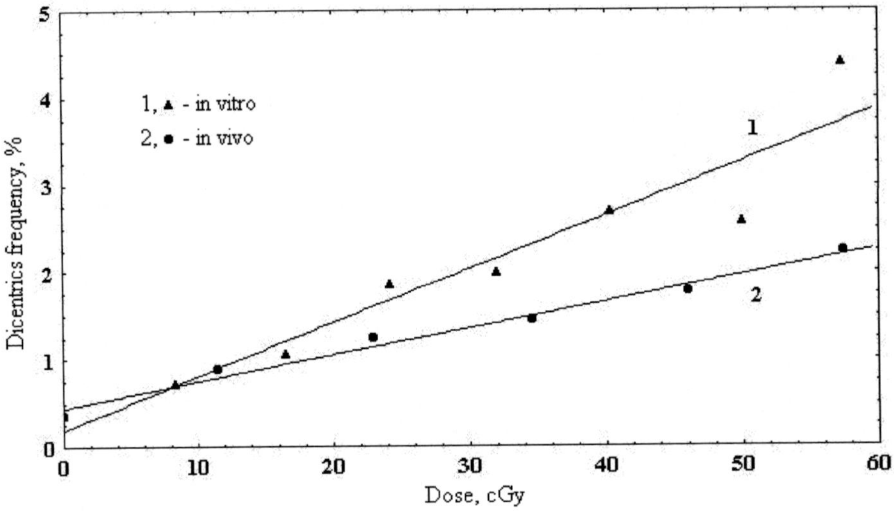

Figure 2. The dose-response curves for dicentrics in lymphocytes irradiated *in vitro* and *in vivo* (FPG)

Table 9. The frequency of dicentrics, observed and predicted by different dose-effect models, in the patient exposed to whole-body irradiation at the dose of 80.5 cGy

Dose, Gy	Dicentrics frequency, %			
	Observed	Predicted by		
		in vivo model	*in vitro* (1) model	*in vitro* (2) model
0.80	3.00 ± 0.70	2.98	4.65	3.21

The model *in vitro* (1) is a linear one, obtained on lymphocytes irradiated *in vitro* (see Figure 2); The model *in vitro* (2) is a laboratory linear-quadratic one, the linear coefficient α was used for calculation the predicted value.

In both cases (*in vivo* and *in vitro*), the linear model described the data obtained in the best way. For translocations the *in vivo* and *in vitro* dose-responses were practically equal, whereas for dicentrics (both in FPG and FISH analysis) the efficiency of *in vivo* irradiation was found to be lower than *in vitro* exposure (Figures 2 and 3). The similarity of translocation responses to fractionated (*in vivo*) and acute (*in vitro*) irradiation indicates that for dicentrics (also exchange type of CA), lower efficiency of *in vivo* irradiation vs. *in vitro* one may hardly explain by the effect of dose fractionation in the case of whole-body irradiation of donors. Most likely, this is due to elimination of some part of lymphocyte precursors with dicentrics, lethal for dividing cells, during 10 – 15 daily courses of whole-body irradiation of the patients. This is also confirmed by the fact that when translocations were scored in the so-called "unstable" cells only (e.g. having dicentrics and/or acentric fragments) their frequency per dose unit, similar to dicentrics, was lower in lymphocytes irradiated *in vivo* than *in vitro* (preliminary data).

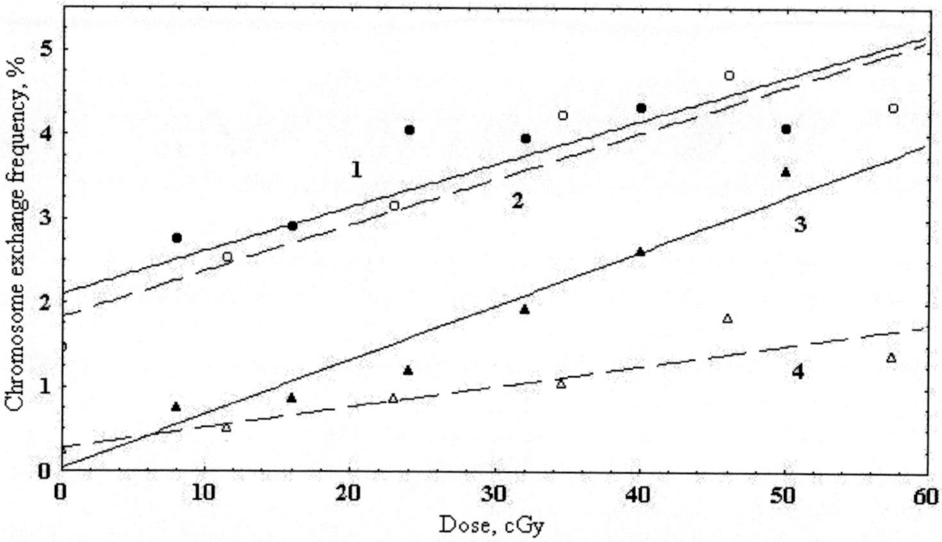

Figure 3. The dose-response curves for translocations and dicentrics in lymphocyte irradiated *in vitro* and *in vivo* (FISH-method): 1, •, continuous line – translocations *in vitro*; 2, ○, dotted line – translocations *in vivo*; 3, ▲, continuous line – dicentrics *in vitro*; 4, Δ, dotted line – dicentrics *in vivo*

On the cancer patient, who received in total 80.5 cGy dose of fractionated whole-body irradiation, an attempt was made to compare the dicentrics frequency observed with that predicted for this dose by different dose-effect models (Table 9). As shown in the Table, the observed frequency of dicentrics coincides most well with that forecasted by the model, obtained at whole-body irradiation of patients. When both *in vitro* calibration curves were used the predicted frequencies of dicentrics were higher than really observed ones. Of course in vivo dose response curves obtained on cancer patients could be applied with great care to the dose estimation of accidentally exposed people. Nevertheless, the data obtained in this study seem to be significant for biodosimetry based on the scoring of dicentrics. Since for dose reconstruction the individual frequency of dicentrics is usually referred to the *in vitro* calibration dose-response curve, the actual absorbed radiation dose in case of protracted exposure is likely to be underestimated.

ACKNOWLEDGMENTS

The authors are grateful to the scientists who participated in the investigation in different times: M.V. Vorob'yeva, A.I. Bogomazova, N.M. Timofeeva, A.Yu. Kanaeva, I.V. Petrova, and A.Yu. Pyukkenen.

The investigation was partly supported by RFFI grants No. 01-04-49118 and 04-04-48816.

References

[1] Sevan'kaev A.V., Anikina M.A., and Zavitaeva T.A., *Radiats. Biol. Radioekol.*, 1995, vol. 35(5), pp. 611 - 617. (Rus)

[2] Vorobtsova I.E., Vorob'yeva M.V., Bogomazova A.N., Timofeeva N.M., Efremova T.N., and Pyukkenen A.Yu., *Radiats. Biol. Radioekol.*, 1995, vol. 37(2), pp. 233 - 239. (Rus)

[3] Vorobtsova I.E., Semenov A.V., Kanaeva A.Yu., Timofeeva N.M., Yakovleva T.K., and Zharinov G.M., *Radiats. Biol. Radioekol.*, 2000, vol. 40(6), pp. 645 - 650. (Rus)

[4] Vorobtsova I., Darroudi F., Semyonov A., Kanaeva A., Timofeeva N., Yakovleva T., Zharinov G., and Natarajan A.T., *Int. J. Radiat. Biol.*, 2001, vol. 77(11), pp. 1123 - 1131.

[5] Perry P. and Wolf S., *Nature*, 1974, vol. 251, pp. 156 - 158.

[6] Kulka U., Huber R., Mueller P., Knehr S., and Bauchinger M., *Int. J. Radiat. Biol.*, 1995, vol. 68, pp. 25 - 27.

[7] Vorobtsova I.E. and Bogomazova A.N., *Radiats. Biol. Radioekol.*, 1995, vol. 35(5), pp. 636 - 639. (Rus)

[8] Jones I.M., Galick H., Kato P., Langlois R.G., Mendelsohn M.L., Murphy G.A., Pleshanov P., Ramsey M.J., Thomas C.B., Tucker J.D., Tureva L., Vorobtsova I., and Nelson D.O., *Radiat. Res.*, 2002, vol. 158, pp. 424 - 442.

[9] Hall S.C. and Wells J.J., *Radiol. Prot.*, 1988, vol. 8(2).

[10] Lucas J.N., Awa A., Straume T., Poggensee M., Kodama Y., Nakano M., Ohtaki K., Weier H.-U., Pinkel D., Gray J., and Littlefield G., *Int. J. Radiat. Biol.*, 1992, vol. 62(1), pp. 53 - 63.

[11] Muss V.F., Pereletov O.N., Larionova L.N., and Vladimirova V.A., *Med. Radiology*, 1992, No. 3, pp. 22 – 25. (Rus)

[12] Zherbin E.A. and Chukhlovin A.B., *Radiation Hematology*, Moscow, Meditsina, 1989, 176 p. (Rus)

[13] Sevan'kaev A.V., Lloyd D.C., and Braselmann H., *Radiat. Prot. Dosim.*, 1995, vol. 58(2), pp. 85 - 91.

[14] Lasutka J.R., *Mutat. Res.*, 1996, vol. 350, pp. 315 - 329.

[15] Rivkind N.B., Tsaritsina L.P., and Skvortsova G.V., *Gemat. i Transfuziologia*, 1996, vol. 68(7), pp. 77 – 79. (Rus)

[16] Neronova E.G., "Cytogenetic parameters of genome instability for the liquidators of the Chernobyl accident consequences in the late period", *Candidate Dissertation Thesis*, St.-Petersburg, 1997. (Rus)

[17] Maznik N.A. and Vinnikov V.A., *Tsitologia i Genetika*, 1997, vol. 31(6), pp. 41 – 47. (Rus)

[18] Vorobtsova I.E., Vorob'yeva M.V., Bogomazova A.N., and Pyukkenen A.Yu., *Radiats. Biol. Radioekol.*, 1995, vol. 35(5), pp. 630 – 635. (Rus)

[19] Mel'nov S.B., "Molecular-genetic and phenotypical effects of anthropogenic mutagenic impacts on a man", *Doctor Dissertation Thesis*, Minsk, Institute of Genetics and Cytology Nat. AS, Belarus, 2004. (Rus)

[20] Semov A.B., Iofa E.L., and Shevchenko V.A., *Radiats. Biol. Radioekol.*, 1994, vol. 34(6), pp. 865 – 871. (Rus)

[21] Vorobtsova I.E., Mikhel'son V.M., Vorob'yeva M.V., Pleskach N.M., Bogomazova A.N., Prokof'eva V.V., and Pyukkenen A.Yu., *Radiats. Biol. Radioekol.*, 1994, vol. 34(6), pp. 798 – 804. (Rus)

[22] Mel'nov S.B., Koryit'ko S.S., Vinnik L.M., and Isaikina Ya.I., In Coll.: *Proc. 4th Intern. Conference "The Chernobyl Catastrophe: Forecasting, Prophylaxis, Treatment and Medicobiological Rehabilitation of the Victims"*, Minsk, 1995, p. 259. (Rus)

[23] Tsyib A.P., Ivanov V.K., and Airapetov S.A., *Med. Radiologia*, 1992, No. 1, pp. 46 – 51. (Rus)

[24] Moiseenko V.V., "The application of biophysical modeling methods to retrospective estimation of doses by chromosomal aberrations in different periods after irradiation", *Candidate Dissertation Thesis*, Obninsk, MRNTs RAMN, 1993. (Rus)

[25] Sevan'kaev A.V., *Radiosensitivity of Human Chromosomes in the Mitotic Cycle*, Moscow, Energoatomizdat, 1987, 160 p. (Rus)

[26] Fedortseva R.F., Kravtsov V.Yu., Belyakov O.V., Vorobtsova I.E., Nekrasov V.N., and Nikiforov A.M., *Int. J. Pediatric Hematol./Onkol.*, 1997, vol. 4(3). pp. 295 - 299.

[27] Vorobtsova I.E., Taker J.D., Timofeeva N.M., Bogomazova A.N., Semenov A.V., and Pleshanov P.G., *Radiats. Biol. Radioekol.*, 2000, vol. 40(2), pp. 142 – 148. (Rus)

[28] Vorobtsova I., Semenov A., Timofeeva N., Kanaeva A., and Zvereva I., *Mech. Ageing and Develop.*, 2001, vol. 122, pp. 1373 - 1382.

[29] Vorobtsova I.E., Lyubimova N.E., Perova A.A., and Semenov A.V., *Ekol. Genetika Cheloveka*, 2004, vol. 2(2), pp. 35 – 40. (Rus)

[30] Snigiryova G., Braselmann H., Salassidis K., Shevchenko V., and Bauchinger M., *Int. J. Radiat. Biol.*, 1997, vol. 71(2), pp. 119 - 127.

[31] Domracheva E.V., Aseeva E.A., Dyachenko L.V., and Rivkind N.B., *Thes. Rep. 3rd Congress on Radiation Investigations*, 14 – 17 October, 1997, Pushchino, 1997, pp. 48 – 49. (Rus)

[32] Littlefield L.G., McFee A.F., and Salomaa S.L., *Radiat. Res.*, 1998, vol. 150(2), pp. 237 - 249.

[33] Granath F., Darroudi F., Auvinen A., Ehrenberg L., Hakulinen T., Natarajan A.T., Rahu M., Rytomaa T., Tekkel M., and Veidebaum T., *Mutat. Res.*, 1996, vol. 369(1-2), pp. 7 - 12.

[34] Bigbee W.L., Jensen R.H., Veidebaum T., Tekkel M., Rahu M., Stengrevics A., Auvinen A., Hakulinen T., Servomaa K., and Boice J.D., *Radiat. Res.*, 1997, vol. 147, pp. 215 - 224.

[35] Saenko A.S., Zamulaeva I.A., Smirnova S.G., Orlova N.V., Selivanova E.I., Matveeva N.P., Kaplan M.F., Nugis V.Y., and Tsyb A.F., *Int. J. Radiat. Biol.*, 1998, vol. 73(6), pp. 613 - 618.

[36] Vorobtsova I.E., Kanaeva A.Yu., Petrova I.A., Semenov A.V., Pleskach N.M., Spivak I.M., Timonina G.A., Prokof'eva V.V., Yartseva N.M., and Mikhel'son V.M., *Tsitologia*, 2004, vol. 46(12), pp. 1030 – 1034. (Rus)

[37] Mazurik V.K. and Mikhailov V.F., *Radiats. Biol. Radioekol.*, 2001, vol. 41(3), pp. 272 – 289. (Rus)

[38] Vorobtsova I.E., Vorob'eva M.V., Korytova L.I., and Shust V.F., *Tsitologia*, 1995, vol. 37(5/6), pp. 449 – 457. (Rus)

[39] Vorobtsova I.E., *Radiobiologia*, 1987, vol. 27(3), pp. 377 – 381. (Rus)

[40] Vorobtsova I.E., *Mutagenesis*, 2000, vol. 15(1), pp. 33 - 38.

[41] Pelevina I.I., Aleshchenko A.V., Antoshchina M.M., Gotlieb V.Ya., Kudryashova O.V., Semenova L.P., and Serebryanyi A.M., *Radiats. Biol. Radioekol.*, 2003, vol. 43(2), pp. 161– 166. (Rus)

[42] Kuz'mina N.S. and Suskov I.I., *Radiats. Biol. Radioekol.*, 2002, vol. 42(6), pp. 741–745. (Rus)

[43] Hagmar L., Brogger A., and Hansteen I.L., *Cancer Res.*, 1994, vol. 54, pp. 2919 - 2922.

[44] Lloyd D.C. and Edwards A.A., *Radiation Induced Chromosome Damage in Man*, Eds. T. Ishihara and M.S. Sasaki, New York: A.R. Liss, 1983, pp. 23 - 49.

[45] *Biological Dosimetry for Persons Irradiated by the Chernobyl Accident*, Eds. D.C. Lloyd and A.V. Sevan'kaev, EUR Report №16532, Brussels: CEC, 1996.

[46] Lloyd D.C., Edwards A.A., Sevan'kaev A.V., Bauchinger M., Braselmann H., Georgiadou-Schumacher V., Salassidis K., Darroudi F., Natarajan A.T., Van der Berg M., Fedortseva R., Fomina Z., Maznik N.A., Melnov S., Palitti F., Pantelias G., Pilinskaya M., and Vorobtsova I.E., "The radiological consequences of the Chernobyl accident", Eds. A. Karaoglou, H. Desmet, G.N. Kelly and H.G. Menzel, *Proc. 1st Int. Conf.*, Minsk, 18-20 March, 1996, EUR Report №16544, Brussels: CEC, 1996.

[47] Buckton K.E., Langlands A.O., Smith P.G., Woodcock G.E., and Looby P.C., *Int. J. Radiat. Biol.*, 1971, vol. 19(4), pp. 369 - 378.

[48] Leonard A., Baltus L., Leonard E.D., Gerber G.B., Richard F., and Wambersie A., *Radiat. Res.*, 1995, vol. 141, pp. 95 - 98.

[49] Schmid E. and Bauchinger M., *Int. J. Radiat. Biol.*, 1974, vol. 26, pp. 31 - 37.

In: 20 Years after the Chernobyl Accident
Editors: E. Burlakova and V. Naidich, pp. 75-85
ISBN 1-60021-249-2

Chapter 6

THE QUESTIONS OF HEALTH MONITORING OF CHILDREN SUBJECT TO RADIATION IMPACT

L. S. Baleva, A. E. Sipyagina, R. N. Terletskaya, E. B. Lavrent'eva, I. N. Yakovleva, T. B. Kuzmina, S. A. Zotova, N. M. Karakhan and N. A. Bondarenko
Children's Scientific-Practical Centre of Radiation Protection, Moscow, Russia

Today, expecting the 20th anniversary of the Chernobyl nuclear accident, the Russian Federation is at the stage of socio-economic transformations, changes and supplementing the basic regulatory documents and laws. Therefore, proper estimation of the population health is of the highest importance, especially radiation-impacted children, living on the territories, polluted with radionuclides. It is also important to indicate the basic reasons currently disturbing health of children and teenagers.

Much has been done during these 20 years, namely:

- the Law "On the social protection of citizens impacted by radiation at the Chernobyl nuclear accident", which includes the 25th article defining social protection, medical care and rehabilitation of children and teenagers has been adopted;
- Federal purpose-oriented programs "Children of Chernobyl" and "Overcoming consequences of the radiation accidents till 2010" has been adopted;
- the three-level health survey defining the scope and order of medical care at all stages of medical observation has been created and adopted;
- the all-year rehabilitation and health promotion system for children, which is the integral and essential part of the health survey system, has been developed and adopted;
- the cohorts of the infant population, which needs long-term health survey, have been determined;
- the level of equipment of childhood and obstetric aid institutions has been increased and medical staff for caring children impacted with radiation has been trained;

- working enfant medical and radiation-monitoring and cancer registers has been created, the system of iodine prophylaxis with centralized purchasing of required drugs has been developed, neonatal screening for inborn hypothyroidism has been adopted, and high-technology methods for diagnosis and curing have been applied.

However, at present the health state of children living on the territories polluted with radionuclides also causes great concern. The basic parameters of the children's health state subject to radiation impact, included in the National Register, were analyzed pursuant to the data of official statistics (statistic forms No. 15, 16, 12, 7 and 32) and the results of 19-year annual health survey and penetrating treatise basing on the All-Russian children's health survey results for 2002.

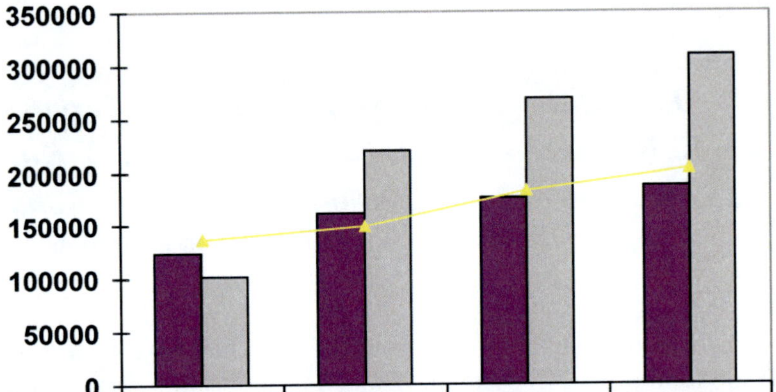

	1990	1995	2000	2003
Bryansk Oblast (form 12)	124197,9	162200,1	175604,9	186916,6
Bryansk Oblast (form 16)	101439,1	219820,2	267951,7	309055,4
RF (form 12)	136419,7	148632,3	181358,4	202040,6

Figure 1. Occurrence of diseases in children (per 100,000 of the children's population)

The comparative analysis of general occurrence of diseases (Figure 1) testifies about its higher level among children from the regions of the Bryanks Oblast polluted with radionuclides since 1991 than in the whole Oblast. This Oblast was higher radiation-polluted the Chernobyl nuclear accident on 26 April, 1986, and now it presents all soil pollution zones with cesium-137. Analogous data were also obtained for other regions of the Russian Federation, higher polluted with radionuclides: Kaluga Oblast and Tula Oblast (228,555.3 and 231,712.1 per 100,000 children, respectively; 202,040.6 for the Russian Federation; the data shown were obtained in 2003).

Altogether, in the period since 1990 till 2003 the occurrence of diseases in polluted regions of the Bryansk Oblast increased three times, whereas in the whole Oblast (as well as in the Russian Federation) – 1.5 times. However, the most intensive increase of the disease occurrence was observed in 1996; in recent three years, the rate of increase reduced and approached their values for other comparable territories (the increase of disease occurrence for children was almost the same – 1.15 times). It is the authors' opinion that this fact clearly illustrates possible decrease of the radiation impact negative consequences for the children's by implementation of adequate rehabilitation and therapeutic measures representing the integral part of developed and carried out health survey.

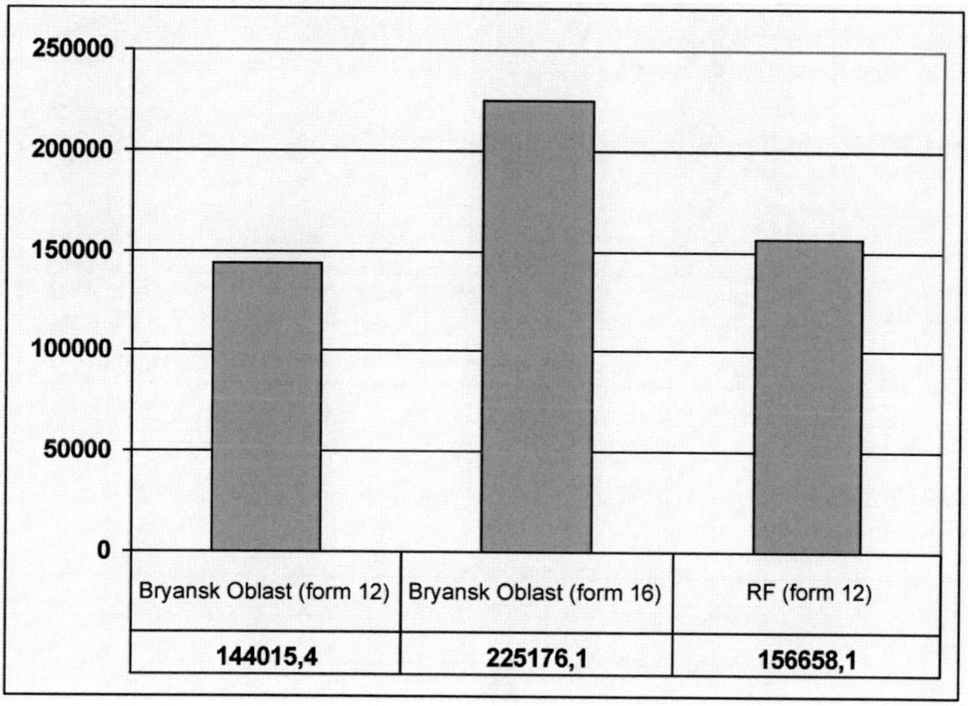

Figure 2. Diseases initially detected in 2003 (per 100,000 of the children's population)

Nevertheless, the preserved excess of total occurrence of diseases in radionuclide-polluted regions and high level of *primary morbidity* (Figure 2) are the reasons for concern for health state formation in the children's living on polluted territories. The following important fact is of special attention that the entire Bryansk Oblast is the endemic region in relation to iodine deficiency (mild case of ioduria) that has the mandatory effect on the morbidity. However, analysis of initially detected morbidity indicates that the health state of children is affected by a combination of endemic disease factor and anthropogenic radiation pollution. That is why, besides the iodine prophylaxis importance, the value of a complex of measures decreasing possible radiation impact on the children's health must be indicated.

The health survey analysis data indicate that dispersion of the child diseases in *both compulsory evacuation zones and with right for compulsory evacuation zones* (282,454 and 330,587 according to form 16) exceeds the general district level (186,916 according to form 12) (Figure 3). Therefore, *the whole* system of health survey measures must be applied to children living in both zones of radionuclide pollution zones mentioned. Twenty years after the radiation accident not so much the radiation pollution level as the fact of dwelling in the regions of any radionuclide contamination becomes of the greater importance.

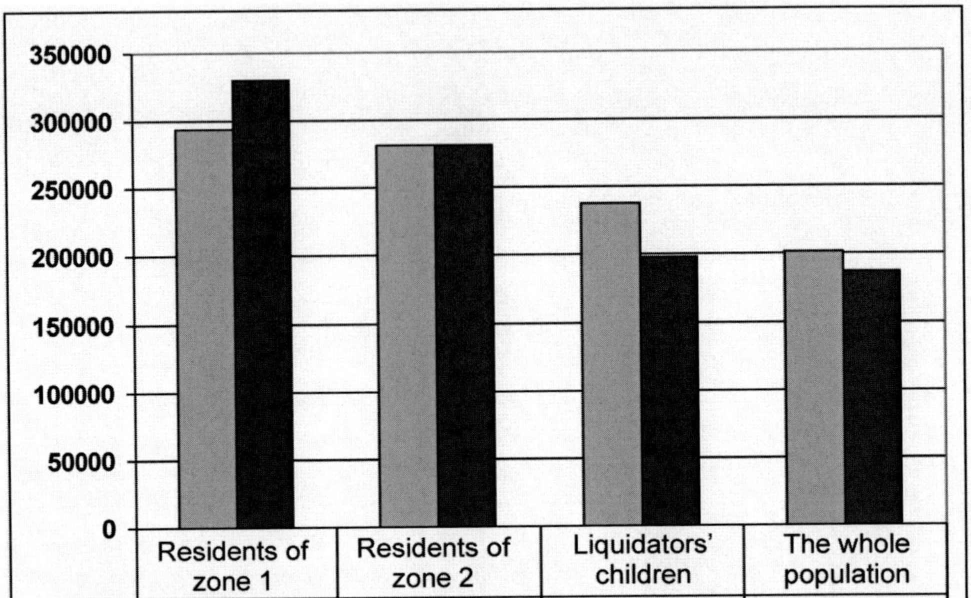

	Residents of zone 1	Residents of zone 2	Liquidators' children	The whole population
BO	293761,1	281454,4	237413,9	202040,6
RF	330587,3	281454,4	197763,6	186916,6

Figure 3. Occurrence of diseases among children from various observation cohorts in 2003 (per 100,000 of the children's population)

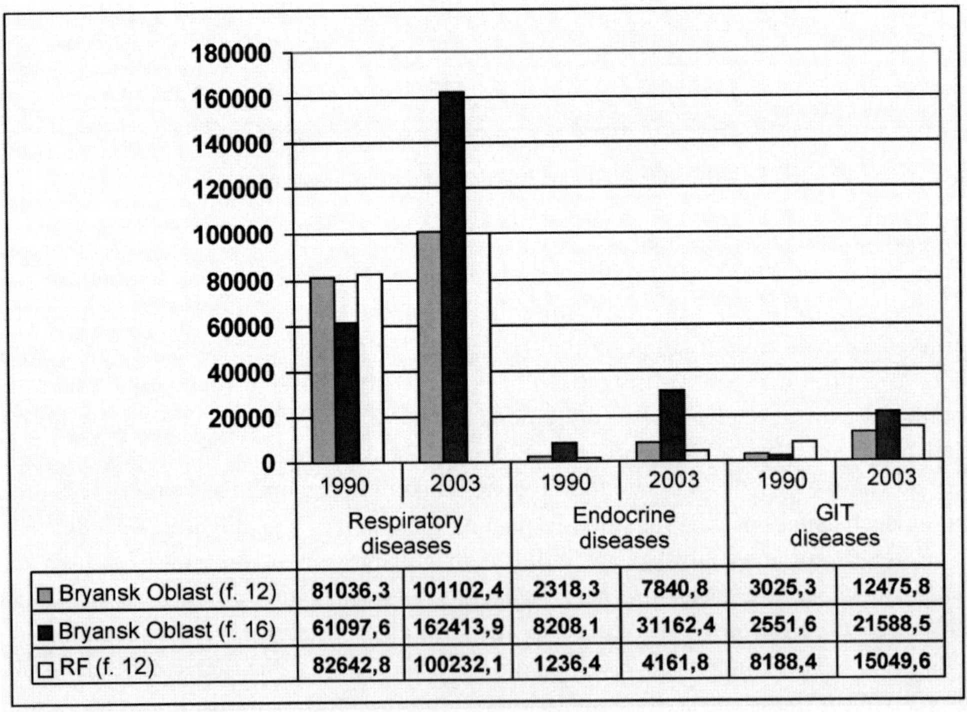

	Respiratory diseases 1990	Respiratory diseases 2003	Endocrine diseases 1990	Endocrine diseases 2003	GIT diseases 1990	GIT diseases 2003
Bryansk Oblast (f. 12)	81036,3	101102,4	2318,3	7840,8	3025,3	12475,8
Bryansk Oblast (f. 16)	61097,6	162413,9	8208,1	31162,4	2551,6	21588,5
RF (f. 12)	82642,8	100232,1	1236,4	4161,8	8188,4	15049,6

Figure 4. Occurrence increase dynamics for some classes of diseases among children (per 100,000 of the children's population)

The analysis implemented shows that children's *morbidity structure* (Figure 4) also possesses structural features, which are:

the first place, as well as in the whole Russia, is occupied by *respiratory diseases* (the 10th class of diseases according to International Classification of Diseases- 10 (ICD-10), to a greater extent, due to high occurrence of pharyngonasal cavity diseases. However, despite the implementation of immunoprophylaxis measures, the level and rate of the occurrence is higher for children and teenagers from radionuclide-polluted regions. For example, in the period between 1990 and 2003, the occurrence of respiratory diseases in children from polluted regions of the Bryansk Oblast was 2.6-fold increase with the maximum observed in 1996, whereas in general for this Oblast and the Russian Federation it showed lower increase (about 1.2 times);

contrary to the All-Russian children's population, in the Bryansk Oblast the second place by occurrence is occupied by *endocrine diseases* (the fourth class of diseases according to ICD-10), accented to predominance of thyroid gland diseases. In the radiation-polluted regions, this pathology increase was observed in the period between 1990 and 2001 (by 3.7 times). For the sake of comparison, for the whole Bryansk Oblast and the Russian Federation this increase equaled 3.3 times. Current rates of increase are lower, but the level of thyroid pathology occurrence remains high (Figure 5).

It is obvious that the questions of prophylactic system measures, aimed at reduction of negative consequences, are preserve their high actuality still. The more so, it should be outlined that the first generation of children from irradiated parents with thyroid gland pathology was born already;

the third place is occupied by *GIT diseases* (liver, stomach and upper intestine diseases). In the all-Russian children's population this class of pathologies takes the second place. This class of diseases (Figure 4) is of special attention because of high increase rate of their occurrence. For example, in 1994, the dispersion index of chronic digestive apparatus pathologies for children living in the radiation-polluted regions increased by 8.4 times (the same parameter averaged by the Bryansk Oblast equaled 4.1 times and 1.8 times for the whole Russia). It is the authors' opinion that of a significant role in this question is the increase of the part of internal radiation due to consumption of food from homestead lands. For solving this problem of the great importance is application of the balanced system of children's feeding at all stages of their development (ontogenesis), first of all, aimed at consumption of ecologically pure products [1].

All the above-mentioned systems of organs relate to the organs – targets for the radiation impact. The emission of radioactive iodine during the Chernobyl nuclear accident made the endocrinal (thyroidal) system the main object for medical measures, aimed at minimization of the accident consequences on the man's health. This allowed for decelerating the increase of thyroidal pathologies. On the other hand, the increase of locally produced food consumption, observed in recent 10 years, may promote for internal radiation increase due to direct or indirect influence of radionuclides on the systems and tissues of internals.

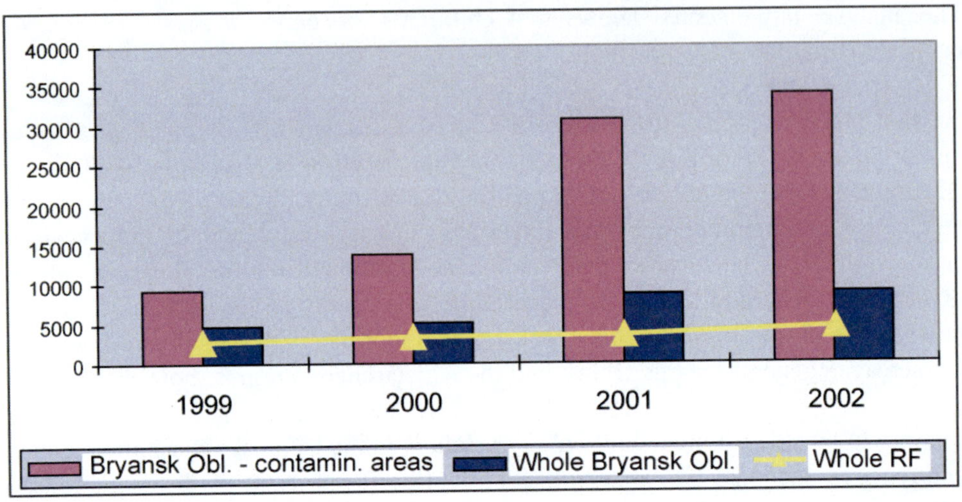

Figure 5. Dynamics of endocrinal system disease occurrence for children **(per 100,000 of the children's population)**

Thus the above data indicate recent assignable interest increasing all over the world to *somatic pathologies* of people subject to the radiation impact [2]. This problem is always urgent for us.

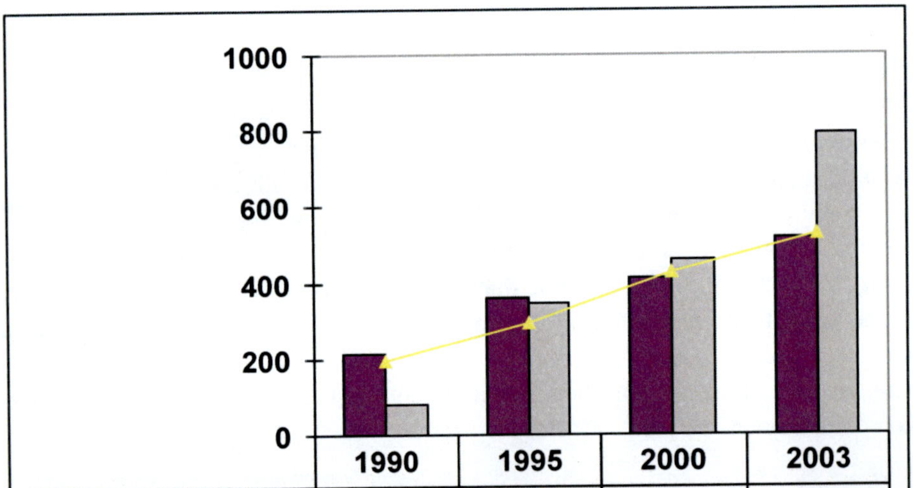

	1990	1995	2000	2003
Bryansk Oblast (form 12)	215,2	362,5	414,5	514,3
Bryansk Oblast (form 16)	80,9	344,4	459,2	793,5
RF (form 12)	193,8	292,2	428,7	523,8

Figure 6. Neoplasm occurrence dynamics for children **(per 100,000 of the children's population)**

Nevertheless, the importance of monitoring of traditionally *classic radiation-induced* diseases [3], such as *neoplasms (especially malignant ones* representing the second class of diseases according to ICD-10) and *congenital abnormalities (deficiency) of development* (the seventeenth class of diseases).

The observation data indicate the increase of *neoplasm* occurrence in radiation-polluted regions. Hence, the rate of this increase was maximal (9.9 times) compared with the whole Bryansk Oblast and the Russian Federation (2.4 and 2.7 times, respectively; Figure 6). For *malignant neoplasms* (Figure 7), the pathology occurrence in the regions polluted by radionuclides increased by 1.9 times (for the sake of comparison, the average indices for the Bryansk Oblast are similar to the all-Russian ones). The maximal level of malignant neoplasms (mostly thyroid gland cancers) was recorded in the period of 1996 – 1999 (with the maximum increase by 2.6 times) with consecutive decrease of the indices. The *structure of malignant neoplasms*, observed for the radiation-polluted regions of the Bryansk Oblast, displays cases of newly obtained localization: in digestive, respiratory and genital organs. Moreover, oncopathology determined is localized in the central nervous system, bones and joints, including malignant neoplasms of lymphoid tissue.

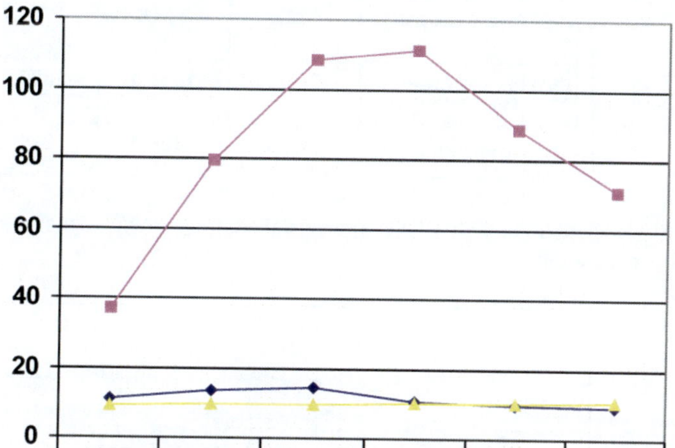

	1991	1995	1997	1998	2000	2003
Bryansk Oblast (form 7)	11,2	13,6	14,3	10,6	9,6	9,1
Bryansk Oblast (form 16)	36,9	79,5	108,2	111,2	88,5	70,7
RF (form 7)	9,2	9,7	9,7	10,1	10,1	10,4

Figure 7. Malignant neoplasm occurrence dynamics for children **(per 100,000 of the children's population)**

As indicated by analysis of *congenital abnormalities (deficiency) of development* (Figure 8), since 1995, children living in radiation-polluted regions of the Bryansk Oblast possess higher concentration of radionuclides compared with unpolluted regions. Hence, the occurrence increase for these class of diseases in children from polluted regions is higher (by 5.1 times), whereas in polluted regions of the Bryansk Oblast and the Russian Federation – by 2.0 and 2.5 times, respectively. *The structure of congenital abnormalities of development*

(Figure 9) displays predominance of deficiency in the blood circulation system development. However, the deficiency in development of the musculoskeletal system, genital organs and the nervous system are also of high significance. In recent five years, examinations of children from radiation-polluted regions of the Bryansk Oblast indicated 3.29-fold increase of congenital cataract and 1.47-fold increase of Down's syndrome occurrences; multiple congenital anomalies and development deficiencies also take place.

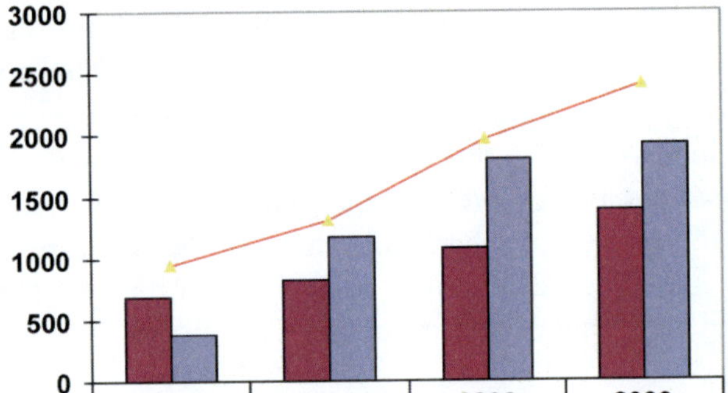

	1990	1995	2000	2003
Bryansk Oblast (form 12)	683,5	814,6	1075,1	1388,1
Bryansk Oblast (form 16)	375,1	1163,9	1800,9	1921,1
RF (form 12)	943,9	1306,3	1970,3	2409,4

Figure 8. Congenital anomaly (development defects) occurrence dynamics for children (per 100,000 of the children's population)

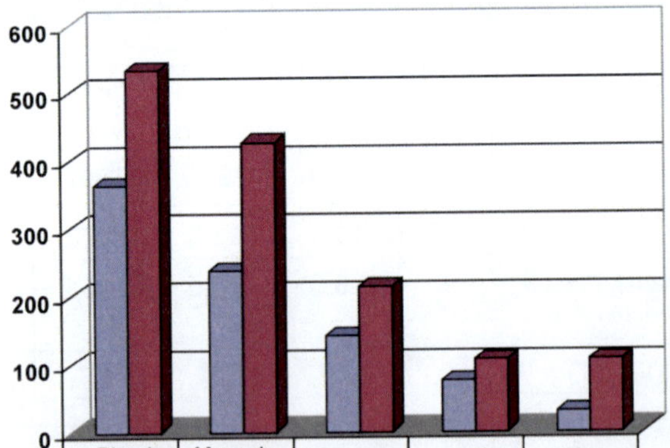

	Blood circulation system	Musculo-skeletal system	Sexual organs	Nervous system	GIT organs
Residents of radiation-contaminated regions	362,7	236,4	142,4	75,2	29,6
Liquidators' children	532,5	426,1	213,1	106,5	106,5

Figure 9. Occurrence of various localizations of congenital defects in the Bryansk Oblast for children in accordance with the cohort belonging in 2003 (per 100,000 of the children's population)

This problem cannot be studied without using high-technology complex *cytogenetic and molecular-genetic methods* of examination and observation of children and their families. In this case, the long-term observation for formation of both genetic (for example, 17th circular chromosome syndrome) and somatic mutations, which lead to genome instability with consecutive formation of a chronic pathology, including malignant neoplasms, congenital anomalies and development deficiencies, are of high significance.

Children subject to radiation impact display the reliable increase of aberrant metaphase level and aberrations of both chromatide (single fragments) and chromosomal (paired fragments, dicentric and circular chromosomes) types, related to unstable chromosomal aberrations. However, an increase of stable chromosomal aberrations (deletions and translocations) was also registered (see Table 1 below).

Table 1. Characterization of cytogenetic indices for children subject to radiation impact and in control (M ± m)

Indices	Children subject to radiation impact	Control
Total number of chromosomal aberrations (frequency, %)	4.68 ± 0.16*	2.54 ± 0.18*
Single fragments (frequency per 100 cells)	3.87 ± 0.21*	1.76 ± 0.12*
Paired fragments (frequency per 100 cells)	0.79 ± 0.15*	0.28 ± 0.06*
Dicentric and circular chromosomes (frequency per 100 cells)	0.34 ± 0.01*	0.06 ± 0.004*
Translocations (frequency per 100 cells)	0.29 ± 0.11	0.14 ± 0.05
Deletions (frequency per 100 cells)	0.28 ± 0.12	0.17 ± 0.07

Note: * Reliable differences from the control ($p < 0.05$).

Examinations of all children subject to radiation impact since prenatal period of development demonstrated much higher changes in the cytogenetic picture: a noticeable broadening of the spectrum of cytogenetic structural defects was observed (the presence of isochromatide fragments, breaks by centromere, and chromatide exchanges). The karyotyping procedure has indicated increasing mosaicism by aneuploidy and trisomy (B, C, D) in children, born in 1988. Children born in 1986 – 1987 showed polyploidy.

The current investigation shows that children subject to radiation impact since prenatal stage of development demonstrate a decrease of individual heterozygosity index that may testify about narrowing the spectrum of adaptive possibilities and, consequently, hypersensitivity to ecopathogenic factor. For children whose prenatal development proceeded under conditions of increased background radiation and born in 1988, the individual heterozygosity index was 2.61 times lower than in control (with reliable differences, $p < 0.05$).

Structural genetic changes in cells may be eliminated by activation of the DNA repair system. Decreasing repair processes, both spontaneous and induced, testify about decreased protective potential and, consequently, increased individual sensitivity to impacts of some ecopathogenic factors. Children born in 1986 – 1987 show the index of gamma-induced repair activity slightly higher than in control, but lower than for dwellers of polluted territories, not prenatally irradiated. The group of children born since 1988 demonstrates reliable ($p < 0.02$) decrease of the repair activity index to (0.97 ± 0.03).

Such investigations, performed with respect to the family aspect, form the basis for proving the estimate of the cause-effect relations between a disease occurrence and the radiation impact factor. This question we assume to extremely urgent. Therefore, in-depth studies of the health of the 1st and the 2nd generation of children born from irradiated parents are carried out in the Children's NP Centre on the Radiation Protection [4, 5].

We also think it important to pay special attention to the cohort of children born from parents – liquidators of the Chernobyl nuclear accident consequences. Figure 9 gives a good illustration of their health. For example, they demonstrate higher pathology occurrence by the initial five positions of the main development deficiency localizations compared with children born not from the liquidators of the Chernobyl nuclear accident consequences, but from parents living on the territories polluted by radiation from Chernobyl NPP.

The study of cytogenetic characteristics, protein polymorphism and DNA radiation activity in children born from parents (fathers) – liquidators of the radiation accident consequences indicates the presence of genome instability in the children's population and narrowing adaptive abilities of descendants, conceived within a year after returning from works on liquidation of the Chernobyl nuclear accident consequences.

Moreover, it should be noted that within recent five years, the infant mortality caused by congenital development defects is 1.35 – 1.45 times higher in the Bryansk Oblast is higher than in the Russian Federation (45.0 and 31.3 per 10,000 children born alive, respectively).

Cooperation with the applied health care organs of the Bryansk Oblast is aimed at the control and prophylaxis of negative consequences of the combined action of ecopathogenic factors on the child population. This is testified about by low increase rate of *children's disability* at high occurrence of chronic diseases compared with the all-Russian indices [6]. In recent years, the increase rate of the number of disabled children in the Bryansk Oblast equaled that in the whole Russia (1.4 and 1.3 times, respectively).

Currently changing economic and social situation imposes the necessity of subsequent development of children's aid, including the victims of ionizing radiation impact. It is very important to implement weighted and elaborated ecological medical and social prophylaxis, which consists of the following aspects:

- the mandatory inclusion of the clauses on the social protection of children and teenagers into coming up law;
- implementation of the state guarantees for children and women support with free medical care and simultaneous preservation of accessibility to expensive medical procedures;
- development and provision for cooperation at different levels of executive power, health care management, mandatory medical insurance funds, various medical insurance departments and public organizations;
- depression of the radionuclide pollution level of both external and internal environment, prophylaxis of goitrous endemia, minor nutrient deficiency, balancing of nutrition rations;

- training of children, adults and new generations using novel informative methods and techniques, simultaneously providing no "*radiophoby*" and occurrence of absolute negativism;
- biosocial adaptation of the child population;
- health survey development (implementation of high-quality examinations, rehabilitation and health promotion), which requires the every possible support for technical outfitting of therapeutic institutions of childhood and obstetric aid with desirable appropriation of required funds for this clause;
- expansion of prophylactic technology application, including the prenatal period, providing for decrease of morbidity and mortality due to congenital development defects (improving methods of prenatal diagnosis and monitoring of congenital development defects);
- increasing the role of scientific centres and research institutes in development and application of effective diagnosis, treatment and rehabilitation methods (including hi-tech), based on the demonstrative medicine and clinical and economic analysis;
- occupational training of medical staff.

The present global development proceeding in the Russian Federation represents the important moment for surveying the population, especially children, from positions of both detection of diseases and chronic pathologies and their prophylaxis.

References

[1] Baleva L.S., Lavrent'eva E.B., and Sokha L.G., 'The role of rational feeding of children living on radiation-polluted territories', *Ecological anthropology. Yearbook*, Ed. T.V. Belookaya, Minsk, 1998, pp. 270 – 272. (Rus)

[2] Shimizu Y., Pierce D.A., Preston D.L., and Mabuchi K., 'Studies of the mortality of atomic bomb survivors. Noncancer mortality: 1950-1990', *Radiation Research*, 1999, vol. 152, pp. 374 – 389.

[3] *MKRZ Recommendations*, Moscow, Energoatomizdat, 1994, Part 1, 191 p.; Part 2, 207 p. (Rus)

[4] Baleva L.S., Kuz'mina T.B., Lavrent'eva E.B., and Sipyagina A.E., 'The health state of the first child generation born from parents subject to ionizing radiation as teenagers and still living on the radiation-polluted territories', *Ecological anthropology. Yearbook*, Ed. T.V. Belookaya, Minsk, Belarus Committee 'Dzetsi Chernobuilya', 2002, p. 90 – 93. (Rus)

[5] Baleva L.S., Sipyagina A.E., Terletskaya R.N., Yakovleva I.N., Lavrent'eva E.B. *et al.*, 'The state of children's health – the progenitors of the first and the second generation from irradiated parents. The principles for medical observation and rehabilitation', *Manual for Doctors*, Moscow, 2004, 12 p. (Rus)

[6] Baleva L.S., Lavrent'eva E.B. Sipyagina A.E., Terletskaya R.N., Yakovleva I.N. *et al.*, 'The features for the causal structure of disability of various age groups of children subject to radiation impact', *Manual for Doctors*, Moscow, 2004, 7 p. (Rus)

In: 20 Years after the Chernobyl Accident
Editors: E. Burlakova and V. Naidich, pp. 87-98
ISBN 1-60021-249-2

Chapter 7

MOLECULAR-GENETIC ANALYSIS OF DNA POLYMORPHISM IN DESCENDANTS OF MICE EXPOSED TO CHRONIC AND ACUTE γ–RADIATION AT DIFFERENT STAGES OF SPERMATOGENESIS

A. E. Myazin**, M. D. Pomerantseva, L. K. Ramaya, G. O. Shaikhaev and V. A. Shevchenko
N.I. Vavilov Institute of General Genetics,
Russian Academy of Sciences, Moscow, Russia

ABSTRACT

Molecular-genetic effects in descendants of mice-males of BALB/c and CBA/lac lines exposed to chronic (with 0.0017 Gy/day dose rate and 0.36 Gy total dose) and acute radiation of 1, 2 and 3 Gy doses were studied. For the basic approach, the analysis of variability of DNA amplification products with a series of random primers (RAPD-analysis) was used. Random primers were applied both single and in mixtures. The work comparatively analyzes genetic radiosensitivity of stem spermatogonia and spermatoblasts. At acute irradiation, the dose–effect dependencies for RAPD-marker polymorphism levels with the mutation maximum at doses of 1 Gy for spermatoblasts and 2 Gy for spermatogonia were obtained. At chronic irradiation, reliable differences with the control were detected for primer 759 and a mixture of primers M1 and M2 used. The comparison analysis of genetic radiosensitivity at different stages of spermatogenesis in mice indicated equal sensitivity of spermatogonia and spermatoblasts for 1 and 3 Gy radiation doses and higher sensitivity of spermatogonia for 2 Gy dose.

Keywords: Genetic effects of radiation, radiosensitivity of spermatogenesis stages, DNA polymorphism, RAPD-analysis

* Main addressee: N.I. Vavilov IGG RAS, A.E. Myazin; 3 Gubkina street, 119991 Moscow, Russia; Tel: (7-095) 135-1139; E-mail: myazin@vigg.ru

The tragedy of the Chernobyl NPP catastrophe happened in 1986 has one more time outlined the urgency of investigations, aimed at genetic effects of ionizing radiation. The study of separate consequences is currently one of the most important directions in the radiation genetics.

In recent decades, the radiation induction of DNA polymorphism was the object of intensive studies. The number of works devoted to analysis of ionizing radiation effects on various markers of DNA polymorphism increases with years. The induction of polymorphism in microsatellite and minisatellite loci is the question most frequently discussed in them. V.G. Bezlepkin and A.I. Gaziev, as well as many other authors consider increasing level of polymorphism in highly variable microsatellite (MCS) and minisatellite (MNS) loci as a manifestation of genome instability (GIS) [1]. A typical feature of highly variable loci is represented by much higher frequency of spontaneous mutations in them compared with encoding loci – by 100 times for MNS and 1,000 times for MCS, approximately [2]. For MCS and MNS, the induction of mutation process due to impacts of physical and chemical factors, ionizing radiation, first of all, has been shown [3 – 14]. In MCS and MNS, comparatively high frequency of spontaneous mutations allow for their using as markers in the studies of radiation consequences manifested in the progeny of irradiated parents, including GIS markers. At the same time, extremely high variability of MCS and MNS is the alerting factor and makes us interpret the results of their use more carefully [3].

Many works implementing another type of markers of DNA polymorphism used for estimation of genetic effects of radiation – random amplified polymorphic DNA (RAPD), were published. The analysis of RAPD-marker polymorphism gives an opportunity for multilocus scanning of hypervariable areas of DNA. The method represents a PCR (polymerase chain reaction) modification and is based on the use of a series of random primers. A singular random primer or a mixture of primers is used in each reaction. Electrophoregrams of the products possessing this amplification produce a picture specific for current genome, the so-called DNA-fingerprints. The specificity of RAPD-marker patterns allows for a broad application of this method to phylogenetic investigations, breeds, lines, grades and strains, as well as to genotyping of microorganisms, including their pathogenic species.

Recently, the RAPD-analysis has been applied to radiation genetic monitoring. Weinberg *et al.* [15] have used amplification with a random primer for examining children of the Chernobyl accident liquidators, for which a reliable increase of non-parent bands e.g. the increase of polymorphic DNA level. The RAPD-analysis was used in the study of transgeneration GIS phenomena, induced by chronic and acute γ–radiation [16 – 21].

The disadvantage of this method is low reproducibility of electrophoregrams related to high sensitivity of the method to the change in amplification parameters and the reaction mixture composition. Of special meaning is stable operation of polymerase and the quantity of matrix DNA e.g. standardization of the reaction requires preliminary specification of its conditions.

The present work was intended to study effects of chronic and acute radiation occurring in the progeny of irradiated mice using random amplified polymorphic DNA assay (the RAPD-analysis).

MATERIALS AND METHODS

Radiation Impact and Crossing

Mice males of BALB/c line were irradiated by 1 – 3 Gy dose on the device GUPOS-1 (producing 4.5 Gy/min dose rate from ^{137}Cs source). During the second and the third week, and three months after the radiation impact, these males were crossed with females of the same line. The interval between the radiation impact and initial coupling was chosen so that spermatozoids participating in fertilization existed at pre-meiotic and post-meiotic stages of maturation during the radiation impact. On the second and the third weeks after irradiation, fertilization proceeded with generative cells, which existed at the spermatoblast stage during irradiation, which is the post-meiotic stage. Irradiated stem spermatogonia (pre-meiotic cells) are converted to spermatozoids ready for fertilization within two-three months. Stem spermatogonia represent the cells most resistant to the ionizing radiation effect. The progeny from control and irradiated males was exposed to standard conditions and killed at 3 – 4 week age. At the end of the experiment, all males and females from all groups were killed for the purpose of sampling from the parent generation.

Male mice of CBA/lac line were irradiated in the device UOG-1 (producing 0.17 cGy/day dose rate from ^{137}Cs source) during 210 days. Total dose applied to mice equaled 35.7 Gy. A week after termination of the radiation impact males from the test and control groups were crossed with hybrid females of CBA/lac X C57/bl line. In both cases, descendants were exposed to standard conditions and killed at the age of 3 – 4 weeks.

Genomic DNA Extraction

DNA was extracted from liver using Diatom DNAprep™ (OOO Izogen Co.) sets in accordance with enclosed instructions. A set is based on application of a sorbing agent, which is selectively linked to nucleic acids, and allows obtaining high-molecular (40,000 – 50,000 b.p.) DNA ready for further use, including the polymerase chain reaction.

DNA Amplification with Random Primers

Table 1 presents sequences of primers, used in the current work. Due to specificity of the method, amplification must proceed under tightly standard conditions. Therefore, we have chosen lyophilized GenePak PCR Core^{tm} mixtures containing components necessary for proceeding of the reaction, including thermally inhibited Taq-polymerase. The prepared reaction mixture was added by 50 ng of DNA and by 10 μM of each primer. The amplification program consisted of 35 cycles: T_d – 95°C, 25 s, T_m – 42 (cycles 1 – 4), then 58°C (cycles 5 – 35), 5 s, T_c – 74°C, 80 s. For PCR, programmable thermostat PT-48 (OOO TDL Co.) was used. A combination of inhibited polymerase and increased temperature of annealing during amplification provided for high reproducibility of PCR results.

Table 1. Sequences of primers used for RAPD-analysis

759	5'-gga cag gga cat aac aat gc-3'
760	5'-gcg cag ggc cac aat aat gg
M1:	
761	5'-gca cag gga cat aat aat gg
763	5'-gct cag ggt tta aac aat gg
765	5'-cga cct aaa gga aac tga tc
M2:	
766	5'-cgt cca aaa gga aac tga tc
768	5'-cgt ccc aaa gga tac tga tc
769	5'-cgt cca agg gga tac tga tc
775	5'-cgt cct aaa gga aac tgg tc

Electrophoresis of Amplification Products

The amplification products were stratified in 1.5% agarous gel, dyed by etidium bromide (10 mg/ml). Gel-recording in transmitting UV-light was implemented with the help of Gel Imager video system (Russia).

RAPD-Pattern Analysis and Statistical Processing of Results

The analysis was based on comparison of electrophoregrams of DNA amplification products, obtained for a descendant and the parent generation for the purpose of detecting alterations, which may be occurrence of new bands or elimination of some of parents' ones. The results of electrophoregram analysis are processed statistically by MS Excel software.

RESULTS

Figures 1 and 2 show results of DNA amplification for mice of different lines and their descendants, obtained with the help of the current primer set. Test and control groups were compared by the following parameters: percentage of newly appeared, "non-parent" bands and average frequency of their occurrence per descendant in each family.

Spontaneous Polymorphism in Descendants of Nonradiated Mice from BALB/C and CBA/Lac Lines

The comparison analysis of RAPD-product polymorphism has determined significant differences in frequencies of new bands for descendants of nonradiation BALB/c line mice and crossing of nonradiated CBA/lac males with hybrid females.

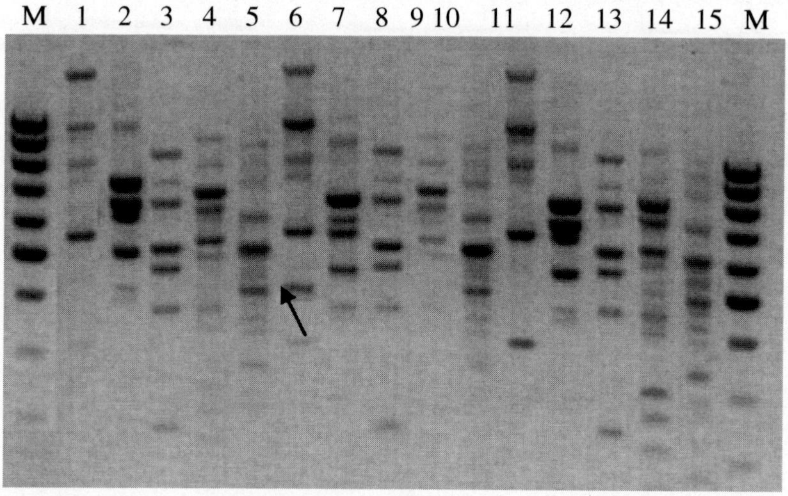

Figure 1. Electrophoregram of RAPD-markers for irradiated male of line BALB/c (1 – 5), descendant (6 – 10) and female (11 – 15); 1, 6, 11 – 759; 2, 7, 12 – 760; 3, 8, 13 – 765; 4, 9, 14 – M1; 5, 9, 15 – M2; M – ladder 1000–100 p.o. Arrows mark nonparent bands

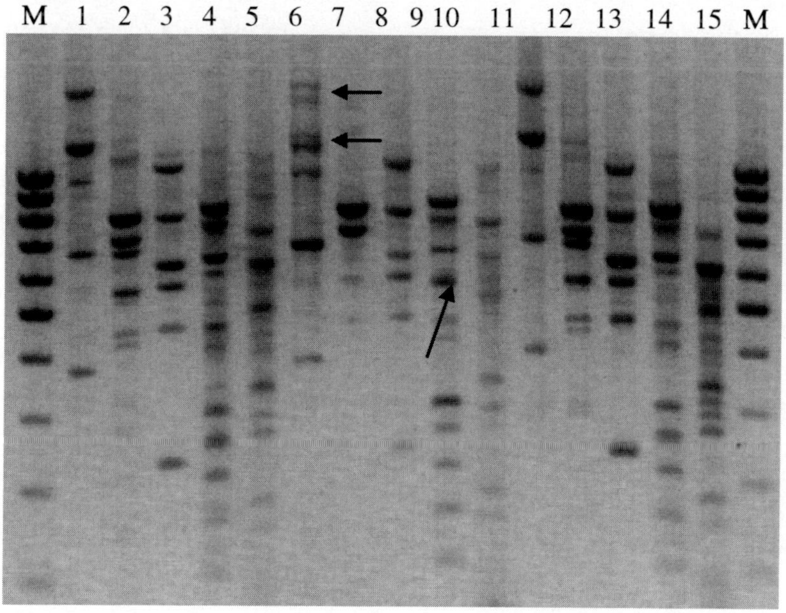

Figure 2. Electrophoregram of RAPD-markers for irradiated male of line CBA/lac (1 – 5), descendant (6 – 10) and female CBA/lac X (11 – 15); 1, 6, 11 – 759; 2, 7, 12 – 760; 3, 8, 13 – 765; 4, 9, 14 – M1; 5, 9, 15 – M2; M – ladder 1000–100 p.o. Arrows mark nonparent bands

The frequencies of "new" bands are presented as histograms in Figure 3. As all primers and primer mixtures are used, the average DNA polymorphism in hybrid crossing descendants significantly exceeded this parameter for F1 BALB/c x BALB/c crossing descendants.

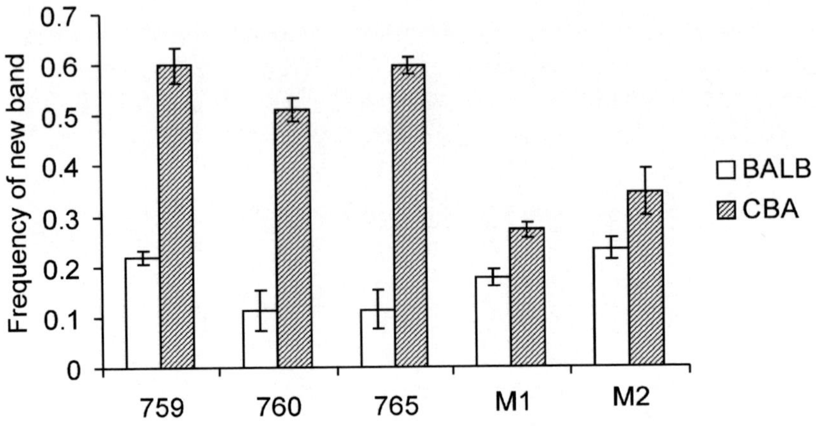

Figure 3. The average frequency of occurrence of new bands for descendants of nonradiated mice

RAPD-Marker Polymorphism in Descendants of Acutely Irradiated Mice of BALB/C Line

Totally, DNA of 98 descendants of mice irradiated by 1 – 3 Gy doses, and 97 descendants from the control group. Ionizing radiation impact increased polymorphism for all radiation doses, used in the experiment. Figure 4 shows dose dependencies for DNA polymorphism for each primer used. Apparently, the differences in the dose–effect curve types may be explained so that different primers annealed different genome areas, differently packed during radiation. Generalized data on all primers are shown in Figure 5. As indicated, the results reliably exceeding polymorphism in the control ($p > 0.05$) were obtained for doses of 1, 2 and 3 Gy. The dose–effect curve has maxima of the number of mutations at 1 Gy dose for spermatoblasts and at 2 Gy for spermatogonia. The comparison analysis of genetic radiosensitivity at different stages of spermatogenesis in mice indicated identical sensitivity of spermatogonia and spermatoblasts at 1 and 3 Gy, and higher sensitivity of spermatogonia at 2 Gy.

RAPD-Marker Polymorphism in Descendants of Chronically Irradiated CBA/Lac Line Male Mice

In a manner similar to acute radiation, DNA of 45 descendants of chronically irradiated mice and 44 descendants of control male mice was analyzed. Figure 6 shows the results of this analysis. A reliable difference in DNA polymorphism of the test group compared with the control was obtained for the primer 759 (33.29%) and mixtures of primers M1 (55.05%) and M2 (23.66%). Note also that analysis of RAPD results for descendants of acutely irradiated mice indicated these primers giving a reliable difference from the control. The intensity of spontaneous polymorphism in CBA/lac and CBA/lac X C57/bl descendants exceeded analogous parameter for descendants of line BALB/c mouse crossing.

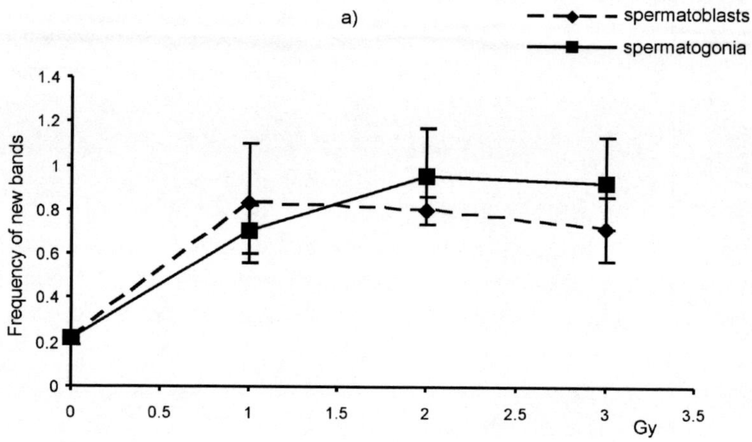

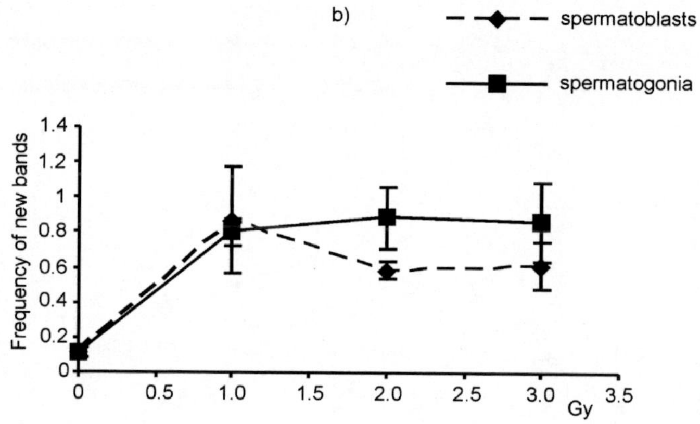

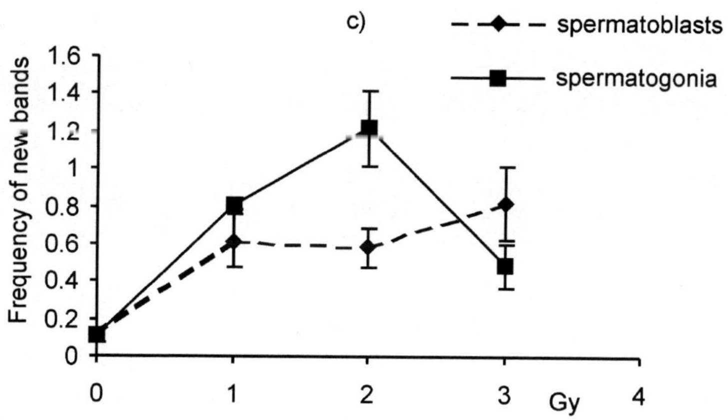

Figure 4. (Continued on next page).

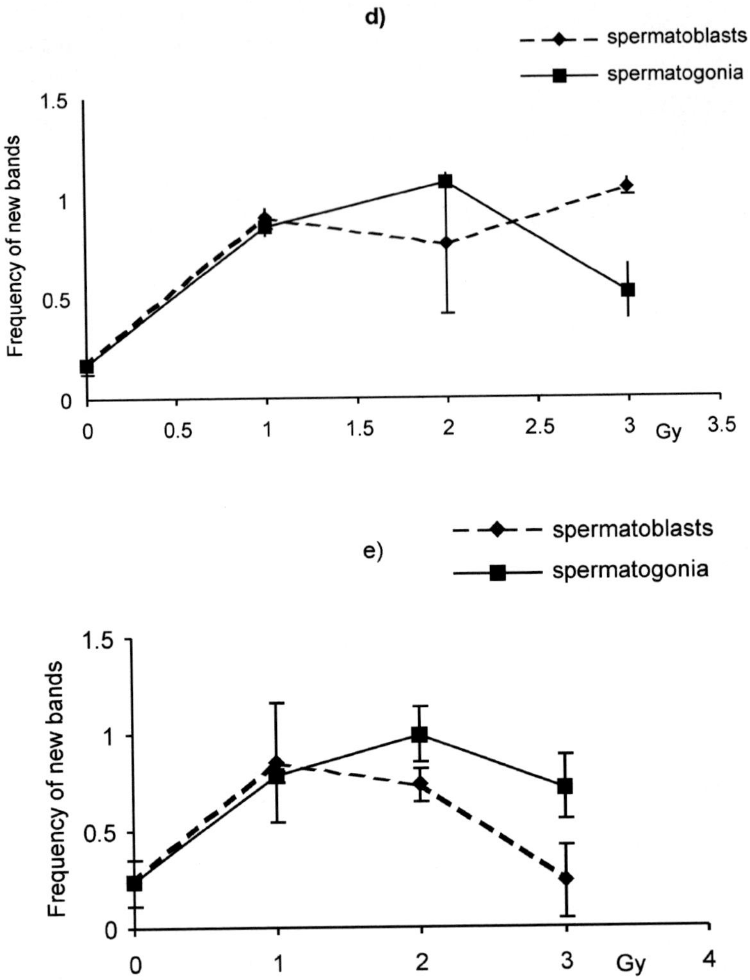

Figure 4. Dose–effect dependencies for frequencies of new bands for the following primers used: a) 759; b) 760; c) 765; d) M1; e) M2 for random amplification with DNA of descendants of irradiated mice.

Discussion

The results obtained in the present work are confirmed by data from the literature. The increase of RAPD-marker polymorphism and mutation frequencies in micro- and minisatellite loci in the descendants of parents exposed to chronic or singular acute irradiation was indicated in many works. Dubrova *et al.* [10] have detected an increased frequency of minisatellite mutations by the method of DNA-fingerprints in descendants of CBA/H line mice exposed to acute or chronic impact of neutrons (0.125 – 0.5 Gy dose) and X-rays (0.5 – 1 Gy). Unexpectedly high level of minisatellite mutations was observed for descendants which parents were exposed to radiation in radioactive contamination zones in Mogilev region after the Chernobyl Accident. The frequency of these mutations in the progeny of irradiated parents noticeably exceeded expected values of point mutations, determined by encoding loci [11]. For example, for the progeny of line BALB/c mice exposed to chronic radiation in 0.25 – 0.5 Gy dose, Bezlepkin *et al.* have observed a reliable increase of nonparent band occurrence frequency compared with the control. Also, acute radiation by 0.5 – 2 Gy dose increased DNA polymorphism intensity in the progeny [16, 17].

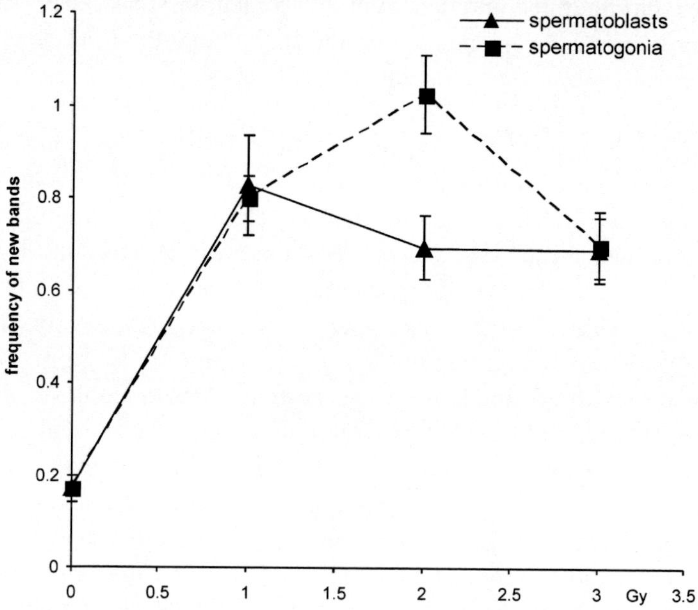

Figure 5. Alteration of average frequencies of "nonparent" bands per a single descendant in the test and the control group in case of acute radiation impact

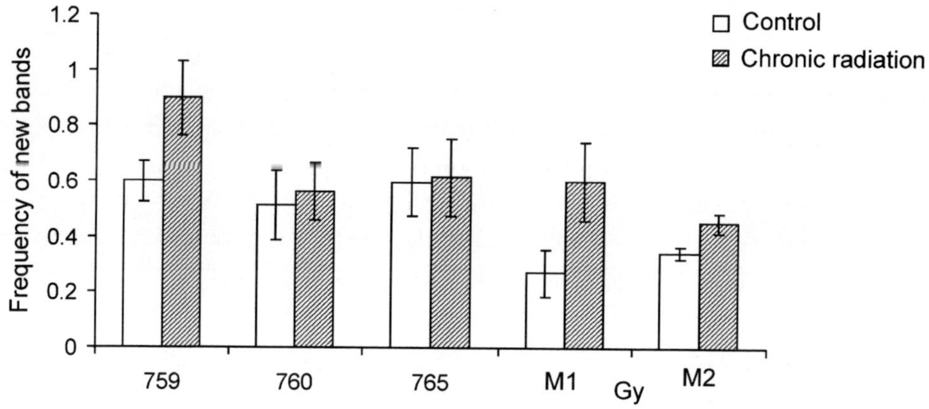

Figure 6. Average frequency of occurrence of "nonparent" bands per a single descendant in the test and the control groups in case of chronic radiation impact

Already in 1950 – 1960ies, the differences in sensitivity to mutagenic action of ionizing radiation of germ cells at different stages of spermatogenesis were shown. According to data from the literature, the lowest genetic radiosensitivity is observed for spermatogonia, whereas spermatoblasts and spermatozoids possess the highest radiosensitivity. It should also be noted that the results of investigations of stem sexual cells and spermatogonia genetic radiosensitivity are fragmentary and contradictory.

In this study the crossing period after radiation impact were selected so that spermatozoa formed from stem cells-precursors, passed through multiple cell divisions in mitosis and meiosis processes during three months after the radiation, may take part in the ovum fertilization. As crossing was implemented on the second and the third week, spermatozoids irradiated at the spermatoblast stage took part in fertilization.

It is common knowledge that compared with other types of spermatogonia stem spermatogonia are more stable to lethal radiation effect. Polymorphism analysis using RAPD indicated a high DNA variability of irradiated mouse descendants, obtained by crossing 3 months after the radiation impact. Note also that, generally, DNA variability for descendants of the "spermatogonium group" e.g. the progeny obtained by crossing 3 months after the radiation impact was higher than for descendants of the "spermatoblast group" (the progeny obtained by crossing within 2 – 3 weeks after the impact). Apparently, this confirms the theory of higher radiation-genetic sensitivity of premeiotic cells, stem spermatogonia, in particular, compared with postmeiotic ones, currently spermatoblasts. For all ionizing radiation doses, used in this study, DNA polymorphism reliably exceeded control values. The specific feature of the results obtained is the presence of a maximum of nonparent band number on the dose–effect curve, most highly expressed for the radiation impact at the stage of stem spermatogonia. Previously, we have analyzed the factors leading to occurrence of mutation maximum in radiobiological experiments [21]. The maximum of the number of mutations may be caused by lethal action of the highest doses of radiation. Taking into account the fact that 2 and 3 Gy doses used in our experiments are quite intensive and decrease viability of sexual cells, this circumstance may really induce the mutation maximum phenomenon.

The experiment on chronic irradiation was performed in cooperation with the laboratory for biological assessment of environmental industry-induced anomalies, GUP MosNPO "Radon". Long-term studies, implemented for the purpose of determining somatic effects of low intensity chronic γ-radiation, indicated that despite the initial alterations of parameters (the percentage of DNA–protein crosslinks, the number of single-stranded breaks, etc.), in future the mouse organism adapts to increase background radiation conditions [23]. For the purpose of studying genetic effects, an experiment on crossing irradiated mice was performed, in which an increase of polymorphism level by 27 – 50% compared with the control was determined. Previous tests, in which the occurrence frequency of anbnormal sperm heads and reciprocal translocations was studied, indicated practical identity of these parameters in irradiated and control mice [24]. The current experiment conditions – total dose and dose rate of radiation – is similar to the radiation conditions at some sites of the 30-km restriction zone around Chernobyl Nuclear Power Plant. At the same time, total annual dose equaled 0.61 Gy, which was 4.12 times higher than the annual dose of natural background radiation [25].

The increased polymorphism level in noncoding loci, registered with the help of RAPD, represented the consequence of mutation frequency induced by radiation. Apparently, this parameter was manifested due to reduced activity of repair systems in the genome areas having no semantic genes.

Apparently, the increased frequency of nonparent bands may occur as a result of both gametal mutations descended from parents and trans-generative (e.g. descending from generation to generation) phenomenon of genome instability in the progeny of irradiated parents. The technical approach applied allowed for detecting damages of premeiotic sexual cells, induced by radiation and realized in the progeny as the increase of RAPD-marker polymorphism. Hence, despite the processes aimed at elimination of cells possessing increased DNA damaging level, which proceed during maturation of sexual cells, high level of polymorphism is detected.

It is known that various areas of DNA are differently packed using histones [26]. It is also known that proteins may definitely shield DNA from free radicals, which are the main damaging factor of the radiation impact [27]. Correspondingly, the difference in polymorphism levels of amplification products with separate primers at different stages may be explained as follows: different genome areas are differently packed and during cell development and division their packing degree changes differently. As a consequence, decrease or increase of polymorphism levels may be explained by different level of shielding of these areas by proteins responsible for genetic material packing.

The results of experiment with chronic irradiation are important in the context of studies of low rate γ-radiation effect on human genome. The conditions of the experiment were approached to radiation conditions in some areas of radiation-contaminated territories, including the 30-km restriction zone of Chernobyl NPP. This allows for extrapolation of data obtained on the model object (the mouse) to the man.

REFERENCES

[1] Bezlepkin V.G. and Gaziev A.I., *Radiats. Biol. Radioekol.*, 2001, vol. 41(5), pp. 475 – 488. (Rus)

[2] Weber J.L. and Wong C., *Hum. Mol. Genet.*, 1993, vol. 2, pp. 1123 – 1128.

[3] Hedrick P.W., *Evolution*, 1999, vol. 53(2), pp. 313 – 318.

[4] Morgan W.F., Day J.P., and Kaplan M.I., *Radiat. Res.*, 1996, pp. 247 – 258.

[5] Weinberg H., Korol A., Kirzhner V. *et al.*, *Proc. R. Soc. Lond.*, 2001, vol. 268, p. 1001 – 1005.

[6] Little J.B., *Int. J. Radiat. Biol.*, 1998, vol. 74, pp. 663 – 671.

[7] Boyd M., Livingstone A., and Wilson L.E. *et al.*, *Int. J. Radiat. Biol.*, 2000, vol. 76(2), pp. 169 – 176.

[8] Paquette B. and Little J.B., *Cancer Res.*, 1994, vol. 54, p. 3173 – 3178.

[9] Dubrova Yu., *Ongene*, 2003, vol. 22, pp. 7087 – 7093.

[10] Dubrova Yu., Plumb M., Brown J. and Jeffreys A., *Int. J. Radiat. Biol.*, 1998, vol. 74(6), pp. 689 – 696.

[11] Dubrova Yu., Nesterov V., Krouchinsky N. *et al.*, *Nature*, April 1996, vol. 380, p. 683 – 686.

[12] Dubrova Yu., Plumb M., Gutierrez B., Boulton E., and Jeffreys A., *Nature*, May 2000, vol. 405, P. 37.

[13] Dubrova Yu. and Plumb M., *Mut. Res.*, 2002, vol. 499, pp. 143 – 150.

[14] Barber R., Plumb M., Boulton E., Roux I., and Dubrova Yu., *PNAS*, May 2002, vol. 99(10), pp. 6877 – 6882.

[15] Vilarino-Guell C., Smith A., and Dubrova Yu., *Mut. Res.*, 2003, vol. 526, pp. 63 – 73.
[16] Bezlepkin V., Vasil'eva G., Lomaeva M., Sirota N., and Gaziev A., *Genetic Consequences of Emergency Radiation Situations. Proceedings*, Moscow, June 10 – 13, 2002, pp. 45 – 46
[17] Bezlepkin V.G., Vasil'eva G.V., Lomaeva M.G., Sirota N.P., and Gaziev A.I., *Radiats. Biol. Radioekol.*, 2000, vol. 40(5), pp. 506 – 512. (Rus)
[18] Sorochinsky B.V., Sokolov N.V., Zelena L.B., Vlasov V.V., Kripka A.V., and Grodzinsky D.M.// *III Congress on Radiation Investigations, Abstracts*, May 2003, Kiev, p. 112.
[19] Hill C.K. and William-Hill D., *J. Radiat. Res.*, 1999, vol. 40, pp. 117 – 127.
[20] Keshava N., Zhou G., Spruill M., Ensell M., and Ong T.M., *Mol. Cell. Biochem.*, 2001, vol. 222, pp. 69 – 76.
[21] Shevchenko V.A. and Pomerantseva M.D., *Genetic Consequences of the Action of Ionizing Radiation*, Moscow, Nauka, 1985, 279 p. (Rus)
[22] Atienzar F.A., Venier P., Jha A.N., and Depledge M.N., *Mut. Res.*, 2002, vol. 521, pp. 151 – 163.
[23] Elakov A.L., Osipov A.N., Pomerantseva M.D., Zaichkina S.I., Syicheva L.P., Puchkov P.V., Rozanova O.M., Ramaya L.K., Akhmadieva A.Kh., Myazin A.E., Kiyatkina M.A., and Syipin V.D., *Thes. Rep. III International Symposium "Action Mechanisms of Superlow Doses"*, 03 – 06 December, 2002, Moscow, Izd. RUDN, p. 78. (Rus)
[24] Elakov A.L., Osipov A.N., Pomerantseva M.D., Ramaya L.K., Myazin A.E., Syipin V.D., and Shevchenko V.A., *Thes. Rep. International Conference "Genetic Consequences of Radiation Emergency Situations"*, 10 – 13 June, 2002, Moscow, Izd. RUDN, pp. 47 – 48. (Rus)
[25] *UNSCEAR 2000. Report to General Assembly, with Scientific Annexes*, vol. II: Effects, 566 p.
[26] Zhimulev I.F., *General and Molecular Genetics*, Novosibirsk, Sib. Univ. Izd., 2003, 479 p. (Rus)
[27] Ryabchenko N.I., *Ionizing Radiation and DNA*, Moscow, Atomizdat, 1979, 192 p. (Rus)

In: 20 Years after the Chernobyl Accident
Editors: E. Burlakova and V. Naidich, pp. 99-118
ISBN 1-60021-249-2

Chapter 8

RADIATION-EPIDEMIOLOGICAL STUDIES OF THYROID CANCER INCIDENCE AMONG CHILDREN AND ADOLESCENTS IN THE BRYANSK OBLAST OF RUSSIA AFTER THE CHERNOBYL ACCIDENT

V. K. Ivanov, A. F. Tsyb, A. I. Gorski, M. A. Maksioutov, K. A. Tumanov and O. K. Vlasov

ABSTRACT

Thyroid cancer incidence between 1991 and 2001 is analyzed for the first time among persons living in the Bryansk Oblast, the worst contaminated area of Russia after the Chernobyl accident, and exposed as children (0-17 years). According to the census of 1989 this population was 375 thousands people. Estimates of thyroid doses were made based on the official document "Reference book of mean thyroid doses in persons of different age resident in the population points of the Bryansk, Tula, Oryol and Kaluga Oblasts contaminated after the Chernobyl accident in 1986.

Between 1991 and 2001 a total of 199 thyroid cancer cases were diagnosed at cancer centers (so-called oncological dispensaries). Diagnoses were confirmed histologically for 95% of the cases. The analysis has revealed statistically significant radiation risk only for those exposed as children of 0 – 14 years old. For this group the standardized incidence ratio (the national incidence rate was a standard) in the considered this time period is estimated to be 5.6 (4.6, 6.8 95% CI) and 10.2 (4.6, 13.6 95% CI) for girls and boys, respectively.

The excess relative risk per 1 Gy (with internal control) for the period 1991-2001 was 9.4 (3.0, 25.4 95% CI) and 29.6 (7.2, 7,460 95% CI) for females and males, respectively. Using external control, the excess relative risk ERR_{1Gy} per unit dose 1 Gy was found to be 10.0 (4.2, 21.6 95% CI) for females and 67.8 (17.1, 5,448 95% CI) for males.

Consideration was given to the dependence of radiation risk on time at exposure, with the focus on two follow-up periods 1991 – 1996 and 1997 – 2001. In 1997 – 2001 the radiation risk is shown to decrease among girls and increase among boys.

Discussion also concerns quality of available information about thyroid cancer cases and screening campaigns carried out after the Chernobyl accident.

INTRODUCTION

In the territories of Belarus, Russia and Ukraine contaminated after the Chernobyl accident a significant increase in thyroid cancer incidence rates was observed after 1991. Among those exposed as children and adolescents in 1986 the excess above the spontaneous incidence level was tenfold and even mor . Exposure of thyroid to radioiodines can be one of possible reasons of the increased incidence rate. At the same time, it may not be excluded that the elevation of incidence rate and improved registration level is explained by increased attention to this issue after the accident (screening effect). Thyroid cancer not being fatal, the indicated effect may play an important role and therefore is discussed among other things in the paper.

By now a large number of publications are available regarding exposure of thyroid gland after the Chernobyl accident [1 – 10].

In works [1 – 6] the incidence analysis is limited to estimating the incidence rates and standardized incidence ratio (SIR). These publications provide evidence to the fact that after 1991 a considerable increase in thyroid cancer incidence has been observed, particularly among those exposed as children and adolescents.

Estimates of dose response coefficients with respect to thyroid cancer during 1991 – 1998 were derived in papers [7 – 9]. As part of analysis in [7, 8] the contaminated territory was divided into zones with assigned doses for which incidence rates were studied. In work [9] doses available for specific population points were used, both for cases and other members of the study population. The estimated excess relative risk per unit dose in these publications ranged from 20 to 40.

All the above studies were concerned with analysis of thyroid cancer incidence in persons exposed as children and adolescents, as the risk of thyroid cancer is the highest in this category.

In work [10] thyroid cancer incidence was studied in adults living in the Bryansk Oblast after the Chernobyl accident. The performed analysis has not revealed any radiation effects among adults.

The present work aims at analysis of thyroid cancer incidence between 1991 and 2001 and risk of radiogenic thyroid cancer among children and adolescents (age at exposure 0 – 17) in the worst contaminated areas of Russia, namely the Bryansk Oblast. The size of the study population is sufficient for performing statistical analysis.

MATERIALS AND METHODS

Population Under Study

The analysis of radiation risks covered the follow-up period 1991 – 2001. The year 1991 was taken as the beginning of the follow-up time with allowance for the minimal latent period of 5 years in induction of radiogenic thyroid cancer.

The subjects of the study were persons exposed as children and adolescents in 1986 (age 0 – 17). Since thyroid dose from radioiodines was accumulating during the first two months after the Chernobyl accident in 1986 and accurate demographic data were not available for specific regions and population points (rayon is a territorial and administrative unit of oblast, there are 28 regions in the Bryansk Oblast), the age structure at the rayon level in 1986 was defined using results of the federal census of 1989. Given this assumption, the study population in 1986 consisted of 373,827 people residing in the Bryansk Oblast. The census data contained only the number of people living in population points and the age structure there was assumed to be the same as in regions they belong to. Since only children and adolescents were studied, the analysis of radiation risks did not take into account changes in the cohort size due to mortality.

Table 1. Key characteristics of the cohort of persons under study exposed as children and adolescents (0 – 17 years old), follow-up period 1991 – 2001

Sex	Females	Males
Population abundance	184,919	188,908
Number of cases	144	55
Average dose per examinant (Gy)	0.08	0.08
Average dose per case (Gy)	0.09	0.22

The risk of radiogenic thyroid cancer is known to depend on sex, age at exposure, thyroid dose and attained age. In turn, thyroid dose due to radioiodines is also a function of age at exposure and increases with a decrease in age [1 – 9]. Therefore, for analysis of the age dependence of incidence rate the study population was divided into 4 categories by age at exposure: children of 0 – 4, 5 – 9, 10 – 14 years, and adolescents of 15 – 17 years plus two categories by sex.

The main demographic characteristics of the study cohort are shown in Table 1.

Medical Information

Procedures for collection and verification of medical information on cancer cases in Russia are described at length in [10, 11].

In the Russian Federation (the former USSR) as regulated by Ministry of Health cancer care involves two levels: cancer centers (dispensaries) at the regional level and consultative services at the rayon level.

At the rayon level, oncological consultations are provided to patients at outpatient departments of rayon hospitals. The responsibility of rayon oncologist is clinical diagnosis of malignant and non-malignant neoplasms, and referral of the patient to the oblast cancer center for specialized diagnosis and treatment.

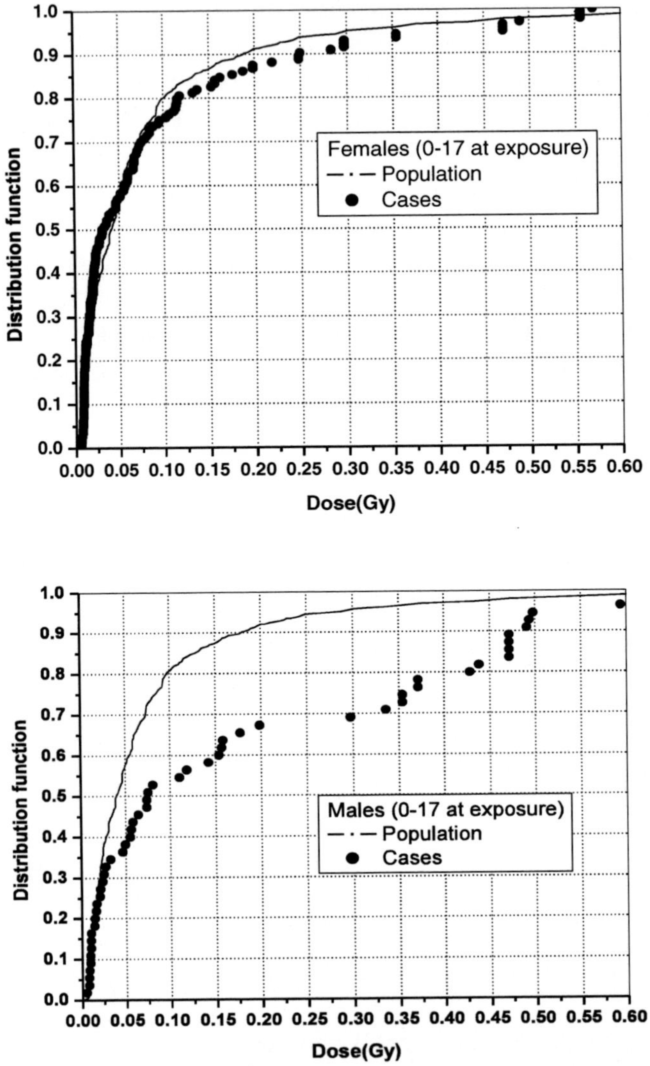

Figure 1. Distribution function of the study population and cases by thyroid dose

The Oblast cancer center is engaged in cancer diagnosis and treatment within its service area (Oblast). In cancer centers there are specialized departments including radiology, chemotherapy, surgery, X-ray, and others. For diagnosis and treatment of difficult tumors, rare tumors (e.g. thyroid cancer), or tumors of uncertain origin, patients are further referred to specialized institutions such as state hospitals and clinics attached to research institutes.

Cancer patients diagnosed and treated at hospitals are reported to the oblast cancer center by sending an abstract from medical record. Following the discharge of a patient, the abstract is forwarded to the oblast cancer center according to the patient's place of residence.

The present study includes 199 cases of thyroid cancer diagnosed between 1991 and 2001 among persons exposed as children (0 – 17 years). Of them, 144 cases are females and 55 males. Information on thyroid cancers used in this analysis is stored in the cancer registry based at the Bryansk Cancer Center. Patients with thyroid cancer were operated on at the hospitals of Bryansk (the majority), or state research centers of Moscow and Obninsk. Diagnoses were confirmed histologically for 95% of the cases.

Dosimetric Information

Thyroid doses from incorporated radioiodines both for cases and the study population were determined using the dose catalogue [12].

The dose reconstruction was based on the official methodology approved by the Russian Scientific Commission on Radiation Protection [13] which is described in detail in [10]. Doses for cases were determined using data on residence place and age at exposure. Doses for males and females living in the same population point were assumed to be equal and dependent on age at exposure only. The distribution of collective dose by age in the study population was derived with data of the census of 1989.

Basic dosimetric data for the studied cohort is shown in Table 1. It is worth noting that the radiation dose for cancer cases among males (0.22 Gy) is significantly higher the mean population dose (0.08 Gy) and the mean dose for females. Hypotheses accounting for these differences are presented in the section "Discussion".

Figure 1 shows the distribution of cancer cases and healthy persons as a function of thyroid dose. As can be seen, the mean dose for those who developed cancer in the Bryansk Oblast was higher that in healthy persons, which may be a qualitative indication of existing dose dependence of incidence.

Method Of Analysis

In the analysis data of 1991 – 2001 for thyroid cancer incidence the cross-tabulation of cases and person years used has dose categories with cut points 0, 0.06, 0.10, 0.2, 0.2+ Gy, along with 5-year intervals of attained age and 1-year intervals of calendar time.

The standard software EPICURE (the AMFIT software) was used for analysis [14]. This analysis was based on a general excess relative risk model which can be stated formally as

$$\lambda(a,t)\cdot(1+\rho(d)\cdot\varepsilon(t,a)), \qquad (1)$$

in which $\lambda(.)$ is the background thyroid cancer incidence rate, $\rho(.)$ is the dose response function, and the function $\varepsilon(.)$ describes dose-effect modification. Background rates are estimated as a function of attained age and calendar time. Calendar time is used to allow for time trends in background rates. The effect modifiers are the covariates, and time since exposure and age at exposure. The linear dose-response function is written as

$$\rho(d) = ERR_{1\mathrm{Gy}} \cdot d\,.$$

In this equation ERR_{1Gy} is the excess relative risk per Gy, and d is the external radiation dose (Gy).

Risk estimates were made using external (spontaneous cancer incidence in Russia as a whole) and internal control group (spontaneous cancer incidence with zero doses).

In calculations with the external control group the risk model takes the form:

$$\lambda^R(a,t) \cdot SIR^u \cdot (1 + \rho(d) \cdot \varepsilon(t,a)), \qquad (2)$$

in which $\lambda^R(a,t)$ is the spontaneous cancer incidence rate in Russia corresponding to the attained age (a) and calendar time (t), SIR^u is the coefficient accounting for a difference between the spontaneous incidence rate in the study cohort and the general population of respective age in the considered time period. In the used model this coefficient is equal to the standardized incidence ratio (SIR) for nonexposed members of the cohort. The value of SIR^u was assumed to be the same for all age groups. The advantage of the selected risk model is that it does not estimate only risk coefficients, but also the difference in spontaneous cancer incidence rate in the followed up cohort and the Russian population.

When risk coefficients were estimated using the internal control, data were stratified by attained age and calendar time.

Projection of temporal changes in the number of thyroid cancers over life time was made. The prognostic model was based on using the risk coefficients derived in this study. Changes in the size of the study population and its age structure due to aging and deaths were taken into account through the survival function which is calculated using Russian age-specific mortality from all causes in the considered period. In the model morbidity and mortality in the study population are assumed to remain unchanged in future and equal to the rates at the end of the considered period.

The dynamics of radiogenic cancers $rad(t)$ by calendar year t was calculated by the formula:

$$rad(t) = SIR^u \cdot AR \cdot CC(t), \qquad (3)$$

where SIR^u is the standardized incidence ratio for unexposed population (national rates are used as control); AR is the attributable risk calculated with dose d as

$$\frac{ERR_{1\mathrm{Gy}}d}{1 + ERR_{1\mathrm{Gy}}d}.$$

For calculating dynamics of spontaneous cases *spon*(t) the following expression was used:

$$spon(t) = SIR^u \cdot (1 - AR) \cdot CC(t). \tag{4}$$

The function $CC(t)$ was determined from the relationship:

$$CC(t) = \int_{t_0}^{t} \int_{u(t_0)}^{u(t)} n(g,\tau) \cdot \lambda^{Th}(g,\tau) \cdot dg \cdot d\tau. \tag{5}$$

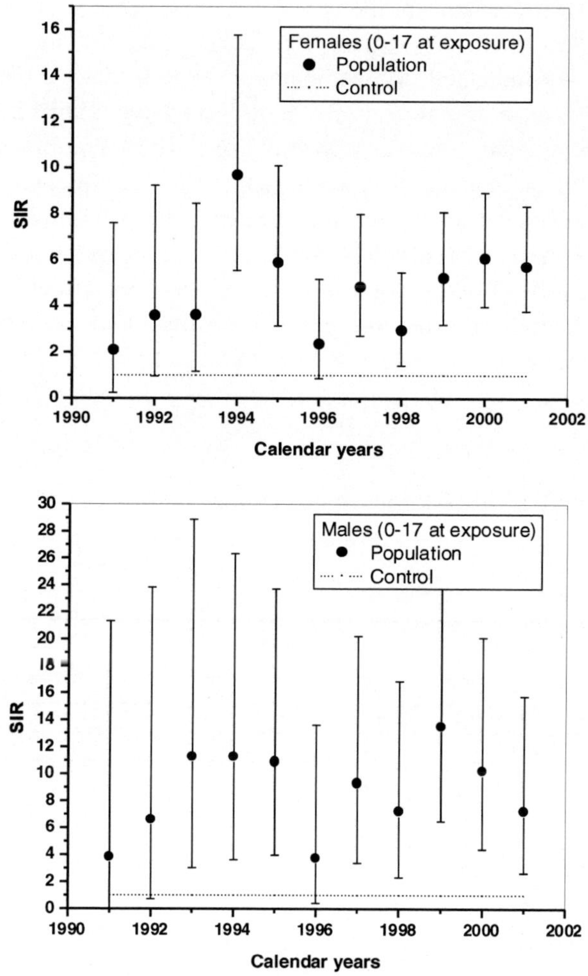

Figure 2. SIR dynamicsThe results of estimating the dynamics of the standardized incidence ratio SIR are shown in Figure 2. As can be seen from the Figure, SIR > 1, i.e. it exceeds the control which is the incidence rate in Russia in general and the value is relatively stable in time.

Here t_0 is the beginning of the follow-up period, $u(t_0)$ is the attained age at the beginning of the follow-up period; $n(g, \tau)$ is the size of the population of age g at time moment t calculated with the survival function and $\lambda^{Th}(g,t)$ are the Russian thyroid cancer rates.

RESULTS

In the first stage of the study focus was placed on thyroid cancer incidence among those exposed as children and adolescents (age 0 – 17 years).

The value of SIR over the whole follow-up period is 4.9 (4.2, 5.8 95% CI) for girls and 8.8 (6.6, 11.5 95% CI) for boys. This excess in SIR can be explained either by exposure to radiation, or regional differences in the thyroid cancer spontaneous incidence rates and registration levels. Main characteristics of this cohort and radiation risk coefficients are presented in Table 2. As indicated, the estimates of excess relative risk per unit dose were derived using internal and external controls. It follows from Table 2 that the relative risk increases with increasing dose, both for girls and boys. The value of the excess relative risk per 1 Gy is positive and statistically significant within 95% confidence intervals. It should be noted that the risk coefficients obtained with different control groups (internal and external) appear to be close for girls (a relatively large number of cases) and different for boys, the number of cases among boys being one third. The subsequent discussion will show that the estimates derived with both control groups are basically close, in other words asymptotic convergence of risk estimates can be seen in both models. It is, therefore, the authors' opinion that in case of a limited number of cases, estimates based on using external control are more reliable.

Table 2. Estimated radiation risk of thyroid cancer among persons exposed as children and adolescents (0 – 17 years), 1991 – 2001 follow-up period

Dose groups	PYR	Mean dose (Gy)	Observed	Expected[a]	RR (95% CI)
Girls					
1	1,328,830	0.026	89	71.9	1
2	300,893	0.079	18	9.1	1.4 (0.9, 2.3)
3	227,620	0.142	19	6.4	2.1 (1.2, 2.6)
4	176,767	0.403	18	3.7	3.2 (1.7, 5.7)
Total	2,034,110	0.080	144	91.1	-
Internal control *ERR* per 1 Gy=9.1 (3.0, 23.3)					
External control *ERR* per 1 Gy=11.2 (5.2, 22.6); SIR=3.1 (2.2, 4.1)[b]					
Boys					
1	1,354,150	0.026	24	19.1	1
2	324,061	0.080	5	3.4	1.0 (0.32, 2.48)
3	234,054	0.143	8	2.7	2.0 (0.83, 3.41)
4	165,727	0.404	18	1.6	6.4 (1.9, 14.0)
Total	2,077,990	0.078	55	26.8	-
Internal control *ERR* per 1 Gy=21.6 (6.0, 151.3)					
External control *ERR* per 1 Gy=43.5 (14.8, 460.8); SIR=2.5 (0.3, 5.1)					

[a] Spontaneous incidence in nonexposed cohort;

[b] SIR for nonexposed members of the cohort

To study the age dependence of radiation risk all cases were divided into three groups by age *a* exposure: 0 – 4, 5 – 9, 10 – 14 and 15 – 17 years. Results of calculations are included in Table 3.

Table 3. Estimated radiation risks of thyroid cancer in girls (1991 – 2001 follow-up period)

Dose groups	PYR	Mean dose (Gy)	Observed	Expected[a]	RR (95% CI)
Girls 0-4 years at exposure					
1	185,673	0.046	5	2	1
2	164,740	0.082	3	1.4	0.8 (0.2, 3.5)
3	118,762	0.145	8	1.1	2.8 (0.9, 9.5)
4	141,948	0.418	16	1.2	5.1 (1.9, 16.1)
Total	611,123	0.161	32	5.6	-
ERR per 1 Gy = 45.3 (5.2, 9953) (internal control)					
ERR per 1 Gy = 28.8 (4.3, 2238) (external control); SIR = 1.7 (NA[c], 6.0)[c]					
Girls 5 – 9 years at exposure					
1	387,428	0.028	16	12.6	1
2	72,670	0.078	3	2.4	1.0 (0.2, 2,9)
3	80,262	0.142	9	2.4	2.8 (1.2, 6.4)
4	25,438	0.342	2	0.7	2.0 (NA, 6.1)
Total	565,799	0.065	30	18.2	-
Internal control *ERR* per 1 Gy = 10.1 (–0.1, 84.7)					
External control *ERR* per 1 Gy = 10.2 (0.1, 75.5); SIR=3.3 (0.9, 6.1)					
Girls 10 – 14 years at exposure					
1	95,042	0.009	12	7.7	1.0
2	179,279	0.013	15	14.21	0.56 (0.22, 1.39)
3	114,475	0.026	7	9.2	0.41 (0.14, 1.12)
4	135,553	0.090	13	10.7	0.68 (0.28, 1.61)
Total	527,347	0.035	47	41.9	-
Internal control *ERR* per 1 Gy = 1.0 (–5.3, 15.0)					
External control *ERR* per 1 Gy = 1.2 (–5.2, 15.4); SIR = 4.4 (2.8, 6.5)					
Girls 15– 17 years at exposure					
1	149,074	0.008	14	9.5	1.0
2	65,913	0.017	9	4.2	1.5 (0.6, 3.3)
3	25,107	0.030	5	1.6	2.3 (0.7, 5.6)
4	89,746	0.065	7	5.8	0.8 (0.3, 2.0)
Total	329,840	0.027	35	21.1	–
Internal control *ERR* per 1 Gy < 0					
External control *ERR* per 1 Gy < 0; SIR = 4.0 (2.3, 6.1)					

[c] NA confidence interval is not defined.

As might be expected, radiation risk increases with a decrease in age at exposure (the highest for girls 0 – 4 years at exposure). Statistically significant risk is observed for those exposed as children of 0 – 14 years, but does not exist for adolescents of 15 – 17 years at exposure. The observed absence of statistical significance for girls of 10 – 14 years in case of using the internal control can be explained by a limited number of cases.

In keeping with the obtained results, analysis of the dose response relationship for thyroid cancer was performed only for persons exposed as children (0 – 14 years). Main characteristics of this population are presented in Table 4.

Table 4. Key characteristics of the study cohort: persons exposed as children (0 – 14 years); 1991 – 2001 follow-up period

Sex	Females	Males
Population size	154,934	159,500
Number of cases	109	47
Mean dose for members of the study population (Gy)	0.09	0.09
Mean dose for cases (Gy)	0.11	0.25

Table 5 includes estimates of the dose response relationship for thyroid cancer. As can be seen from the Table, the values of the excess relative risk are positive and statistically significant. The relative risk increases with increasing dose. The values of the excess relative risk ERR_{1Gy} for boys is higher that for girls.

Table 5. Estimated radiation risks of thyroid cancer persons exposed as children (0 – 14 years old) (1991 – 2001 follow-up period)

Dose groups	PYR	Mean dose (Gy)	Observed	Expected	RR (95% CI)
Girls					
1	1,020,780	0.028	59	52.4	1
2	287,387	0.080	13	9.7	1.1 (0.6, 2.0)
3	221,024	0.143	19	6.8	2.2 (1.3, 3.9)
4	175,079	0.404	18	4.2	3.2 (1.7, 5.9)
Total	1,704,270	0.090	109	73.1	–
Internal control *ERR* per 1 Gy = 9.4 (3.0, 25.4)					
External control *ERR* per 1 Gy = 10.0 (4.2, 21.6); SIR = 3.4 (2.3, 4.7)					
Boys					
1	1,053,430	0.029	16	11.4	1
2	309,908	0.080	5	2.8	1.1 (0.34, 2.97)
3	227,245	0.143	8	2.2	2.1 (0.8, 5.2)
4	163,912	0.404	18	1.4	7.1 (3.2, 16.2)
Total	1,754,500	0.088	47	17.8	–
Internal control *ERR* per 1 Gy = 29.6 (7.2, 7,460)					
External control *ERR* per 1 Gy = 67.8 (17.1, 5,448); SIR = 1.7 (NA, 4.8)					

The value of SIR for this group over the whole follow-up period from 1991 to 2001 was 5.6 (4.6, 6.8 95% CI) and 10.2 (4.6, 13.6 95% CI) for girls and boys, respectively. If SIR = 5.6 for girls is compared with the value 3.4 for unexposed persons from Table 5, it may be concluded that the gain in incidence for girls as a result of exposure is 5.5/3.4 = 1.6-fold, i.e. the regional differences in incidence rates and registration levels are the main contributors to the excess above national level. For boys such an excess is equal to 10.2/1.7 = 6.

Time dependence of radiation risk was studied for two five-year follow-up periods: 1991-1996 and 1997 – 2001. Results of estimating the dose response coefficients are presented in Table 6. As can be seen from the table, the radiation risk in 1997 – 2001 decreased, which leads us to assume that the peak of radiation risk of thyroid cancer for girls has already occurred. The derived decrease of risk with time is qualitatively confirmed by the distribution of the population and number of cases with respect to two factors of radiation risk, namely radiation dose and age at exposure.

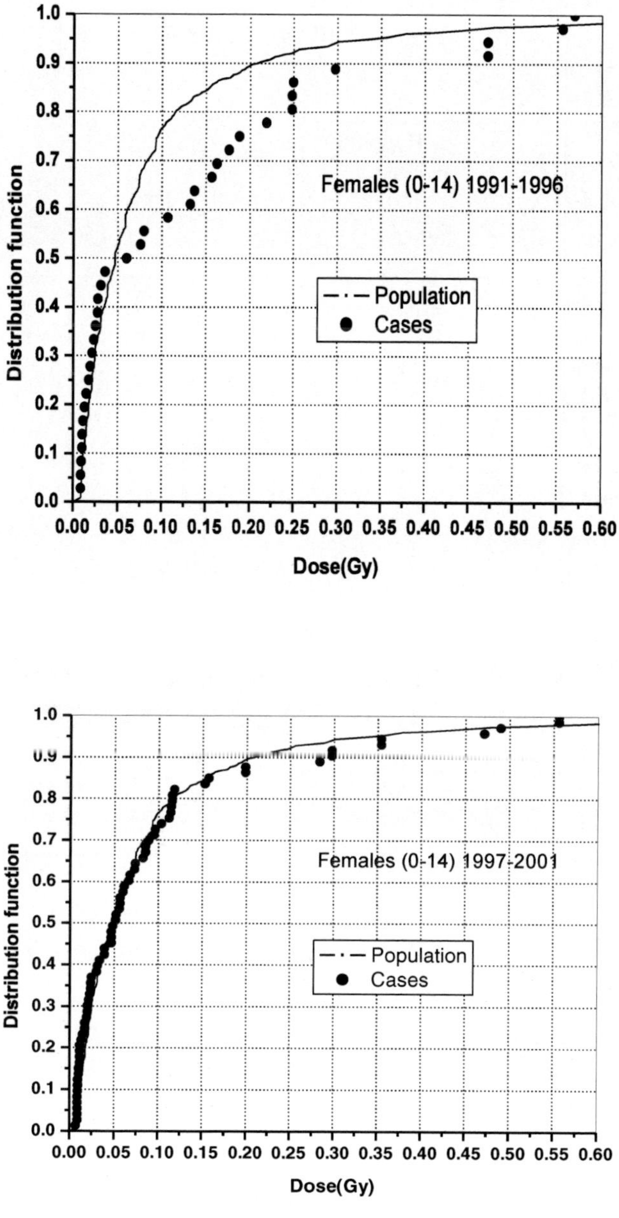

Figure 3. The distribution function of the population and observed cases by thyroid dose for girls

Figure 3 presents the distribution of the study population and observed cases by thyroid dose. For the unexposed persons for whom incidence rates are proportional to size of age groups, these distributions are identical. Comparison of the left and right hand side plots shows that the rates are different in 1991 – 1997, but are virtually the same in 1997 – 2001. Note that the doses for the study population and cases start to differ from 0.05 Gy between 1991 and 1996, whereas in 1997 – 2001 the difference is observed starting from 0.15 – 0.20 Gy, which may be explained by higher radiosensitivity of girls who were the first to fall ill.]

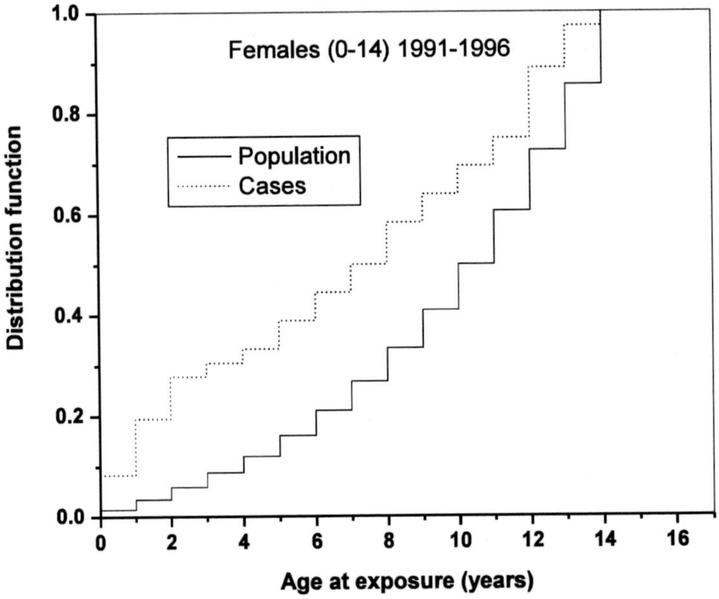

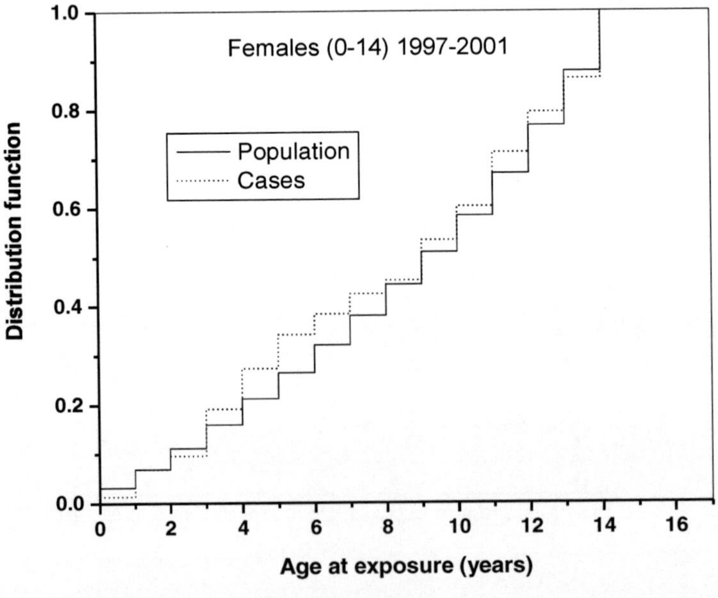

Figure 4. The distribution function of the study population and observed cases by age for girls

Table 6. Estimates of time dependence of radiation risk of thyroid cancer in girls (0 – 14 years old at exposure)

Dose groups	PYR	Mean dose (Gy)	Observed	Expected	RR (95% CI)
Follow-up period 1991 – 1996					
1	556,790	0.028	18	14	1
2	156,756	0.079	2	2.8	0.5 (NA, 0.7)
3	120,559	0.143	7	1.9	2.3 (0.8, 5.8)
4	95,497	0.404	9	1.1	4.3 (1.5, 4.7)
Total	929,602	0.090	36	19.8	
Internal control *ERR* per 1 Gy = 12.6 (2.1, 78.3)					
External control *ERR* per 1 Gy = 34.4 (9.4, 6,74.1)					
Follow-up period 1997 – 2001					
1	463,991	0.028	41	35.8	1
2	130,630	0.079	11	6.1	1.5 (0.7, 2.8)
3	100,465	0.143	12	4.3	2.2 (1.1, 4.3)
4	79,581	0.404	9	2.6	2.6 (1.1, 5.5)
Total	774,668	0.090	73	48.8	
Internal control *ERR* per 1 Gy = 7.6 (1.1, 27.9)					
External control *ERR* per 1 Gy = 4.8 (0.7, 13.6)					

Table 7. Estimates of time dependence of radiation risk of thyroid cancer in boys (0 – 14 years at exposure)

Dose groups	PYR	Mean dose (Gy)	Observed	Expected	RR (95% CI)
Follow-up period 1991 – 1996					
1	574,601	0.028	7	4.2	1
2	169,041	0.079	3	0.7	2.1 (0.4, 8.1)
3	123,952	0.142	2	0.6	1.6 (NA, 6.2)
4	89,406	0.404	5	0.3	5.3 (1.6, 23.7)
Total	957,000	0.088	17	5.8	-
Internal control *ERR* per 1 Gy = 19.7 (NA, 13,790)					
External control *ERR* per 1 Gy = 21.9 (2.8, 5,180)					
Follow-up period 1997 – 2001					
1	478,834	0.028	9	5.5	1
2	140,867	0.079	2	1.6	0.6 (NA, 2.6)
3	103,293	0.142	6	1.2	2.3 (0.7, 7.3)
4	74,505	0.404	13	0.8	7.4 (2.6, 21.6)
Total	796,500	0.088	30	9.1	-
Internal control *ERR* per 1 Gy = 39.4 (NA, 3,231)					
External control *ERR* per 1 Gy = 290 (NA, 65,100)					

Figure 4 shows the distribution of the population and cases by age at exposure. What is peculiar to this characteristic is that for thyroid cancer due to radioiodines it makes possible qualitative assessment of exposure without knowing thyroid doses. In fact, both thyroid dose and radiation risk here is a function of age at exposure and both of them tend to increase with decreasing age at exposure (children). In other words, this distribution is more responsive to exposure than the risk dependence on dose. Results presented in Figure 4 confirm of the risk assessments of Table 6. Between 1997 and 2001 the radiation risk per unit dose decreased and the distributions by age at exposure, both for the cases and the study population, are close.

The results of calculating the indicated characteristics for boys are shown in Tables 7 and Figures 5 and 6. As shown by data presented, the opposite effect is observed for boys, the risk increasing from 1997 to 2001 along with the mean dose for the cases.

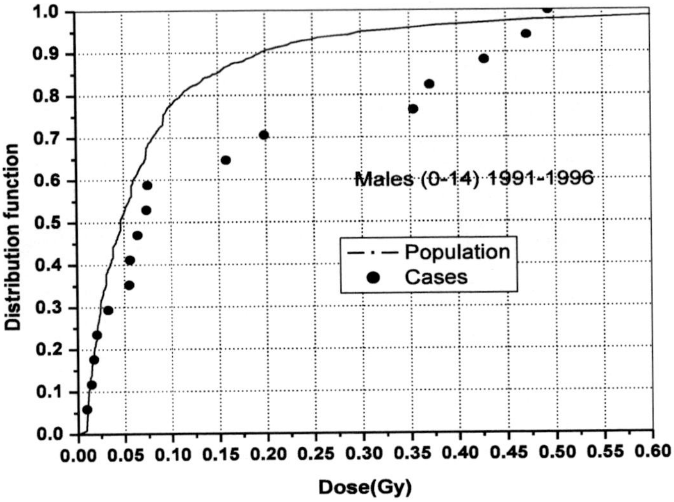

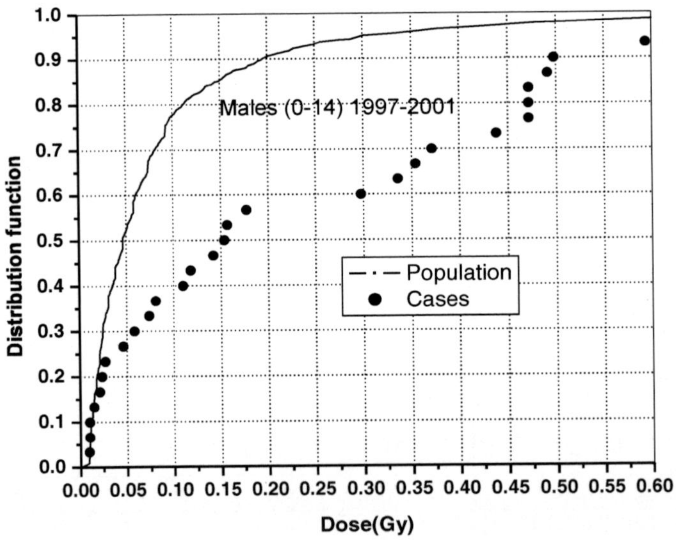

Figure 5. The distribution function of the population and observed cases by thyroid dose for boys

Figures 7 and 8 present results of the projection of thyroid cancer lifetime risk.

Figure 7 shows temporal changes in the cumulative incidence rate for thyroid cancer. It can be seen that the incidence of thyroid cancer will be observed till 2050. Over the whole period from 1991 to 2050 a total of 2822 cases of thyroid cancer are expected among females, of them 1293 radiogenic cancers and 1529 spontaneous cancers. For males, these values are

392 and 147 cases, respectively. Thus, altogether 1685 radiogenic and 1676 spontaneous thyroid cancers can be expected to occur.

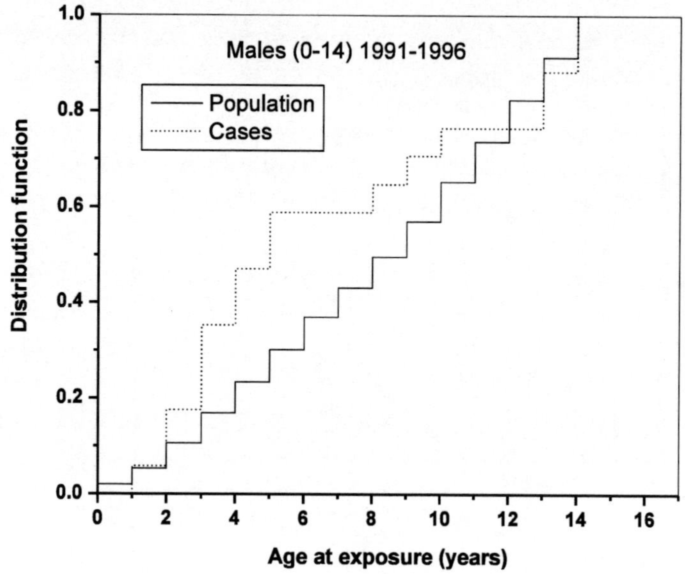

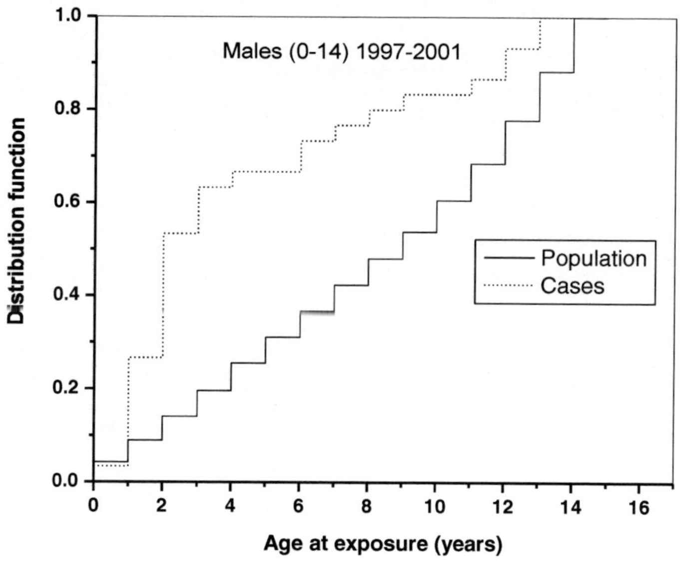

Figure 6. The distribution function of the population and observed cases by age for boys

Figure 8 shows the density distribution for cases by calendar time (density is normalized over total number of cases). According to this distribution, the peak of the annual incidence rate, both for males and females, is expected to occur in 2025 – 2030.

In conclusion it should be pointed out that thyroid cancer incidence rates in adults were also analyzed within the discussed study. As in [10], no positive trend was found in the dose dependence of incidence rate.

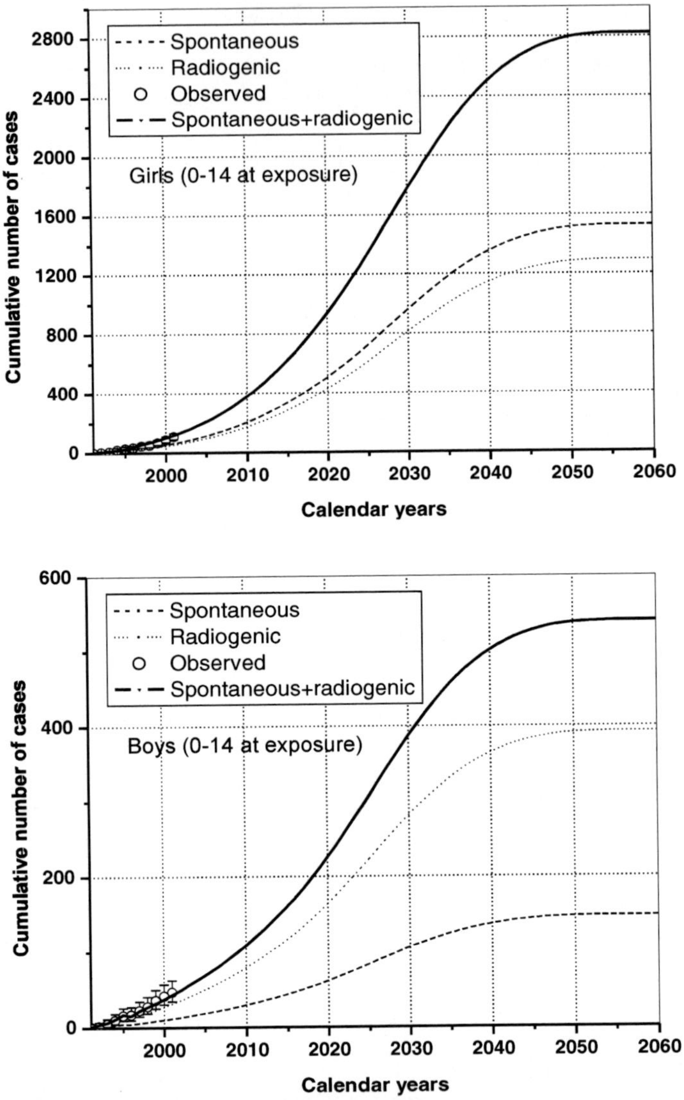

Figure 7. Forecasted cumulative number of thyroid cancer cases

DISCUSSION

The estimates of radiation risk derived in the study, within confidence intervals, are in agreement with other published data [7-9]. As in other works, a significant radiation risk for those exposed as children has been confirmed.

Outcomes of the post-Chernobyl studies of thyroid cancer incidence suggest that the concepts of the latency of radiogenic cancers and dose response coefficients should be revised. The derived risk coefficients appear to be much higher those in works [15-17] obtained primarily for external exposure of thyroid gland. Besides, these coefficients are significantly different from the ICRP estimates (the ICRP 60 publication) in which effects from internal irradiation are three times lower those due to external exposure.

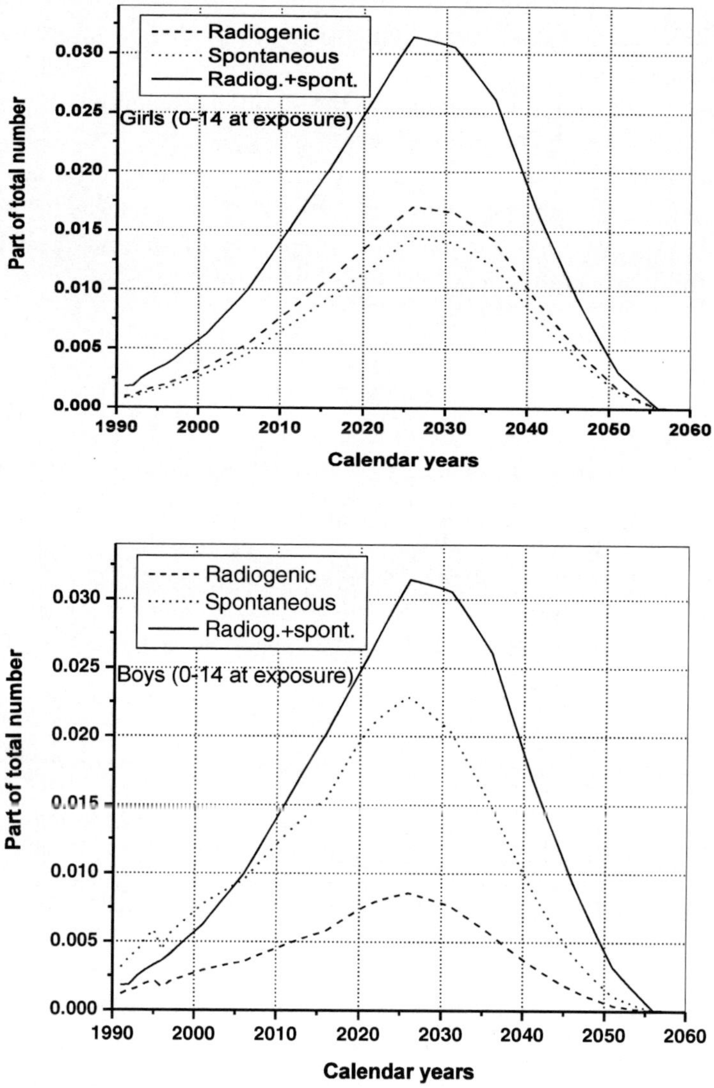

Figure 8. The distribution density for the proportion of cases as a function of calendar time

Of interest are the following of the obtained results: a significant excess of the mean dose for boys with thyroid cancer above that for girls; an increase in the mean dose (risk) among boys between 1997 and 2001 and a decrease in the mean dose (risk) in girls. The decreasing radiation risk for girls leads us to a hypothesis that the risk coefficient varies with time (is decreasing currently) and the peak of risk seems to be passed. The opposite phenomenon of a risk increase is observed for boys. The increase in the mean dose among boys with thyroid cancer may mean that the currently occurring radiogenic cancers are due to a high radiation dose, rather than high radiosensitivity. In the first period from 1991 to 1996, for both girls and boys, the effect of high radiosensitivity was playing the lead role.

For assessing health effects of thyroid exposure, projections of thyroid cancer dynamics were made using the risk coefficients derived in the present study.

Needless to say that the performed projections are approximate, given the assumption that the morbidity and mortality in the study population will remain unchanged in the forthcoming period. Yet, despite existing limitations such predictions make possible assessing health effects of exposure in terms of thyroid cancer incidence. The estimates of 1685 cases of radiogenic cancers also provide evidence to the significance and relevance of the problem and the need for special health care with respect to this group.

The obtained results should be treated as tentative, since the performed analysis shows that risk coefficients tend to change with time. Moreover, the analysis of the dose response relationship does not take into account the component of thyroid dose due to external exposure, while this consideration may lead to different estimates.

Conclusion

Analysis of thyroid cancer incidence between 1991 and 2001 was performed for persons exposed as children and adolescents (0 – 17 years) and living in the Bryansk Oblast, the worst contaminated after the Chernobyl accident.

The results of the performed statistical analysis lead us t conclude:

The incidence rate for thyroid cancer in the study population is higher that for Russia in general. The standardized incidence ratio SIR is estimated to be 4.9 (4.2, 5.8 95% CI) for girls and 8.8 (6.6, 11.5 95% CI) for boys.

The statistically significant radiation risk of radiogenic cancers is observed only for persons exposed as children (0 – 14 years). The excess relative risk per 1 Gy is 9.4 (3.0, 25.4 95% CI) and 29.6 (7.2, 7460 95% CI) for girls and boys, respectively. The value SIR for this group over the entire follow-up period from 1991 to 2001 is 5.6 (4.6, 6.8 95% CI) and 10.2 (4.6, 13.6 95% CI) for girls and boys, respectively.

The comparison of risk coefficients for the time periods 1991-1996 and 1997 – 2001 reveals that the temporal trend of ERR per unit dose in time is negative for girls (risk in the first period is higher than in the latter) and positive for boys.

The projections based on using the risk coefficients derived in the present study show that in the studied population the thyroid cancer incidence will be observed till 2050. Over the entire period from 1991 to 2050 a total of 2822 cases of thyroid cancer are expected, of them 1293 are radiogenic and 1529 spontaneous cases. For males, the projections are 392 and 147 cases respectively. Thus a total of 1685 radiogenic and 1676 spontaneous cancer cases can be expected to occur. The maximum incidence (number of cases per year) should be expected in 2025 – 2030.

REFERENCES

[1] Likhtarev I.A., Sobolev B.G., Kairo I.A. *et al.*, 'Thyroid cancer in the Ukraine', *Nature*, 1995, vol. 365, p. 365.

[2] Jacob J., Goulko G., Heidenreich W.F. *et al.*, 'Thyroid cancer risk to children calculated', *Nature*, 1998, vol. 392, pp. 31.

[3] *Proceeding of an International Seminar on Radiation and Thyroid Cancer*, Eds. Thomas G., Karaoglou A., Willliams E.D., Brussels-Luxembourg, World Scientific Publishing, 1999.

[4] Ivanov V.K., Gorski A.I., Tsyb A.F., Maksioutov M.A., and Rastopchin E.M., 'Dynamics of thyroid cancer incidence in Russia following the Chernobyl accident', *J. Radiol. Prot.*, 1999, vol. 19(4), pp. 305 – 318.

[5] Ivanov V.K., Gorski A.I., Pitkevitch V.A., and Tsyb A.F., 'Risk of radiogenic thyroid cancer in Russia following the Chernobyl accident', In: *Radiation and Thyroid Cancer. Proc. International Seminar on Radiation and Thyroid Cancer*, Eds. Thomas G., Karaoglou A., Willliams E.D., Brussels-Luxembourg, World Scientific Publishing, 1999, pp. 89 – 96.

[6] Heidenreich W.F., Kenigsberg Y., Jacob P., Buglova E., Gulko G., Paretzke H.G., Demidchik E.P., and Golovneva A., 'Time trends of thyroid cancer incidence in Belarus after Chernobyl accident', *Radiation Res.*, 1999, vol. 151, pp. 617 – 625.

[7] Jacob P., Kenigsberg Y., Zvonova I., Gulko G., Buglova E., Heidenreich W.F., Golovneva A., Bratilova A.A., Drozdovitch V., Kruk J., Pochtennaja G.T., Balonov M., Demidchik E.P., and Paretzke H.G., 'Childhood exposure due to the Chernobyl accident and thyroid cancer risk in contaminated areas of Belarus and Russia', *Brit J. Cancer*, 1999, vol. 80(9), pp. 1461 – 1469.

[8] Likhtarev I.A., Kairo I.A., Shpak V.M. *et al.*, 'Radiation-induced and background thyroid cancer of Ukranian children (dosimetric approach)', *Int. J. Radiat. Med.*, 1999, vol. 3-4, pp. 51 – 66.

[9] Ivanov V.K., Gorski A.I, Tsyb A.F., Maksioutov M.A., Vlasov O.K., and Godko A.M., *The Risk of Radiogenic Cancer in the Population of the Bryansk and Oryol Regions of Russia after the Chernobyl Accident. Chernobyl: Message for the 21st Century. Proceeding of the Sixth Chernobyl Sasakava Medical Cooperation Symposium*, Moscow, Russia, 30-31 May 2001, Elsevier, 2002, pp. 85 – 93.

[10] Ivanov V.K., Gorski A.I., Maksioutov M.A., Vlasov O.K., Godko A.M., Tsyb A.F., Tirmarche M., Valenty M. and Verger P., 'Thyroid cancer among adolescents and adults in the Bryansk Region of Russia following the Chernobyl Accident', *Health Phys.*, January 2003, vol. 84, Number 1.

[11] Winkelmann R.A., Okeanov A., Gulak L., Remennik L., Rahu M., and Storm H.H., 'Cancer registration techniques in the New Independent States of the former Soviet Union', *IARC Technical Report No. 35*, Lyon, 1998, pp. 22 – 43.

[12] *Reference Book of Mean Thyroid Doses in Persons of Different Age Residents in the Population Points of Bryansk, Tula, Oryol and Kaluga Oblasts, Contaminated After the Chernobyl Accident in 1986*, Ed. M.I. Balonov and I.A. Zvonova, *Bulletin of the National Radiation Epidemiology Registry*, Special issue, 2002. (Rus)

[13] 'Methodology for reconstruction of thyroid doses from iodine radioisotopes in residents of the Russian Federation exposed to radioactive contamination as a result of the Chernobyl accident in 1986', *Guidelines MU-2.6.1-00b*, 2000. (Rus)

[14] Preston D.L., Lubin J.H., Pierce D.A. and McConney M.E., *EPICURE*, Hirosoft International Corporation, Seatle, USA, 1993.

[15] Thompson D.E., Mabuchi K., Ron E., Soda M., Tokunaga M., Oshikubo S., Sugomoto S., Ikeda T., Terasaki M., Izumi S., and Preston D.L., 'Cancer incidence in atomic bomb survivors. Part II: Solid tumors, 1958-1987', *Radiation Res.*, 1994, vol. 137, S17 – S67.

[16] Shore R.E., 'Issues and epidemiological evidence regarding radiation-induced thyroid cancer', *Radiation Res.*, 1992, vol. 131, pp. 98 – 117.

[17] Ron E., Lubin J.Y., Shore R.E., Mabuchi K., Modan B., Pottern L.M., Shneider A., Tucker M., and Boice J.D., 'Thyroid cancer after exposure to external radiation: a pooled analysis of seven studies', *Radiation Res.*, 1995, vol. 141, pp. 259 – 277.

In: 20 Years after the Chernobyl Accident
Editors: E. Burlakova and V. Naidich, pp. 119-170
ISBN 1-60021-249-2

Chapter 9

THE HEALTH STATE OF THE CHERNOBYL ACCIDENT LIQUIDATORS: IMMUNOLOGIST'S OPINION 20 YEARS AFTER

I. V. Oradovskaya
State Scientific Center the Institute of Immunology,
Ministry of Health, Moscow, Russia

ABBREVIATIONS

ID	immune deficiency
IMIN	immune insufficiency
MO	Moscow Oblast
LO	Leningrad Oblast
NW	Northwest
LAC	Liquidation of the accident consequences
Chernobyl NPP	Chernobyl Nuclear Power Plant
IPS	immunopathological syndrome
ARVI	acute respiratory-viral infections
AD	allergic diseases
ChD	chronic diseases
CMSD	Central Medico-Sanitary Division
RG	risk group
CRHVI	chronic recidivating herpes-viral infection
ARW	accident recovery works
SAS	severe atopic syndrome
MCP	mine-chemical plant
BCD	blood circulation system diseases
IHD	ischemic heart disease
CVD:DEP	dyscirculatory encephalopathy
GIT	gastrointestinal tract

HD	hypertension disease
VD	vascular dystonia
NCD	neurocirculatory dystonia
UD	ulcer disease
VC	vesical calcification
CRI	chronic renal insufficiency
SRCI	Scientific Research Clinical Institute
ARCERM	All-Russian Center of Extremal and Radiation Medicine

PART 1. CLINICAL IMMUNOLOGICAL PARAMETERS

The industrial catastrophe at the Chernobyl Nuclear Power Plant (NPP) has induced to revise our opinion on low-dose irradiation consequences. The radiation impact was accompanied by a strong effect of other negative factors, which always appear at a radiation accident: the effects of radionuclide's, heavy metals, toxicants, working under extremely adverse labor conditions, high physical, emotional and psychic stress, enough strong and intensive. These adverse risk factors impacted the accident liquidators – a cohort of people manifesting different individual reactivity and psychophysiological conditions, an ability to adapt to unusual, extremely complicated conditions of the 30-kilometer restriction zone. The effect of these adverse factors induced a stress of compensatory-adaptive mechanisms, which in the case of their overloading promoted a breakdown of adaptive reactions and development of disadaptation syndrome, based on the immunological mechanisms and lasted for long time in many cases. The whole complex of these factors has necessarily effected on the formation of remote consequences unusual for classical radiobiology and not thoroughly studied yet. One of the manifestations of remote effects is increasing frequency of the clinical signs of immunity disturbance and chronic somatic diseases.

MATERIALS AND METHODS

Since 1986, the experts from Institute of Immunology, Ministry of Health of the Russian Federation are performing a long-term monitoring of the health state and immune status of people who have taken part in liquidation of Chernobyl Accident consequences. In remote time of the third-fourth five-year period after the accident the main groups for monitoring were formed from liquidators from the Moscow Oblast (MO), North-west (NW) region of Leningrad Oblast (LO) and Krasnoyarsky krai (Figure 1).

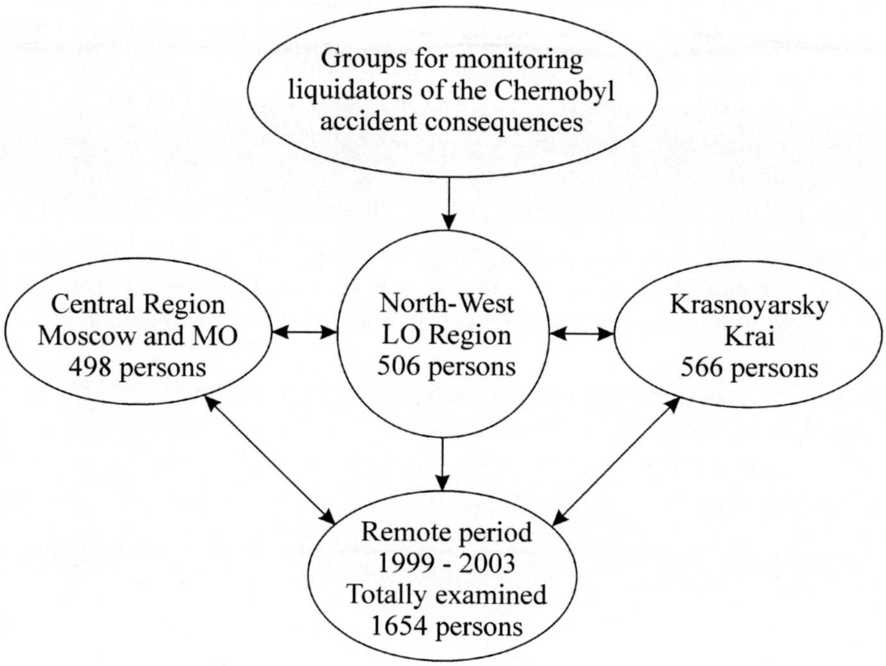

Figure 1.

Immunological monitoring was performed according to the general technique based on a specialized clinical immunological and allergological examination using standardized test card, designed in the Institute of Immunology for detecting the immune insufficiency (IMIN) and other dysfunctions of the immune system prior to laboratory stage [13, 16 – 18]. The main section of screening represents IMIN clinical symptoms subdivided by the leading immunopathological syndrome (IPS): infection, allergic (atopic), infection-allergic (a combination of infection and allergic syndromes), and autoimmune syndromes. The infection syndrome comprises frequently repeating, chronic (with frequent aggravations), recrudescent pathological processes of bacterial, viral and fungal etiology, frequent acute respiratory-viral infections (ARVI) happening 3 – 4 times a year or more (Table 1). Other clinical IPS comprise allergic diseases (AD) and responses, autoimmune diseases, and immunoproliferative states. Allergological examination comprised collection of allergic case history, skin test performance with domestic, epidermal, fungal and, according to indications, food allergens, and determination of general IgE. The card also comprises data on frequency of the most widespread chronic diseases (ChD).

RESULTS AND DISCUSSION

Dynamics of IMIN Clinical Symptom Occurrence for Liquidators from Moscow and Moscow Oblast

In the period of 1997 – 2003, over 500 persons – liquidators of the Chernobyl accident consequences from Moscow and MO were monitored. Annually, up to 130 persons are involved in specialized clinical immunological and allergological examination, about 30 – 40 of whom are examined 2 – 3 times. In the case of low dynamics, the immunoepidemiological analysis is supplemented with data of a single examination only. Already by the beginning of the third five-year period after the Chernobyl catastrophe, the liquidators from Moscow and MO (Figure 2) manifested high frequency of clinical symptoms of the immunity disorders equal 84.88% (73:86). In the period of 1999 – 2000, the occurrence dynamics of IMIN clinical symptoms varied from 78.31% (69:87) to 88.54% (85:96), characterized by increased occurrence in 199 and 2001 with a tendency to slight decrease in 2000 and 2002. The presence of close RG values for a six-year period allowed for calculating the average parameter, which for the period of 1997 – 2002 equaled 83.49% (435:521) or 834.9:1000. This is very high occurrence of clinical symptoms for the immune system dysfunction, exceeding that for the personnel of a nuclear park [5, 14, 15, 19, 24]. In 2003, a considerable increase of this parameter to 97.3% (107:110) of persons, i.e. they were observed almost in all passed the examination.

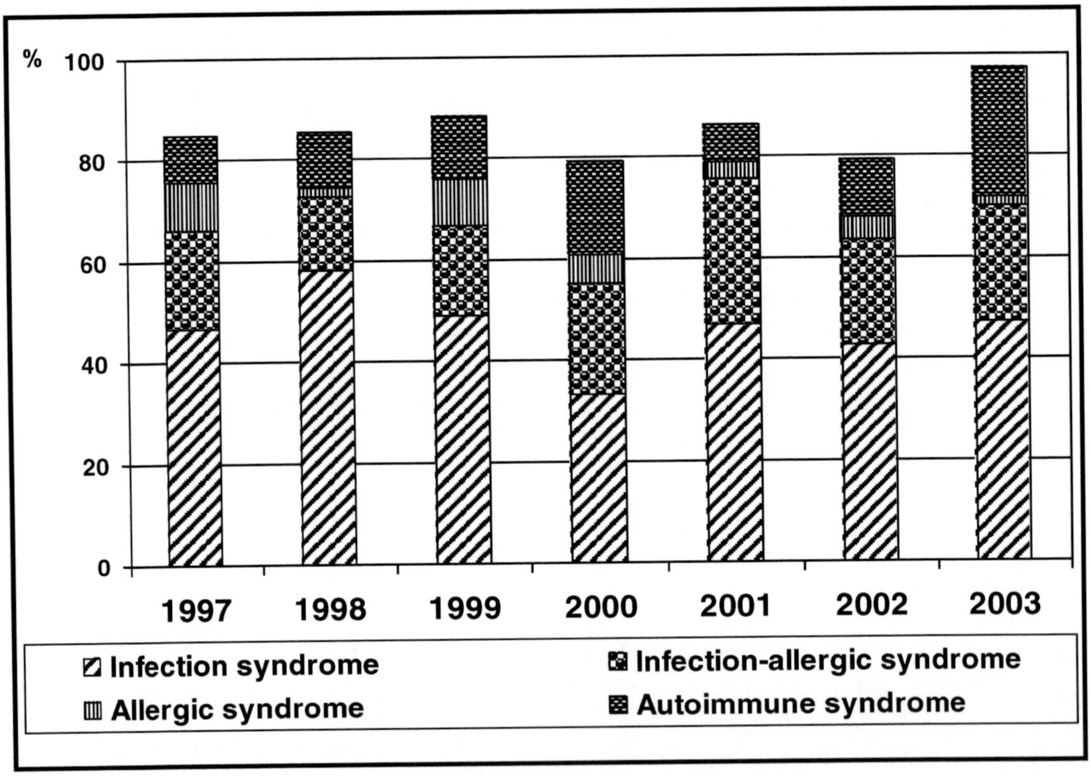

Figure 2. Dynamics of GR values for IMIN and distribution of immunopathological syndromes in the studied groups of liquidators from Moscow and MO, 1997 – 2003

Table 1. Structure and dynamics of occurrence frequency of IMIN clinical symptoms in groups of persons – the liquidators of the Chernobyl accident consequences from Moscow and MO, according to data of monitoring performed in 2000 – 2003

No.	IMIN clinical manifestations	2003 $n = 110$	2002 $n = 131$	2001	2000 $n = 87$
	I. Infection syndrome:				
1	Chronic bronchitis, frequent	10.0	**15.27**	10.61	12.64
2	Frequent bronchitis with chronic infection of ENT-organs	–	1.53	1.52	2.30
3	Bronchitis with bronchial component and increased sensitivity to ARVI	–	**15.27**	15.15	11.49
4	Bronchitis, obstructive chronic	**49.09**	**14.50**	13.64	6.90
5	Chronic bronchitis (total)	**59.09**	**46.56**	**40.91**	**33.33**
6	Frequent anginas, including at the chronic tonsillitis background	10.90	5.34	4.55	1.15
7	Frequent ENT-organ infections	**24.54**	12.98	12.12	**19.54**
8	Bacterial skin and hypoderm infections	12.72	7.63	7.58	9.20
9	Fungal infections	10.90	8.40	13.64	9.20
10	Frequent ARVI (3 – 4 times a year or more)	**65.45**	**39.69**	**40.91**	**43.68**
11	Frequent ARVI combined with CRHVI	**60.91**	**21.37**	**19.70**	**21.84**
12	CRHVI without frequent ARVI	13.60	3.05	0	0
13	Gastroenteropathy with dysbacteriosis	12.72	16.79	13.64	10.34
14	Chronic pyelonephritis	10.0	5.34	6.06	17.24
15	Durable subfever	6.36	7.63	12.12	6.90
16	Parodontosis with frequent aggravations	12.72	3.82	15.15	6.90
17	Lymphadenopathy	4.54	0.76	3.03	0
	IMIN. Allergic syndrome				
1	Skin-atopic manifestations (atopic dermatitis, neurodermatitis, eczema)	2.72	3.05	6.06	3.45
2	Severe atopic syndrome (SAS)	5.45	–	–	–
3	Allergic conjunctivitis	–	3.05	3.03	0
4	Allergic rhinitis	6.36	5.34	6.06	3.45
5	Rhinoconjunctivitis	7.26	3.05	0	2.30
6	Recrudescent hives	1.81	1.53	0	1.15
7	Pollinosis	8.18	9.92	4.55	9.20
8	Bronchial asthma, pre-asthma	6.36	3.82	7.58	5.75
9	Sensibilization to allergens (domestic dust, drugs, etc.)	19.09	16.03	19.70	18.39
10	Food intolerance	2.72	2.29	4.55	5.75
	IMINI. Autoimmune diseases	**25.45**	11.45	**7.58**	**18.39**
1	Autoimmune thyroiditis	**20.90**	6.11	4.55	**12.64**
2	Sugar diabetes (type 2 – demanding insulin)	0.9	4.58	1.52	5.75
3	Rheumatoid arthritis	0.9	1.53	1.52	1.15
4	Rheumatism (rheumatic fever)	1.81	–	–	–
5	Chronic glomerulonephritis	0.9	–	–	–
6	Psoriasis	0.9	0.75	1.52	1.15

In the structure of IMIN clinical symptoms for liquidators from the Moscow Region, the infection syndrome is predominant (Figure 2, Table 1). Its frequency was comparable for 1997, 1999, 2001 and 2002, varying within the range of *42.75 – 48.96%.* In 1998, the frequency increased sharply to 58.18%, whereas in 2000 it decreased to 33.33%. Hence, a decrease of frequency of IMIN clinical symptoms for infection syndrome was accompanied by increasing occurrence of combined pathology of the infection-allergic genesis. In the group of liquidators examined in 2001, the occurrence of IMIN clinical symptoms of both syndromes and the highest total index of them were observed. In the period of 1998 – 2001, increasing occurrence of IMIN clinical symptoms for combined pathology of the infection-allergic syndrome *(14.55% → 17.7% → 21.84% → 28.79%)* indicated increasing severity of the pathology. In 2002 – 2003, the occurrence frequency of infection-allergic syndrome combined pathology had similar values. For the liquidators from Moscow and MO the occurrence of AD without infectious symptoms is low. Total occurrence of AD and combined pathology of infection-allergic syndrome is high due to increasing frequency of infection-allergic diseases. Among liquidators examined in 2000, *the frequency of autoimmune diseases was increased*, which was observed in 18.4% of the liquidators. This is the expected effect of the radiation impact for the remote period. Autoimmune diseases were observed both isolated and combined with another chronic pathology and/or at the background of clinical symptoms of the immune system dysfunctions. In 2000, the main clinical entities were autoimmune thyroiditis and sugar diabetes of type two demanding insulin (Table 1).

In the fourth five-year period after the Chernobyl accident, the structure of IMIN clinical symptoms of the infection syndrome possessed high occurrence of chronic, frequently repeating and obstructive bronchitis. The dynamics of their frequency testifies about a considerable increase of occurrence of this pathology, specifically in 2003 (Table 1), obstructive bronchitis dominating in the structure. The occurrence frequencies of ARVI and ARVI combined chronic recrudescent herpes viral infection (CRHVI) are very high: *654.5:1000* and *609.1:1000*, respectively. High frequency of respiratory apparatus diseases in liquidators was promoted by inhalation damage of airways during implementation of works on liquidation of the accident consequences, the so-called "irritation syndrome of upper airways". The average frequency of this syndrome observed for the liquidators of the accident consequences equaled 47.64%, summed up from 37.05 – 81.54% with respect to external γ–irradiation dose perceived and 30.3 – 60.0% with respect to periods of presence in the zones with elevated background radiation e.g. increasing with the radiation dose and periods of presence in the zones of accident consequence liquidation. This was also promoted by decreasing cellular immunity already in the initial 1 – 1.5 years after leaving the zones of accident recovery works (ARW) and the presence of imbalance in the system of both cellular and humoral immunity, observed during many years.

The data obtained on high occurrence of chronic bronchitis are comparable with these obtained in Research Institute of Pulmonology at single-step epidemiological examination, performed in 1994 for liquidators from Ryazan' (44.3%) that exceeded for liquidators from Vladimir (34.9%) [2, 8]. Generally, the cohort of examined liquidators was composed of people drafted by military state departments. For this group, the experts from Research Institute of Pulmonology have also observed high frequency of ARVI occurrence (37.9%).

ARVI cases as frequent as 2 – 3 times per year were determined for 36.6% of the liquidators [2]. According to the authors' observations, higher ARVI and chronic bronchitis frequencies are reasoned by more remote observation time and the difference in the persons forming the cohort to be examined. In remote time periods after the accident frequent infections represent an unfavorable symptom of deteriorating state of the immune reactivity in liquidators and the *risk factor and ChD severity increase*, specifically dominating cardiovascular pathology, because they shorten reserves of the vitally important organs leading to pervert responses of the immune system, formation of antibodies to self tissues of the organism and autoimmune pathology. Viruses of Herpesviridae family may promote intensification of atherogenesis processes and *early development of elderly age sicknesses*: atherosclerosis and related diseases in patients below 50 that induces *shortening of the life duration*.

Table 2. Quantitative distribution of GR values by IMIN and leading IPS in groups of Chernobyl accident liquidators from NW Region of Leningrad Oblast, 1999 – 2003

No.	IMIN clinical syndromes	Frequency of occurrence (%)				
		1999 *n* = 177	**2001** *n* = 112	**2002** *n* = 108	**2003** *n* = 104	**Average for** 2002 - 2003
1	Infection	17.51% (31:177)	21.43% (24:112)	13.21% (14:108)	44.23% (46:104)	28.30% (60:212)
2	Infection-allergic	14.12% (25:177)	10.71% (12:112)	17.59% (19:108)	18.27% (19:104)	17.92% (38:212)
3	Allergic	5.08% (9:177)	13.39% (15:112)	13.21% (14:108)	5.77% (6:104)	9.43% (20:212)
4	Autoimmune diseases	7.34% (13:177) *	8.93% (10:112) *	23.15% (25:108) *	13.46% (14:104) *	18.40% (39:212)
5	Presence of IPS clinical symptoms	44.07% (78:177)	54.46% (61:112)	66.67% (72:108)	81.73% (85:104)	74.06% (157:212)

* The frequency of occurrence is indicated by the leading IPS; autoimmune diseases were determined in combination with other IMIN clinical symptoms.

According to data of the period of 1997 – 2002, the average index of *autoimmune pathology* occurrence in the liquidators from the Moscow region equaled 119.0:1000. In 2003, a considerable increase of the frequency of immune-type pathology to 254.5:1000 was observed, mostly due to increasing frequency of autoimmune thyroiditis. The frequency dynamics of autoimmune diseases manifested gradual increase in the period of 1997 – 2000 (183.9:1000), a decrease in 2001 and reverse increase to the level of 1998 – 1999 in 2002, and then a considerable increase in 2003. Differences in the frequency of autoimmune pathology in the groups of the accident consequence liquidators were also strictly dependent on heterogeneity of groups under examination.

Dynamics of IMIN Clinical Symptom Occurrence Among Liquidators from LO Northwest Region

Clinical symptoms of the immune pathology among liquidators of the Chernobyl accident consequences, recruited from NW Region of Leningrad Oblast in 1999, were observed in 44.07% of examinants, which was much lower compared with the liquidators from Moscow Oblast for the period under consideration (Table 2). For the period of 1999 – 2003, the frequency of IPS clinical symptoms for liquidators from NW Region of Leningrad Oblast steadily increased: *44.07% → 54.46% → 66.67% → 81.73%* (Figure 3). In 2003, it reached the level of the liquidators from Moscow Oblast, determined in 2002.

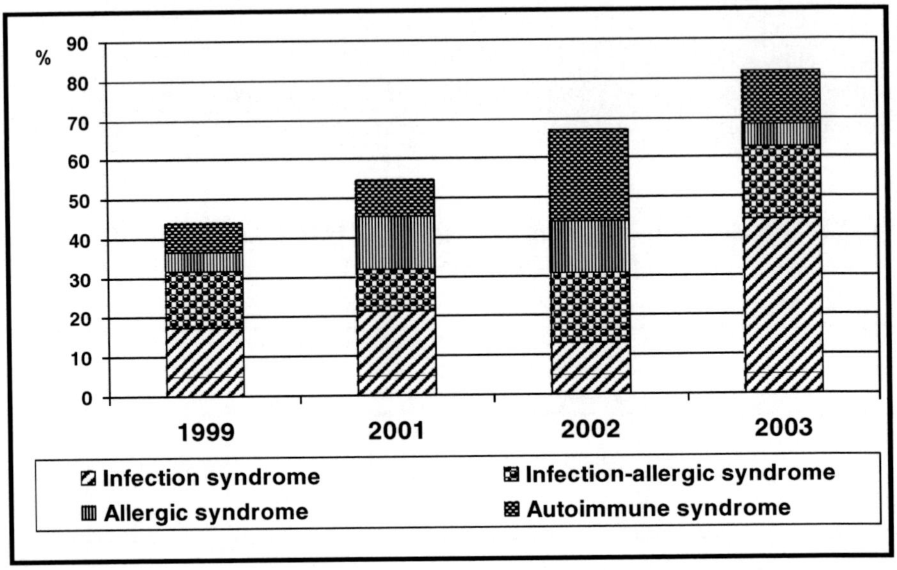

Figure 3. GR dynamics with respect to IMIN and distribution of main IPS among examined groups of Chernobyl accident liquidators from NW Region of Leningrad Oblast according to monitoring data in the period of 1999 – 2003

In 1999 and 2001, the main IMIN clinical symptom was also infection index, but its frequency in this period was much lower: 17.51% → 21.43%. However, in 2003, this value was considerably increased reaching the level, observed for liquidators from the Moscow Region in 2001 – 2003. In 2003, the occurrence of combined pathology of the infection-allergic syndrome was found slightly lower, but also comparable with the Moscow Region level: 18.27% (2001) – 20.61% (2002) – 22.72% (2003). The AD frequency in liquidators from NW Region of Leningrad Oblast exceeded this parameter for MO. Total frequency of AD and combined pathology of infection-allergic syndrome occurrence is high. The presence of allergic pathology is observed for 1/3 of the liquidators (30.8%, 33:108), examined in 2002, and 1/4 (24.04%), examined in 2003.

Lower frequency of IMIN clinical symptoms, observed for liquidators from NW Region of Leningrad Oblast before 2001, may be caused by two reasons underlining importance of the immunological monitoring:

- since 1986, the liquidators from NW Region of Leningrad Oblast were subject to specialized examination, which was of high prophylactic importance, and until 1996 were examined at annual dispensary immunocorrection procedures;
- diagnostic capabilities of MSD in a small town and the absence of special immunological service in it.

At the fourth five-year period after the Chernobyl accident, the structure of IMIN clinical symptoms with the leading infection syndrome manifested airway diseases (frequent obstructive and chronic bronchitis), frequent ENT-organ infections, fungal infections, chronic recrudescent herpes-viral infections combined with ARVI or most frequently CRHVI without ARVI as the most frequently occurring sicknesses (Table 3). As compared with the previous data, in 2003 an increased occurrence of obstructive bronchitis, CRHVI, fungal infections and chronic pyelonephritis was observed. Occurrence of frequent ENT infections, 2-fold elevated as comparing data of 2002 with 2001, remains high.

Table 3. Dynamics of IMIN clinical symptom frequency and structure in groups of liquidators from NW Region of Leningrad Oblast according to monitoring data of 1999 – 2003

No.	IMIN clinical symptoms	2003	2002	2001	1999
		n = 104	*n* = 108	*n* = 112	*n* = 177
	I. Infection syndrome:				
1	Chronic bronchitis, frequent	12.50	7.41	8.93	4.52
2	Frequent bronchitis with chronic infection of ENT-organs	0.96	0.93	2.68	–
3	Bronchitis with bronchial component and increased sensitivity to ARVI	0.96	1.86	1.79	0.56
4	Bronchitis, obstructive chronic	**22.12**	5.56	3.58	–
5	Bronchitis (total)	**32.77**	**15.74**	**16.96**	5.08
6	Chronic pneumonias, frequent	0	5.56	0	0
8	Frequent chronic infections of ENT-organs	**25.96**	**25.0**	12.50	0.56
9	Bacterial skin and hypodermal infections	1.92	2.79	2.68	1.13
10	Fungal infections	**23.08**	**15.74**	13.39	6.21
11	Frequent ARVI (3-4 times a year of more)	9.62	10.19	**19.64**	12.99
12	Frequent ARVI combined with CRHVI	6.73	4.63	2.68	4.52
13	Herpes not accompanied by frequent ARBI	**20.19**	2.79	0.89	–
14	Gastroenteropathy with dysbacteriosis	5.77	2.79	1.79	1.69
15	Chronic pyelonephritis	**13.46**	2.79	2.68	–
16	Lymphoadenopathy	0	0.93	0	0.56
17	Durable subfever	0.96	1.86	0.89	0.56
18	Parodontosis with frequent aggravations	0.96	3.70	2.68	2.26

No.	IMIN. Allergic syndrome:				
1	Skin-atopic manifestations (atopic dermatitis, neurodermatitis, eczema)	0	**12.04**	4.46	2.26
2	Severe atopic syndrome (SAS)	0	0.93	–	–
2	Allergic conjunctivitis	0	**10.19**	0.89	
3	Allergic rhinitis	2.88	3.72	2.68	4.52
4	Rhinoconjunctivitis	5.77	**10.19**	1.79	1.69
5	Recrudescent hives	6.73	2.78	5.36	1.13
6	Pollinosis	**12.50**	**10.19**	7.14	7.34
7	Bronchial asthma	5.77	6.48	2.68	2.83
8	Pre-asthma	0	1.86	1.79	0.56
9	Sensibilization to allergens (domestic dust, drugs, etc.)	11.54	18.52	16.07	18.08
10	Food intolerance	0.96	7.41	2.68	–
11	Latent sensibilization	5.77	9.26	8.93	–
13	Insect allergy	0.96	0.93	0.89	–
	IMINI. Autoimmune diseases:				
1	Autoimmune thyroiditis	11.54	14.81	8.04	5.65
2	Sugar diabetes (type 1, type 2 – demanding insulin)	0	6.48	1.79	3.39
3	Rheumatoid arthritis	0.96	1.86	1.79	–
4	Widespread vasculitis	0	0.93	–	–
5	Psoriasis	0.96	3.72	3.57	1.73

The dynamics of AD structure indicates atopic pathology increase in 2002: skin-atopic manifestations, atopic bronchial asthma, pollinosis, and allergic pathology with impairment of mucous linings (allergic conjunctivitis, rhinitis, rhinoconjunctivitis), which is typical of patients exposed to low-dose radiation [6]. Expansion of the nosological spectrum and increasing occurrence frequency by all nosological forms was observed. In patients examined in 2003, pollinosis dominated in the structure of allergic pathology.

The dynamics of IMIN clinical symptoms by leading IPS testifies about increasing occurrence frequency of autoimmune pathology in the liquidators. Hence, the most considerable elevation of the frequency was observed in the period of 2001 – 2002, when autoimmune diseases were detected in each fourth examined liquidators. As compared with other groups (from Moscow and MO, and Krasnoyarsky krai) in this period, liquidators from NW Region of Leningrad Oblast demonstrated the highest occurrence frequency for autoimmune diseases. The elevation of frequency of autoimmune diseases was also observed in 2000 for the group from Moscow region but with slightly lower occurrence frequency. The elevation of autoimmune pathology occurrence in the liquidators from Moscow and MO, observed in 2003, is comparable with this index determined in 2002 for the liquidators from NW Region of Leningrad Oblast. The elevated frequency of autoimmune pathology is the expected effect of radiation damage, which for the liquidators from NW Region of Leningrad Oblast obtained maximal manifestation 15 – 17 years after their participation in the accident recovery works in the zones with increased radiation, whereas for the liquidators from Moscow region this happened two years earlier. In the structure of autoimmune diseases allergic thyroiditis dominates.

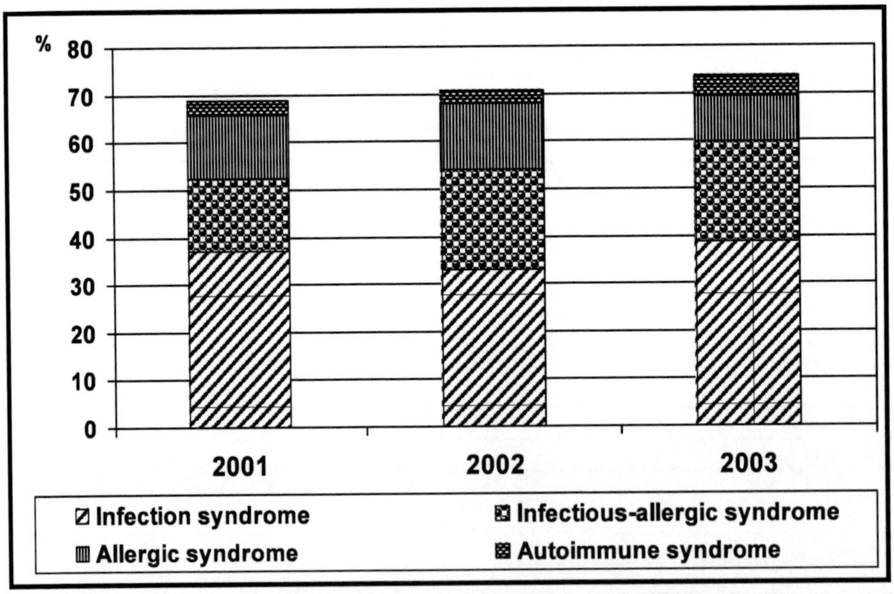

Figure 4. Dynamics of GR values by IMIN and distribution of main IPS among examined groups of liquidators from Krasnoyarsky krai according to monitoring data of 2001 – 2003

Dynamics of IMIN Clinical Symptom Occurrence for the Liquidators from Krasnoyarsky krai

By the fourth five-year period after the Chernobyl accident, for 2/3 part of the group of liquidators from Krasnoyarsky krai was determined by the presence of clinical symptoms of immunopathological states.

As compared with the examination groups from other regions, dynamics of slightly varying clinical immunological indices manifested relatively close values of the IMIN risk group during 2001 – 2003 tending to increase (*68.83% → 70.62% → 73.61%*), and similar values of quantitative distribution of the main immunopathological syndromes: infectious (2001, 2003), infectious-allergic (2002 – 2003) and AD (2001, 2003). In the period of 2001 – 2003, dynamics of the allergic pathology manifested decreasing AD frequency (*13.42% → 13.92% → 9.72%*) and increasing frequency of combined pathology of infectious-allergic syndrome (*15.15% → 21.13% → 20.83%*) (Figure 4). Compared with two other groups of liquidators, the liquidators from Krasnoyarsky krai manifested much lower frequency of occurrence of the autoimmune pathology with detected nosological forms of diseases: *3.03% → 2.58% → 4.17%.*

The structure of IMIN clinical symptoms of the infectious syndrome (Table 4) manifested frequent chronic bronchitis as a monosign or in combination with ENT pathology, chronic obstructive bronchitis, frequent ARVI, ARVI combined with CRHVI, and chronic pyelonephritis. Similar to other groups of observation, the cohort of liquidators from Krasnoyarsky krai demonstrate increasing frequency of obstructive bronchitis forms, this value being 2.14 times higher in the liquidators from Moscow and MO and 2.22 times higher in the liquidators from NW Region of Leningrad Oblast, simultaneously.

In the course of allergic pathologies, no considerable differences from other groups was observed, which varied with respect to the composition of examined groups. A pathology damaging mucous linings dominates: allergic rhinitis, rhinoconjunctivitis, pollinosis, and skin-atopic manifestations (2001 – 2002). In the one fifth of examined liquidators, sensibilization to domestic and epidermal allergens, drugs, etc. is observed.

Basing on regional features, among liquidators of the Chernobyl accident from Krasnoyarsky krai total index of immune pathology clinical symptoms is comparable for the personnel of Mining Chemical Plant (MCP), which is also located in Krasnoyarsky krai. Quantitative distributions of main IPS are different. The liquidators possess IMIN clinical symptoms with the leading infectious syndrome and combined pathology of infectious-allergic syndrome (54.12%) dominating in the IPS structure. Hence, MCP personnel indicated relatively uniform distribution of IMIN clinical syndromes with a sight tendency to increase of infectious genesis symptoms.

Table 4. Dynamics of IMIN clinical symptom frequency and structure among liquidators from Krasnoyarsky krai according to monitoring data for 2001 – 2003

No.	IMIN clinical manifestations	2003	2002	2001
		n = 74	*n* = 194	*n* = 231
	I. Infectious syndrome:			
1	Chronic bronchitis, frequent	12.16	8.25	12.12
2	Chronic bronchitis, infrequent	4.05	10.31	–
3	Frequent bronchitis with chronic ENT infection	2.70	3.61	5.19
4	Bronchitis with bronchial component and increased sensitivity to ARVI	5.41	4.12	2.16
5	Chronic obstructive bronchitis	**22.97**	7.22	9.52
6	Bronchitis (total)	**36.49**	**33.51**	**29.0**
7	Chronic pneumonias	1.35	–	–
8	Frequent ENT infections	**2.70**	**6.70**	4.33
9	Bacterial infections of skin and hypodermis	1.35	3.09	3.03
10	Fungal infections	6.76	**7.22**	5.19
11	Frequent ARVI (3-4 times or more)	**27.03**	**25.26**	**25.11**
12	Frequent ARVI combined with CRHVI	**14.86**	**7.22**	**9.09**
13	CRHVI without frequent ARVI	4.05	1.55	–
14	Gastroenteropathy with dysbacteriosis	4.05	1.55	3.03
15	Chronic pyelonephritis, cystitis	**5.41**	3.09	6.06
16	Durable subfever н/э	6.76	5.15	2.16
17	Parodontosis with frequent aggravations	–	2.06	2.60
18	Lymphoadenopathy	–	1.03	1.30
	Allergic syndrome			
1	Atopic dermatitis	1.35	3.09	7.79
2	Eczema	1.35	5.15	2.60
2	Allergic conjunctivitis	–	2.58	0.43
3	Allergic rhinitis	–	4.64	5.19
4	Rhinoconjunctivitis	5.41	6.70	7.79
5	Pollinosis	5.41	5.67	7.79
6	Bronchial asthma	**6.76**	3.09	2.16
7	Recrudescent hives	–	2.06	0.43
9	Sensibilization to allergens from domestic dust, drugs, etc.	**24.32**	**19.59**	**21.65**
10	Food intolerance	1.35	–	–
11	Latent sensibilization	6.76	5.15	3.46
	Autoimmune diseases			
1	Autoimmune thyroiditis	2.70	2.06	0.87
2	Sugar diabetes (type 1, type 2 – demanding insulin)	0*	1.03*	0*
3	Rheumatic arthritis	–	0.52	0.43
4	Rheumatism	–	–	0.43
5	Autoimmune glomerulonephritis	1.35	–	0.87
6	Psoriasis	1.35	0.52	1.30

In 2001 and 2003, 2 cases of type 2 sugar diabetes in each year were observed, whereas in 2002 – 4 cases.

Table 5. The frequency of CD occurrence in liquidators of the Chernobyl accident consequences from Moscow and MO according to data of monitoring performed in 2001 – 2003

No.	Diseases	Moscow and MO		NW Region of Leningrad Obl.		Krasnoyarsky krai	
		2003	**2002**	**2003**	**2002**	**2003**	**2002**
		n = 110	*n* = 133	*n* = 104	*n* = 108	*n* = 74	*n* = 194
	Blood circulatory disease	**98,18**	**85.71**	**85.58**	**72.22**	**85.14**	**81.44**
1.	Atherosclerosis + HD	**82.72**	**56.39**	**63.46**	**47.22**	**58.11**	**44.32**
2.	Ischemic heart disease	**71.81**	**48.87**	40.38	**43.52**	**36.49**	**26.80**
3.	VD, NCD	13.64	25.56	14.42	13.89	21.62	**25.77**
4.	CVD: Discirculatory encephalopathy	**86.36**	**49.62**	**59.62**	**42.59**	**71.62**	**51.03**
	Neuropsychic sphere pathology	**80.0**	**41.99**	**34.62**	**25.93**	**82.43**	**40.72**
5	Asthenic syndrome, neurasthenia	**53.63**	20.30	15.38	9.26	13.51	9.28
6	Increased fatigability syndrome	**62.72**	19.55	17.31	18.52	**78.38**	**22,68**
7	Organic diseases of cerebrum, psychoorganic syndrome	**14.54**	**12.78**	5.77	2.78	**25.68**	**14.43**
8	Polyneuropathy	9.09	3.03	**0.96**	–	2.70	4.12
	Digestive system diseases (total)	**96.36**	**72.18**	**66.35**	**52.78**	**64.87**	**63.92**
9	**Diseases of gastrointestinal tract** (chronic gastritis, gastroduodenitis, stomach ulcer disease, and dodecadactylon), **total**	**95.45**	**51.88**	**56.73**	**47.22**	47.30	**42.27**
10	Chronic cholecystitis, cholecystopancreatitis	**49.09**	**40.60**	**25.96**	20.59	45.95	**41.24**
11	Fatty hepatosis, fatty degeneration of liver	17.27	5.26	**9.62**	5.56	16.22	8.69
	Musculoskeletal system diseases	**100.0**	**66.17**	**53.85**	**53.7**	**52.70**	**52.06**
12	Deforming vertebral osteochondrosis	**91.81**	**56.39**	**48.08**	**46.30**	**48.65**	**44.32**
13	Chronic polyarthritis, osteoarthrosis	39.09	14.29	1.92	6.48	8.11	7.22
14	Osteoarthrosis	**31.82**	**12.78**	**10.58**	6.48	1.35	1.56
	Other chronic pathology						
15	Vein diseases	8.17	6.79	15.39	10.19	4.05	4.12
16	Multiple caries	10.0	6.02	0	3.72	1.35	4.12
17	Thyroid diseases with AT	**42.72**	**30.08**	**26.92**	**26.85**	24.39	22.16
18	Visual organ pathology of uninformative etiology	**50.0**	9.77	8.65	8.33	16.22	6.19
19	Radiation cataract	8.18	3.76	0	4.63	2.70	2.06
20	Acoustic apparatus pathology of noninfectious etiology	6.36	1.50	**0.96**	**0.93**	**4.05**	1.55
21	Dyshidrotic (nonallergic) eczema	3.63	0.75	–	–	–	–
22	VC	**26.36**	16.03	10.58	11.11	**13.51**	**7.22**
23	CRI	4.54	–	–	–	1.35	0.52
24	Sugar diabetes, type 2	4.54	3.01	0	0.93	2.70	2.06
25	Benign tumors	**35.45**	14.10	12.50	16.67	13.51	4.12
26	States after cancer elimination	2.72	1.50	3.85	2.78	4.05	4.12
27	Healthy	0	0	0	0	0	2.06
28	**presence of 3 or more diseases**	**100.0**	**84.21**	**86.54**	**72.22**	**77.03**	**74.23**
29	**Chronic diseases** (total)	**100.0**	**96.99**	**100.0**	**89.81**	**86.49**	**97.94**
30	Disability due to diseases caused by works as liquidators of Chernobyl accident consequences	35.45	33.08	26.92	25.93	43.24	38.66
	including: group I	0.9	0.75	0	0.93	4.05	1.55
	group IMIN	25.45	23.31	20.19	13.21	18.92	18.04
	group IMINI	9.09	8.27	5.77	12.04	20.27	19.07
31	By general disease :	19.09	19.08	5.77	3.72	5.41	3.61
32	Total	54.55	51.13	32.69	29.63	48.65	42.27

Dynamics of Occurrence of Chronic Somatic Diseases among Liquidators from Moscow and Moscow Oblast

Already by the third five-year period after the Chernobyl accident, high frequency of chronic pathology was detected among the liquidators from Moscow region (Figure 6) [12]. In the period of 1997 – 2002 and especially in 2003 year of the fourth five-year period, analysis of the frequency dynamics of the most widespread ChD has determined a considerable elevation by all classes of diseases (Table 5, Figure 5). Chronic diseases and polymorbidity syndrome were observed in all (100%) examined liquidators. A considerable increase of blood circulatory disease (BCD) frequency is observed for all nosological forms accompanied by atherosclerosis. BCD`s are determined for 98.2% of liquidators from the observation group. Compared with 2002, the frequency of atherosclerosis with hypertensive disease (HD) and ischemic heart disease (IHD) increased by 1.47 times and cerebrovascular disease clinically expressed as dyscirculatory encephalopathy (CVD:DEP) – by 1.74 times; hence, 61.8% of liquidators demonstrated systemic atherosclerosis with HD. Cardiovascular system diseases are accompanied by dysfunction of neuropsychic sphere of various severity: clinical manifestations of asthenic syndrome, increased fatigability syndrome, psychoorganic syndrome at the background of cerebrum organic damages and CVD:DEP. For 96.36% (106:110) of liquidators, digestive system diseases were observed. The most considerable elevation is observed for GIT diseases: *51.88% → 95.45%.* Hence, all examined liquidators had diseases of the musculoskeletal system, comprising 91.8% of deforming osteochondrosis, among which 39.09% were deforming osteoarthrosis and/or chronic polyarthritis. Besides vertebral osteochondrosis, the liquidators show signs of the systemic skeletal disease – metabolic osteopathy, most often observed in persons participating in liquidation of the accident consequences in 1986 – 1987. Data on the frequency of bone and joint diseases correlate with the results by other investigators. Currently existing data do not allow for unambiguous reasoning of skeleton damaging of the liquidators, who had no such pathology before participation in liquidation of the accident consequences. Apparently, the systemic damage of bones and joints follows from incorporation of radionuclides of the osteotropic group [10, 11].

In 2003, among other nosological forms of diseases increased occurrence of thyroid diseases is observed, mostly due to autoimmune thyroiditis, noninfectious pathology of visual organs, VC and benign tumors (papillary tumor, basalioma, adenoid tumor, etc.), observed for more than 1/3 of the examined group. The data presented indicate an abrupt health deterioration for the liquidators in 2003 compared with the period of 2001 – 2001. Hence, increasing pathology is observed by both IMIN clinical manifestations and chronic somatic diseases.

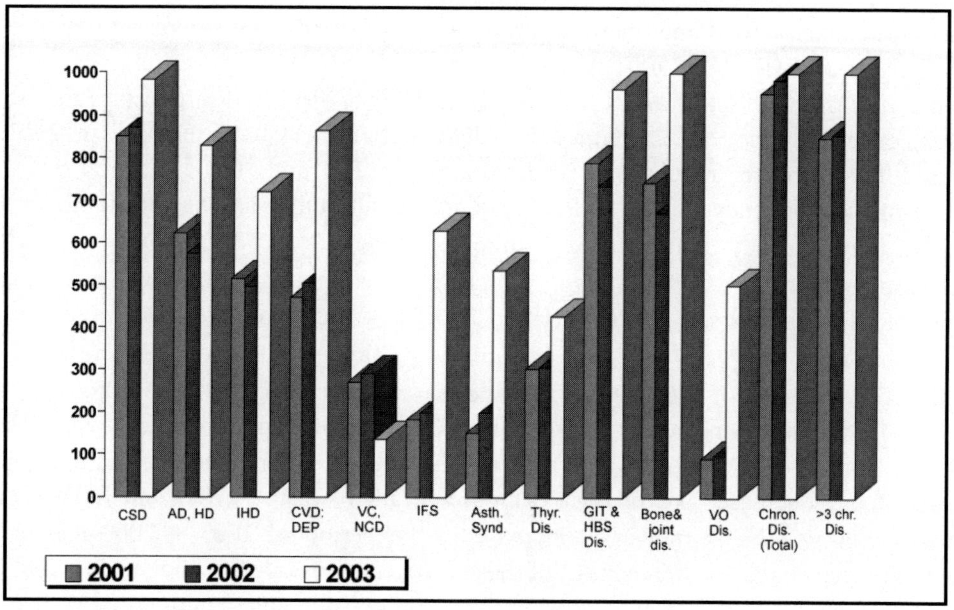

Figure 5. The occurrence frequency of the most widespread chronic diseases among liquidators of the Chernobyl accident from Moscow and MO (per 1000 persons)

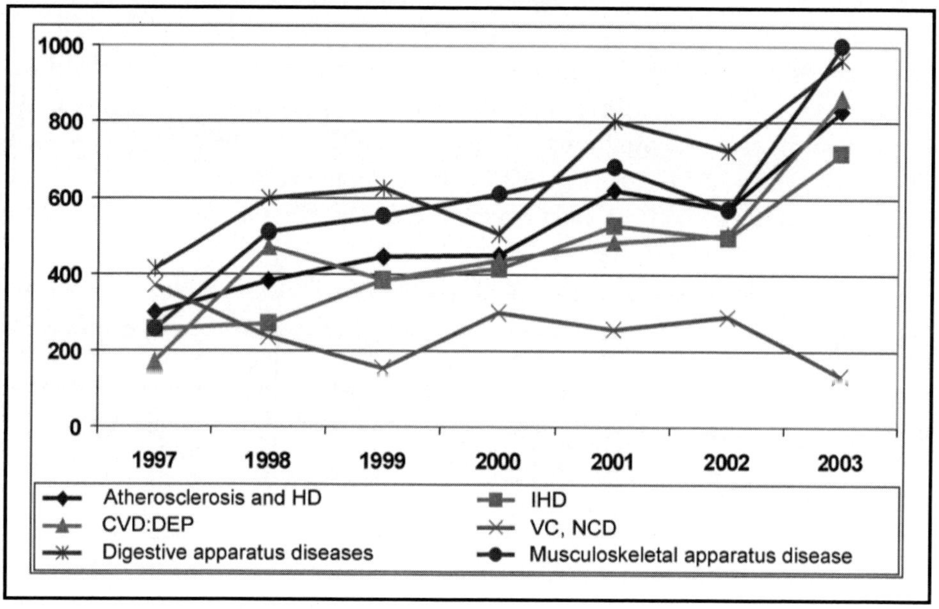

Figure 6. Dynamics of the occurrence frequency of CSD, digestive system and musculoskeletal apparatus of the Chernobyl accident liquidators from Moscow and MO in the remote period 1997 – 2003 (per 1000 persons)

Dynamics of CSD occurrence frequency variation for the liquidators from Moscow region generally shows the increase of atherosclerosis and HD in the period of 1997 – 1999 with a plateau in 1999 – 2000 and two peaks in 1999 and 2001, followed by slight decrease in 2002 relative to previous value for 2000 (Figure 6). In the same periods IHD peak is observed at general tendency to some increase from 1999 to 2000 and decrease together with atherosclerosis and HD from 2001 to 2002. Dynamics of CVD:DEP was characterized by a sharp increase from 1997 to 1998, a decrease from 1998 to 1999 and further uniform increase to 2002, exceeding the value obtained in 1998, and further sharp increase in 2003. The decrease of occurrence frequency for atherosclerosis, HD and IHD is related not to real decrease in the frequency of this pathology, but with mortality among the liquidators. The change in the frequency of functional disorders of IHD – VD, NCD shows a considerable decrease in the period of 1997 – 1999, which happens simultaneously with the frequency increase of atherosclerosis, HD, NCD, IHD, CVD:DEP, and then to 2000 an increase of VC, NCD frequency is observed with some variability in 2001 – 2002 and considerable decrease in 2003. The increase of VC, NCD frequency in the period of 1999 – 2000 happens at relatively close values for atherosclerosis, HD, IHD and a tendency to increase of CVD:DEP frequency. Further on, the peak of atherosclerosis, HD and IHD frequency coincides with VC, NCD frequency decrease, as well as a decrease of atherosclerosis, HD and IHD frequency coincides with VC, NCD frequency increase in the period of 2001 – 2002 e.g. showing regularly inverse character of dynamics of these diseases, most clearly observed in 2003.

The presented data indicate the adverse state and considerable deterioration of the health of the liquidators of the Chernobyl accident consequences. The absence of practically healthy people among examined liquidators, high frequency of CSD accompanied by atherosclerosis and age-dependent, as well as high frequency of polymorbidity confirms progressing aging of the liquidators.

The dynamics of digestive system chromic diseases possessed a considerable increase in the period of 1997 – 1998, relatively similar parameters in the period of 1998 – 1999 with a tendency a small increase, two minima in 2000 and 2002 and a maximum in 2003. For the musculoskeletal system, the change of ChD frequency in the period of 1997 – 1999 coincides with that for digestive system ChD and then manifests a continuous increase till 2001, a slight decrease in 2002 a maximum in 2003.

As follows from the data obtained, in 2002 a tendency to decrease of CSD frequency was observed: atherosclerosis, HD, IHD, as well as chronic diseases of GIT, bones and joints. One of the reasons for decreasing frequency of chronic pathology is mortality among the liquidators due to presence of polyorganic pathology, the *polymorbidity syndrome*, in them, i.e. the presence of three of more chronic diseases and early development of such socially valuable diseases as CSD, which show a progressive tendency for the liquidators: first of all, CVD:DEP, cardiovascular pathology, and early aging. The accelerated proceeding of chronic diseases is promoted by high frequency of IMIN clinical manifestations of the infectious syndrome, frequent respiratory-viral infections, increased fatigability syndrome, and depressive syndrome that predisposes activation of slow viral infections, among which CSHVI should be mentioned and, as mentioned above, the frequency of which in the liquidators from Moscow and MO is very high.

Dynamics of Chronic Somatic Disease Occurrence among Liquidators from NW Region of Leningrad Oblast

In the third five-year period, the occurrence of chronic diseases and IMIN clinical manifestations in the Chernobyl accident liquidators from NW Region of Leningrad Oblast was lower compared with these from Moscow and MO [12]. In the fourth five-year period, the liquidators from NW Region of Leningrad Oblast possess a considerably increased occurrence frequency of chronic pathologies. Chronic diseases are detected in all liquidators clinically examined in 2003; for 86.54% of liquidators the presence of polymorbidity e.g. 3 or more chronic diseases was determined (Table 5, Figure 7). A tendency to permanent increase of the number of patients with three or more chronic diseases was observed – the factor indicating deterioration of the health of liquidators. In 2003, the occurrence of circulatory system diseases also increased reaching 858.8:1000 that corresponded to CSD frequency among liquidators from Moscow megalopolis in 2002. The structure of CSD also possesses dominating atherosclerosis, HD and CVD:DEP, the frequency of which in 2002 is by 1.34 and 1.40 times higher, respectively. The highest increase rate is observed for cerebrovascular pathology.

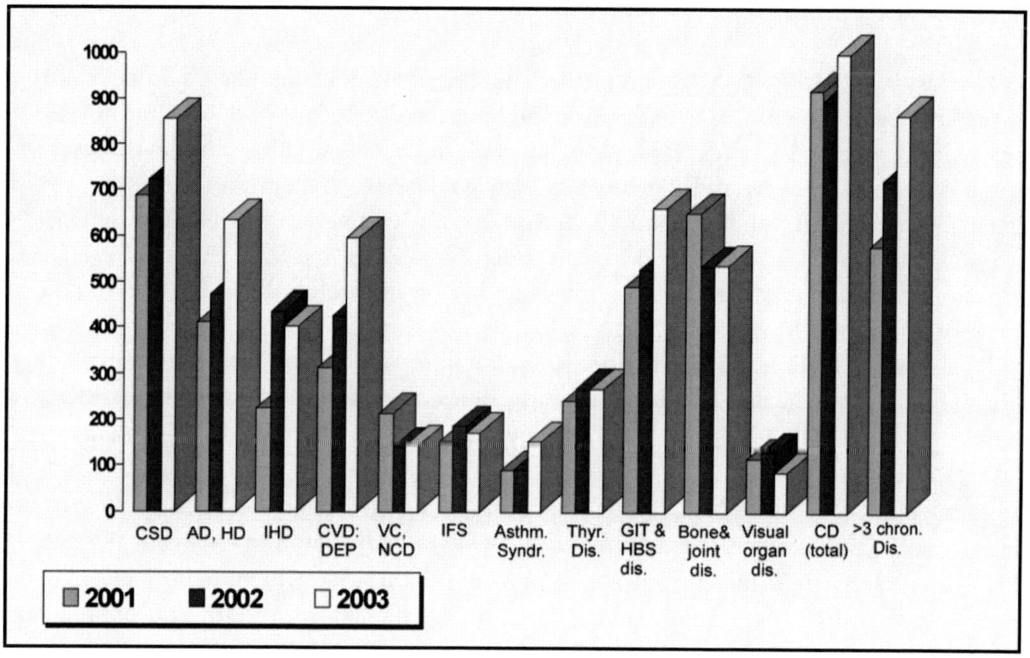

Figure 7. The occurrence frequency of the predominant chronic diseases among the Chernobyl accident liquidators from NW Region (per 1000 patients)

A tendency to increase for GIT ChD and hepatobiliary system (HBS) is observed, but lesser than for CSD, specifically for atherosclerosis, HD and CVD:DEP. More than a half of liquidators examined in 2002 – 2003 had bone and joint diseases: deforming osteochondrosis, osteoarthritis, pain muscular-joint syndrome, etc.

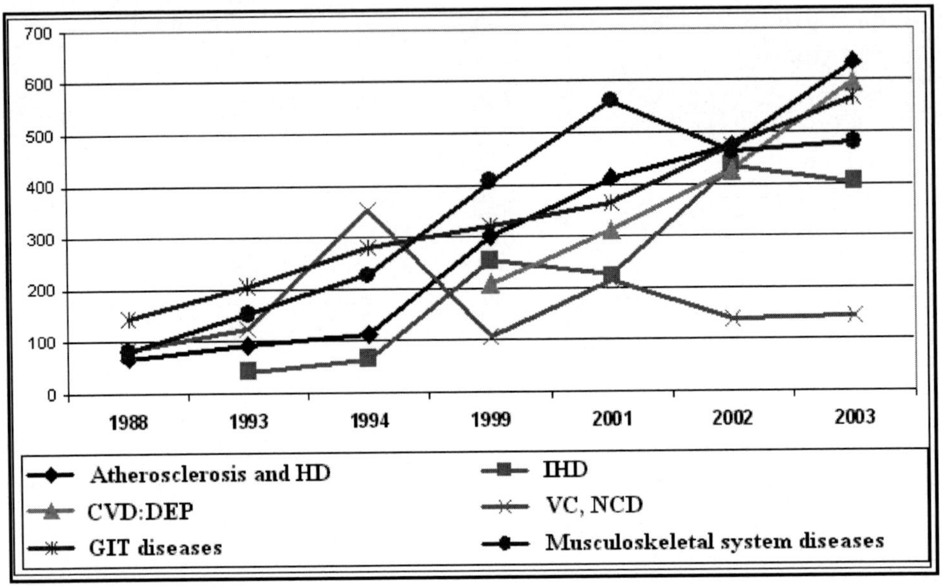

Figure 8. Dynamics of the occurrence frequency of chronic pathologies among liquidators from NW Region of Leningrad Oblast (per 1000 patients), 1988 – 2003

Generally, in 1999 – 2003, among the Chernobyl accident liquidators from NW region the occurrence frequency of CSD is the highest, in 2003 followed by digestive system diseases, and bone and joint diseases at the third place. Hence, in 2002 chronic diseases of the musculoskeletal apparatus and digestive system possessed very close occurrences, in 2003, digestive system ChD show much higher frequency compared with the musculoskeletal apparatus. For the liquidators from Moscow and MO, in 2001 and 2002 digestive system ChD were more frequent than ChD of the musculoskeletal apparatus. In 2003, the situation for liquidators from NW Region of Leningrad Oblast was similar.

Dynamics of CSD occurrence variation in the liquidators from NW Region of Leningrad Oblast (Figure 8) indicated relatively uniform, but considerable and continuous increase of atherosclerosis and HD in 1994 – 2003 and CVD in 1999 – 2003. The dependence shows two peaks of IHD frequency in 1999 and 2002 and some decrease in 2003 compared with 2002. The frequencies of VC and NCD increased in the period of 1988 – 1994 e.g. in the early time after the accident, showing the maximum in 1994, followed by a decrease in the period of 1994 – 1999 and 2-fold increase in 2001 compared with 1999, which, nevertheless, is much lower than in 1994. Then the frequency decreased from 2001 to 2002 and approaching a uniform plateau in the period of 2002 – 2003.

For the liquidators from NW Region of Leningrad Oblast, the dynamics of digestive system disease occurrence manifested continuous increase in the period of 1988 – 2003 with a peak in 2003. The bone and joint diseases also showed continuous and even much faster increase of the occurrence till 2001, some decrease in 2002 and a plateau in 2003. Both examined groups of liquidators from NW Region and Moscow region demonstrated increasing occurrence frequency of digestive system ChD in 2003, for the Moscow region group this value being 1.45 times higher. The occurrence of skeletal apparatus diseases in the liquidators from Moscow was 2 times higher, simultaneously.

Dynamics of the Occurrence of Chromic Somatic Diseases Among Liquidators from Krasnoyarsky krai

The study of chronic pathology occurrence among liquidators from Krasnoyarsky krai (Figure 9, Table 5) also indicated their high frequency: 95.24% (220:231) in 2001 and 97.94% (190:194) in 2002. The cases of occurring 3 ChD or more showed different frequencies: 93.1% in 2001 and 74.23% in 2002 that was generally related to different composition of the groups examined. In 2001, the average age of liquidators obtaining ChD equaled (50.63 ± 0.63), and for 3 ChD or more – (50.76 ± 0.64), e.g. no dependence of ChD occurrence on the liquidator age is observed. In 2002, the ChD occurrence frequency in the liquidators from Krasnoyarsky krai was similar to that for the Moscow region, whereas the occurrence of 3 ChD or more for the Moscow region fell within the range of 84.21% – 84.85%, which much higher than for the liquidators from Krasnoyarsky krai. Hence, this parameter is practically equal for Krasnoyarsky krai and NW Region: 74.23% (742.23:1000) and 72.22% (722.22:1000), respectively. Te group of liquidators examined in 2003 demonstrated 86.49% occurrence of chronic pathologies, which is lower value compared with previous years. Most likely, the reason for such difference is much shorter group of patients examined in 2003 compared with 2001 – 2002. According to data on 2003, the polymorbidity frequency is comparable with the data obtained in 2002 tending to increase.

The structure of chronic pathologies among liquidators from Krasnoyarsky krai, as well as in other groups, showed CSD predominance: atherosclerosis, HD, CVD:DEP, IHD. According to observations in 2001 and 2002, CSD occurrence equaled 787.88:1000 and 814.43:1000, respectively. The average age of the liquidators possessing atherosclerosis, arterial hypertension, and HD was (51.75 ± 1.02) and for the case of CVD:DEP – (50.12 ± 1.30). Of course, clinical manifestations of these diseases appear much earlier. They relate to the elderly age diseases and are predominant in all groups of the Chernobyl accident liquidators that also indicated noncoincidence of biological and official ages of the patients with a tendency to premature aging.

Premature aging of the liquidators already 7 – 10 years after the accident was also confirmed by the experts from Kharkov Research Institute of Radiology of Ukraine. An increase of biological age compared with population standards was determined. A reverse dependence between biological aging rate and real age of irradiated liquidators. The highest indices of biological and cardio-pulmonary age were observed among liquidators, exposed to radiation impact in the age before 25. Ten years after the accident, 66% of the liquidators aged 30 – 39 manifested a noticeable increase of the levels of cholesterol and atherogenic lipoproteins both low and high density, and triglycerides at the background of antiatherogenic lipoprotein cholesterol level decrease. These atherogenic shifts were accompanied by highly expressed intensification of peroxide oxidation of lipids at the background of decreasing activity of antioxidant enzymatic systems. The determined totality of changes is typical of elder age groups [21].

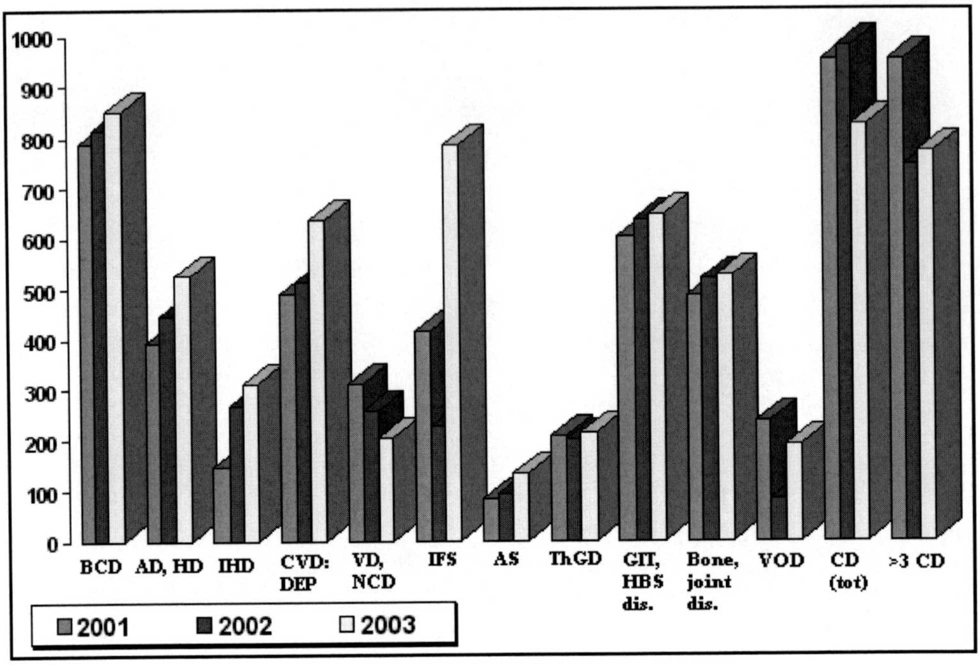

Figure 9. The occurrence frequency of chronic pathologies among accident liquidators from Krasnoyarsky krai (per 1000 patients)

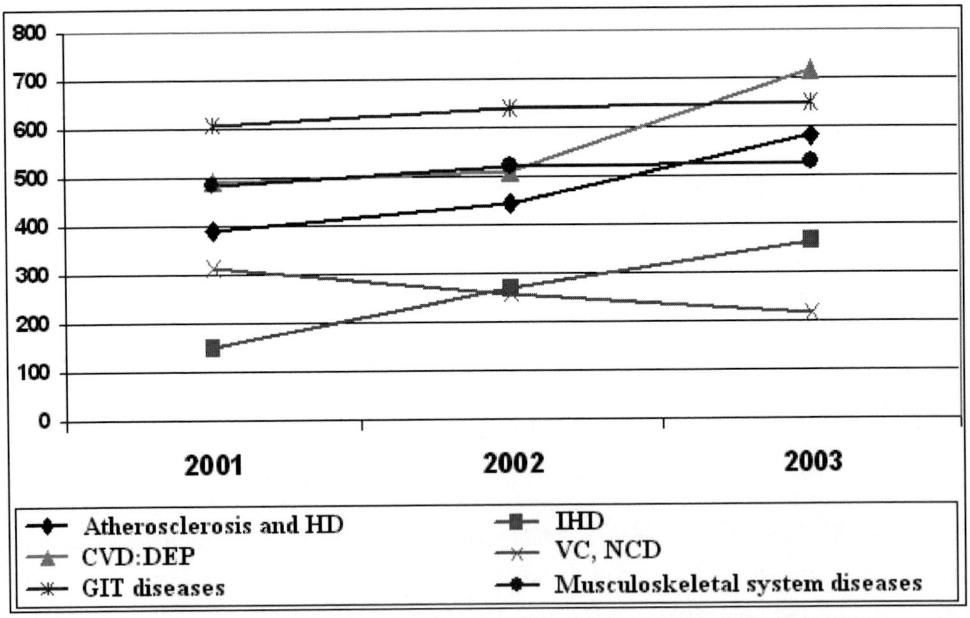

Figure 10. The occurrence frequency of chronic pathologies among liquidators from Krasnoyarsky krai (per 1000 patients), 2001 – 2003

Thus, the younger the liquidators of the Chernobyl accident were working in the zone in 1986 – 1987, the higher is the aging rate, first of all, of the cardiovascular system. It is known that the low dose radiation induces development of sclerotic and necrotic changes and a decrease of functional capabilities of the hypothalamus– hypophysis– adrenal cortex system. Among liquidators this induced reduction of compensatory abilities and the state of extended dysadaptation syndrome. This was also promoted by disorders in the immune system: the presence of secondary IMIN and imbalance in the cellular immunity system. Yu.G. Grigoriev [4] has noted early manifestation of the age pathology as cerebral atherosclerosis and HB increasing the frequency of infarctions and cerebral insults, induced by low dose impact of both ionizing and nonionizing radiation. The development of premature aging syndrome is also notified by the investigators from Belarus [3].

Early aging of the liquidators is also confirmed by changes of fundus of the eye: early phacosclerosis, retina angiodystrophy, early cataracts, and both radiation-induced and age-related early presbyopia. Intensification of aging processes is also manifested by higher cortical functions. The liquidators possess increased level psychoemotional stress, alarm reactions, decreasing adaptive capabilities and, at the same time, inadequate assessment of the self conditions. The main reasons for these dysfunctions are progressing deterioration of the health state and anxiety of occurring more severe diseases. Health deterioration, inability to perform usual scope of working operations at common rate, and unemployment induce psychological discomfort, perplexity, pessimism and socially disorderly life that initiated a coterie of problems, which can hardly be solved in the deconditioning state, provided for imbalance in functioning of the main vitally important systems of the organism and faster aging.

Compared with 2001 – 2002, in 2003 among liquidators from Krasnoyarsky krai a tendency to CSD frequency increase was preserved, accompanied by atherosclerosis, specifically CVD:DEP with decreasing frequency of VC, which also confirmed aging of the liquidators (Table 5). In 2003, the frequency of cerebrovascular pathology increased by 1.46 times reaching 716.2:1000. The observations performed in the mentioned three years of the remote period indicate permanent predominance of CVD:DEP in the CSD structure at significantly high frequency of atherosclerosis and HD and lower frequency of IHD occurrence compared with other groups. Before 2003, the CSD structure of liquidators from Moscow region showed predominance of atherosclerosis and HD. Only in 2003, CVD:DEP becomes predominant showing very close occurrence frequency to atherosclerosis and HD: 863.6:1000 and 827.2:1000, respectively. Though all three groups of observation show annual increase of the occurrence frequency of some nosological forms of diseases, accompanied by atherosclerosis, the occurrence of this pathology among liquidators from the Moscow region is extremely higher (Table 5).

High frequency of digestive system, bones and joints diseases with pain syndrome is observed. In 2001, the average age of liquidators with long-term pain syndrome in bones, muscles, joints, and spinal column equaled (49.08 ± 1.03) and with fixed diagnosis of deforming osteochondrosis – (50.75 ± 0.86). In 2003, high frequency of digestive system diseases is preserved (648.7:1000), including GIT (47.3%) and hepatobiliary system (51.4%). By the occurrence frequency digestive system diseases are ranked second, whereas as by the rate of increase for the latest three years approach the first place value (Figure 10). The occurrence frequency of digestive system diseases in the groups of liquidators from Krasnoyarsky krai and NW Region of Leningrad Oblast are similar, whereas for Moscow and MO they are much higher. The third place in the ChD structure is occupied by bone and joint diseases, the frequency of which in the period of 2002 – 2003 possesses identical value. As compared with other groups of the occurrence of chronic pathologies of the musculoskeletal system among liquidators from Krasnoyarsky krai and NW Region of Leningrad Oblast are equal, whereas in the Moscow region it is twice higher.

As follows from the authors' data, in the fourth five-year period the three groups of accident liquidators possess high frequency of CVD:DEP occurrence. For liquidators from Krasnoyarsky krai, it is most frequently observed at the background of cerebrum organic damages. In 25.7% of cases clinical manifestations of psychoorganic syndrome, most frequently of the asthenodepressive type, as well as increased fatigability syndrome, very frequently (78.38%) observed in the liquidators, examined in 2003. Development of the psychoorganic syndrome and depressions at the background of organic vascular damages of the central nervous system is mostly related to the presence of many somatic diseases and thyroid pathologies, remote consequences of psychic trauma, caused by participation in liquidation of the accident consequences – the secondary processing factor typical of the liquidators [20]. Among the liquidators, cerebrovascular pathology is combined with clinical manifestations of the immune system disorders. For example, patients from Moscow and MO, and NW Region possess high frequency ChD combined with IMIN clinical symptoms: infectious and infectious-allergic syndromes, AD and practically all cases of autoimmune diseases. According to data of 2002, the overwhelming majority (93.44%) of liquidators possessed a combination of IMIN clinical symptoms with chronic pathologies, and in 2003 100% of examined liquidators possessing clinical manifestations of the immune system disorders also showed occurrence of chronic pathologies.

Thus, by the age of 50 the liquidators obtained a broad spectrum of chronic pathologies damaging many organs and systems and, therefore, manifested polysystemic e.g. polymorbidity character. The authors' data on the high frequency of chronic diseases, continuous increase of their occurrence and CSD predominance among the liquidators coincide with the results obtained by experts from Belarus, Ukraine and Latvia. According to data obtained in Minsk SRCI of Radiation Medicine and Endocrinology, since 1997 CSD became predominant diseases among liquidators, the second place is occupied by the digestive system diseases and the third place – by nervous system and sensory organ diseases [1]. The presence of polyorganic pathology among liquidators is also note by the experts from All-Russian Center of Extremal and Radiation Medicine. Generally, 76% of liquidators obtained multiple vegetovascular disorders with NCD and VC symptoms; 63% of patients with IHD manifest GIT diseases; in 75% of cases, the respiratory function disorder or chronic bronchitis is detected. Over 95% of the liquidators possess pathologies of the musculoskeletal system, joint diseases, and for 98% disorders of the psychic status are observed [7].

The following data show the level of ChD frequency among liquidators exceeding their occurrence among other cohorts of people. For example, MCP personnel located in Krasnoyarsky krai shows CSD occurrence frequency equal 281.9:1000 [24, 25]. Compared with the cohort of liquidators, it is 3.02 ties lower (851.4:1000). Among CSD, VC and atherosclerosis complicated by HD are the most frequent diseases: 137.3:1000 and 127.7:1000, respectively. The liquidators from this region show atherosclerosis and HD occurrence frequency equal 581.1:1000, which is 4.55 times higher. Among personnel of Chepetsk Uranium Chemical and Metallurgical Processing Plant having contacts with the low dose radiation factor, the frequency of CSD was found equal 246.3:1000, which is comparable with this parameter detected for MCP personnel [5]. Among examined MCP personnel, the occurrence of digestive system chronic pathology equaled 424.1:1000, whereas among examined group of common population from this region – 344.0:1000. Among the liquidators from Krasnoyarsky krai, this parameter equaled 648.7:1000, which is 1.55-fold higher than for the personnel mentioned and 1.9-fold higher than for common population. Among liquidators the frequency of GIT chronic diseases among liquidators is 1.77-fold higher than for MCP personnel and 1.8-fold higher than for common population. Moreover, the frequency of deforming osteochondrosis is 2.35 and 2.64 times higher, respectively.

The comparison of the occurrence frequency of cerebrovascular pathology among the liquidators, MCP personnel and common population demonstrates drastic differences. The frequency of CVD:DEP is the following: 716.2:1000 for the liquidators; 31.3:1000 for MCP personnel and 72.0:1000 for common population, which manifests 22.9-fold and 9.95-fold exceeding of the liquidator index compared with others, respectively. The data need no comment, because they give a convincing proof of horrible damages of the liquidators and progressive deterioration of their health.

The extending occurrence frequency of chronic diseases has induced a considerable increase of annual incapacitating of the liquidators, most of all, in three groups examined in Krasnoyarsky krai. According to data of monitoring in 2001 – 2003, this parameter fell within the range of 386.6 – 432.4:1000. Meanwhile, just few people (3.6 – 5.4%) are incapacitated due to common diseases. For the liquidators from Moscow region, incapacitating due to radiation impact in Chernobyl is lower: 330.8 – 354.5:1000, at considerably higher disability due to common reasons: 190.9:1000. The main reasons for incapacitating liquidators are somatic diseases, which usually deteriorate workability of cause complete disability, decreases quality of life, reduces the life duration, and even exceeds damage cause to the health state by neoplasms. According to data by Central Institute of Workability Examination, ten years after the accident the main incapacitating pathologies were nervous system diseases (70%), mostly cerebrovascular ones, circulatory system pathologies (23%), musculoskeletal apparatus diseases (6.2%), and the rest given by GIT and respiratory diseases, and thyroid pathologies [9].

Total disability index is the highest among liquidators from the Moscow Region that corresponds to higher occurrence of chronic somatic pathology and clinical symptoms of the immune system insufficiency. In the Moscow Region, they dominate by all nosology types and classes of diseases present. Apparently, here an additional effect of “megalopolis syndrome” factors takes place. Similar data were obtained in the analysis of the disease frequency among the liquidators from St. Petersburg and Leningrad Oblast, according to which morbidity rate in St. Petersburg is 1.4 – 3.3 times higher than in other areas of NW Region: Leningrad, Novgorod, Pskov, and Tver’ Oblasts [22].

The data on disability in 2003 in Moscow and MO are comparable with the previous monitoring results for 2002 (Table 5). Despite a significant increase of frequency and broadening of chronic pathology spectrum, no considerable increase of disability is observed. This is related to both mortality among liquidators with state disability and limitation of the list of diseases, by which the liquidators are related to Chernobyl NPP.

In the period of 2001 – 2003, characteristics of disability in the group of liquidators from NW Region of Leningrad Oblast is comparable with the data mentioned in a letter of information by I.I. Shantyir’ *et al.* as of 1999. The frequency varied by years of participation in liquidation of the Chernobyl accident consequences: 32.1:100 for the liquidators of 1986,

24.8:100 for 1987, 16.2:100 for 1988, and 9.6:100 for 1989. Disability shows linear dependence on time of participation in liquidation works and is reliably (1.3 – 3 times) higher for the liquidators of 1986 compared with others participated in the liquidation in the following years. As of 1999, the examined group of liquidators includes disabled patients giving much lower frequency (10.2%), induced by participation in liquidation of the Chernobyl accident consequences and general diseases (1.12%). As of 2003, a group of disabled patients gives 32.69%, including 26.92% related by Chernobyl NPP accident liquidation works and 5.77% – to general diseases.

CONCLUSION

[1] The highest frequency of IMIN clinical manifestations and other symptoms of the immune system dysfunctions in the remote period after the Chernobyl accident was observed for the liquidators from Moscow and MO. The extremely high frequency was detected in 2003 – 97.3%. IMIN clinical symptoms are mostly represented by frequent respiratory-viral infections, ARVI combined with CRHVI, obstructive bronchitis, and frequent ENT infections. In 2000 and mostly in 2003, the increase of autoimmune disease frequency was observed (18.39% → 25.45%).

[2] In the period of 1999 – 2003, dynamics of IMIN clinical manifestation frequency among liquidators from NW Region of Leningrad Oblast was characterized by annual increase: 44.07% → 54.46% → 66.67% → 81.73% giving the average 74.06% for the last two years. In 2002, the structure of immunopathological syndromes showed increasing frequency of autoimmune diseases (23.15%); in 2003, the infectious syndrome dominated. In the structure of IMIN clinical manifestations of the infectious syndrome an increase of bronchitis is observed, mostly obstructive, fungal and herpes-viral infections; the frequency of ENT infections is also high (25.0 – 25.96%).

[3] In the period of 2001 – 2003, the liquidators from Krasnoyarsky krai had similar general indices of immunopathology clinical symptoms: 68.83% → 70.62% → 73.61%, comparable with data for MCP personnel. The structure of IMIN clinical manifestations show predominant the same symptoms, but their frequency is much lower as compared with the Moscow Region. As compared with rest two groups, the liquidators from Krasnoyarsky krai possess much lower frequency of autoimmune diseases.

[4] In the fourth five-year period after the Chernobyl accident, the liquidators of consequences possess high frequency of chronic diseases. All liquidators from Moscow and MO and NW Region possess chronic somatic diseases demonstrating high frequency and rate of polymorbidity: 100.0% for Moscow region, 86.54% for NW region, 77.03% – 95.24% for Krasnoyarsky krai. These indices testify about significant deterioration of the liquidators' health state.

[5] In the remote period CSD is predominant pathology among liquidators. The increase of occurrence frequency and predominance of elderly age diseases (atherosclerosis, HD and diseases related to atherosclerosis, such as IHD, CVD:DEP) testify about premature aging of the liquidators. Among CSD, CVD:DEP and atherosclerosis with HD are predominant. High frequency of chronic diseases related to atherosclerosis, including CSD, was promoted by high occurrence of infectious IMIN clinical manifestations,

frequent respiratory-viral infections with chronic recrudescent herpes-viral infections, as well as extended deconditioning syndrome.

[6] The comparison of examined groups of liquidators, MCP personnel and population of Krasnoyarsky krai indicated high exceeding of chronic pathology frequency among liquidators of the Chernobyl accident consequences: 3.02 and 2.6 times by CSD, 4.55 and 3.30 times by atherosclerosis with HD, 22.9 and 9.95 times by CVD:DEP; compared with MCP personnel and the population, respectively. For other classes of diseases and separate nosological forms of diseases, a significant excess of frequency among liquidators is also observed. These data persuasively confirm abrupt deterioration of the liquidators' health.

[7] Among liquidators from NW Region, the second place by the occurrence frequency is occupied by digestive system diseases; at the third place are musculoskeletal system diseases. For the liquidators from Moscow and MO, the first place is occupied by diseases of bones and joints (100%), followed by chronic diseases of the digestive system (96.4%). They compete with one another by the occurrence frequency. The highest frequency of thyroid gland diseases is observed among liquidators from Moscow region. Besides the risk factors of the radiation catastrophe, the increase of chronic disease occurrence frequency is also affected by the "megalopolis syndrome".

[8] The highest incapacitating was also determined among liquidators from Moscow and MO due to increasing frequencies of common diseases and in Krasnoyarsky krai – due to diseases, induced by the Chernobyl accident.

REFERENCES

[1] Aderiho C.N., *Intern. J. Radiat. Med.*, Special issue, vol. 3(1-2), p. 148.

[2] Antonov N.S., Yakushin S.P., Stulova O.Yu. *et al.*, In: *The Pathology of Respiratory Organs Among Liquidators of The Chernobyl NPP Accident*, Moscow, GRANT, pp. 44 – 64. (Rus)

[3] Vishnevskaya V.P., "Psychological aspects of premature aging of the liquidators", *Proc. 2nd Intern. Conference on "Remote Medical Consequences of the Chernobyl Catastrophe"*, Kiev, 1998, p. 198. (Rus)

[4] Grigor'ev Yu.G., *Radiats. Biol. Radioekol.*, 1997, vol. 37(4), pp. 695 – 702. (Rus)

[5] Zhuravleva A.I., "Clinical-immunological assessment of the health state of chemicometallurgical production of uranium processing personnel", *Candidate Dissertation Thesis*, Moscow, 2000, 27 p. (Rus)

[6] Erokhina S.M., "The features of allergic reactivity in people exposed to radiation impact", *Candidate Dissertation Thesis*, Moscow, 1996, 24 p. (Rus)

[7] Katashkova G.D., Bonitenko Yu.Yu., Nikiforov A.M. *et al.*, In: *Pathology of the Remote Period Among Liquidators of the Chernobyl Accident Consequences*, Ed. Prof. A.M. Nikiforov, St. Petersburg, BINOM, 2002, pp. 165 – 175. (Rus)

[8] Kopyilev I.D., Ammosova S.P., Maracheva A.V. *et al.*, In: *The Pathology of Respiratory Organs Among Liquidators of The Chernobyl NPP Accident*, Moscow, GRANT, pp. 165 – 176. (Rus)

[9] Lavrova I.D., "Criteria of estimation of the health state and rehabilitation of people incapacitated due to radiation catastrophes", *Proc. Conf.*, November 12, 1997, pp. 35 – 36. (Rus)

[10] Nikiforova I.D., Shantyir' I.I., Tyutin L.A. *et al.*, *Med. Radiol. Radiats. Bezop.*, 2000, vol. 60(6), pp. 14 – 20. (Rus)

[11] Nikiforova I.D., Bonitenko Yu.Yu., and Bakhtin M.Yu., In: *Pathology of the Remote Period Among Liquidators of the Chernobyl Accident Consequences*, Ed. Prof. A.M. Nikiforov, St. Petersburg, BINOM, 2002, pp. 242 – 255. (Rus)

[12] Oradovskaya I.V., *Fiziologia i Patologia Immunnoi Sistemy*, 2004, No. 4, pp. 8 – 23. (Rus)

[13] Oradovskaya I.V., "Immunological monitoring of big groups of population, including the cohort of people participated in liquidation of the Chernobyl NPP accident consequences", *Candidate Dissertation Thesis*, Moscow, 1991, 48 p. (Rus)

[14] Oradovskaya I.V., Luss L.V., Leiko I.A. *et al.*, *Allergia, Astma i Klin. Immunologia*, No. 12, pp. 15 – 23. (Rus)

[15] Oradovskaya I.V., Oprishchenko M.A., Leiko I.A. *et al.*, *Radiats. Biol. Radioekol.*, 2000, vol. 40(2), pp. 177 – 187. (Rus)

[16] Petrov R.V., Khaitov R.M., Pinegin B.V. *et al.*, *Immunologia*, 1992, No. 6, pp. 51 – 62. (Rus)

[17] Petrov R.V., Khaitov R.M., Oradovskaya I.V. *et al.*, *Immunologia*, 1992, No. 4, pp. 43 – 53. (Rus)

[18] Petrov R.V. and Oradovskaya I.V., *Proc. Rep. All-Union Conf. "Methodology of Arrangement and Results of Large-Scale Immunological Examinations"*, Moscow – Angarsk, 1987, pp. 215 – 234. (Rus)

[19] Radzivil T.T., "Immune status of people being in professional contact with plutonium-239", *Candidate Dissertation Thesis*, Moscow, 1995, 24 p. (Rus)

[20] Rumyantseva G.M., Chinkina O.V., and Plyiplina D.V., *Proc. 2nd Intern. Conference on "Remote Medical Consequences of the Chernobyl Catastrophe"*, Kiev, 1998, p. 128. (Rus)

[21] Simonova L.I., Gertman V.Z., and Abramova L.P., *Ibid*, p. 370. (Rus)

[22] Shantyir' I.I., Astaf'ev O.M., and Makarova N.V., *The Health State of the Chernobyl NPP Accident Consequences*, St, Petersburg, 2002, 115 p. (Rus)

[23] Shantyir' I.I., Makarova N.V. *et al.*, *Disability and Mortality Among Liquidators of the Chernobyl NPP Accident Consequences. Inform. Letter*, St. Petersburg, 2001, 34 p. (Rus)

[24] Shuvatova E.V., Oradovskaya I.V., Leiko I.A. *et al.*, *Fiziologia i Patologia Immunnoi Sistemy*, 2004, No. 9, pp. 3 – 16. (Rus)

[25] Shuvatova E.V., "Clinical-immunological characteristics of Krasnoyarsky krai Mine-Chemical Plant personnel and the population dwelling close to production", *Candidate Dissertation Thesis*, Moscow, 2005, 24 p. (Rus)

ABBREVIATIONS

abs. un. – absolute units

$CD3^+$	T-lymphocytes
$CD4^+$	T-helpers
CTL–$CD8^+$	cytotoxic T-lymphocytes/killers
$CD4^+/CD8^+$	immunoregulation index
$CD16^+$, NK-cells	natural killers
NPA	neutrophil phagocytic activity
$CD19^+$	B-lymphocytes
$CD25^+$	lymphocyte early activation marker
$CD95^+$	marker activation and apoptosis
HLA-DR^+	delayed activation marker
IgM	class M serum immunoglobulin
IgG	class G serum immunoglobulin
IgA	class A serum immunoglobulin
IgE	total immunoglobulin E
SRD	special risk divisions

PART 2. DYNAMICS OF THE IMMUNE STATUS PARAMETERS

The second component of immunological monitoring in the remote period, as well as in the earlier times, is analysis of laboratory immunological examination results with assessment of the immune status of liquidators, dynamics of an alteration of it by years at main immunopathological syndromes (IPS), most frequently occurring chronic diseases (ChD), the dependence of the immune status variation on the time of participation in liquidation of the accident consequences, the zone of implementation of accident recovery works (ARW), doses of external radiation and the age of liquidators. This part of the report presents analysis of the dynamics of immune status indices in the same groups of observation: liquidators from Moscow and Moscow Oblast (MO), Northwest (NW) region of Leningrad Oblast (LO) and Krasnoyarsky Krai, among which clinical immunological monitoring was performed.

The laboratory immunological examination included determination of the quantity of leukocytes, relative and absolute content of lymphocytes, general population of T-lymphocytes ($CD3^+$), subpopulations of T-helpers ($CD4^+$) and cytotoxic T-lymphocytes (CTL–$CD8^+$), immunoregulation index ($CD4^+/CD8^+$), natural killers ($CD16^+$), and neutrophil phagocytic activity (NPA) with latex. Assessment of the humoral unit of the immune status included determination of B-lymphocytes ($CD19^+$), the level of level immune serum globulins of three main classes M, G and A by radial immunodiffusion and general IgE methods. Activation markers CD25, CD95, HLA-DR, and activated T-lymphocytes were determined.

For laboratory verification of the immune insufficiency (IMIN) and determination of persons with reduced and increased immune status parameters the principle of sigmatic deviations from the control was used. For the control, data obtained by Prof. B.V. Pinegin *et al.* in the laboratory of clinical immunology and immunodiagnosis of the Institute of

Immunology for a group of practically healthy persons and primary donors of Moscow were taken. These data were used many times for implementing high scope immunological examinations and monitoring in the framework of the program on "The Immune status of a man" [1 – 5].

Dynamics of Changes in the Immune Status Indices for Liquidators from NW Region According to Monitoring Data of 1995 – 2003

In 1995, assessment of the immune status of the Chernobyl accident liquidators in the early period indicated considerable changes with T-cell unit insufficiency ($CD4^+$%, absolute value, $CD8^+$%), imbalance in parameters of cytological cells with CTL insufficiency and increased concentration of natural killers, decreased concentration of B-lymphocytes and insufficient serum IgG (Table 1). Monitoring data obtained in 1999 indicated noticeable variations of parameters due to imbalance in cellular, phagocytic and humoral unit of the immune status. Imbalanced state of the cellular unit was observed by a reliable increase of lymphocyte concentration, dissociation in T-cell immunity indices with insufficient T-lymphocytes(%) and CTL-$CD8^+$(%), increasing absolute values of $CD3^+$, $CD4^+$, immunoregulation index, and considerable decrease of natural killer concentration compared with 1995. The presence of disimmunoglobulinemia was determined by serum IgM (↑), IgG (↑) and general IgE (↑↑) indices. The average values of phagocytic activity of neutrophils were reliably higher. As compared with earlier periods after the Chernobyl accident, the increased NPA level was observed in 1992, whereas in 1993 – 1995 it indicated stable and reliable decrease.

As observed from the immune status assessment results for a group of liquidators examined in 2001, absolute values of lymphocytes, percentage of T-lymphocytes, T-helpers, and immunoregulation index were reliably lower than for the control, whereas cytotoxic cell concentration increased: relative and absolute values of CTL – $CD8^+$ and NK-cells, NPA and serum IgM, IgG (Tables 1 and 2, Figure 1). As compared to 1999, dynamics of the immune status indices showed a reliable decrease of relative values of lymphocytes, T-helpers, immunoregulation index, NPA (which remained elevated compared with the control), serum IgA, and reliable increase of cytotoxic cells – relative values of $CD8^+$, $CD16^+$ and B-lymphocytes, and concentration of serum IgM, IgG. Highly expressed imbalance in all units of the immune system, valuable variations in cytotoxic cell parameters, T-cell unit insufficiency by T-lymphocyte, T-helper and immunoregulation indices, disimmunoglobulinemia with a tendency to increase the level of IgM, IgG, IgE and decrease the level of serum IgA.

As compared with the examination performed in 2001, the monitoring results of 2002 indicated changes in parameters of T-cell unit showing reliable increase of concentrations of T-lymphocytes, T-helpers up to the control level, immunoregulation index and decreasing relative and absolute values of CTL. A reliable change in activation marker parameters was found: the valuable decrease of early activation markers – activated $CD25^+$ lymphocytes, which testifies about decreasing functional activity of the cellular unit by interleukin-2 reception level, and elevation of lymphocytes expressing delayed activation markers – HLA-DR^+, as well as activated T-lymphocytes at the background of decreasing parameters of B-lymphocytes. The presence of disimmunoglobulinemia was indicated by differently directed change of the humoral unit parameters, observed at the comparison of serum Ig changes, monitored in 2001 – 2002, with parameters of 1999 – 2001. The patients examined in 2002 demonstrated reliable decrease of serum IgG and general IgE, and elevation of serum IgA. The presence of stable disimmunoglobulinemia is one of the symptoms of the immune system aging. As compared with parameters obtained in 1999 – 2001 and the control level, NPA in the patients examined in 2002 was decreased.

Table 1. Average values of the immune status parameters of Chernobyl accident consequence liquidators from NW Region according to monitoring data obtained in 1995 – 2001

No.	Immune status indices	Average values, $M \pm m$								
		2001 (n = 112)			**1999** (n = 177)			**1995** (n = 287)		
1	Leukocytes	6.54	±	0.18	6.10	±	0.14	**6.08**	±	**0.11°↓**
2	Lymphocytes, %	**30.18**	±	**0.88^{++}↓**	**34.42**	±	**0.92 °↑√√↑**	30.8	±	0.47
3	Lymphocytes, abs	**1.90**	±	**0.06°↓**	2.07	±	0.076	**1.86**	±	**0.04°°↓**
4	CD3$^+$, %	**65.31**	±	**1.06°°↓**	**63.80**	±	**0.94°°°↓**	62.49	±	0.65 (CD5$^+$)
5	CD3$^+$, abs	1.23	±	0.04	**1.34**	±	**0.057√↑**	1.16	±	0.03
6	CD4$^+$, %	**38.28**	±	**0.85$^+$↓°↓**	40.69	±	0.69	**38.37**	±	**0.58°↓**
7	CD4$^+$, abs	**0.72**	±	**0.03$^+$↓**	**0.85**	±	**0.035√√↑**	**0.71**	±	**0.02°↓**
8	CD8$^+$, %	**34.93**	±	**0.83^{+++}↑°°↑**	**26.60**	±	**0.68**	**25.42**	±	**0.49 °↓**
9	CD8, abs	**0.66**	±	**0.03$^+$↑°°↑**	0.56	±	0.03	0.48	±	0.02
10	CD4$^+$/CD8$^+$	**1.21**	±	**0.05^{+++}↓°°↓**	**1.81**	±	**0.09 °↑**	**1.68**	±	**0.04°↑**
11	CD16$^+$, %	**19.35**	±	**0.84^{+++}↑°°°↑**	**13.33**	±	**0.60 °↓√√√↓**	**20.98**	±	**0.51°°°↑**
12	CD16$^+$, abs	**0.36**	±	**0.02^{++}↑°↑**	**0.27**	±	**0.017°↓√√√↓**	**0.39**	±	**0.01°°°↑**
13	CD25$^+$, %	3.96	±	0.26	–	±	–	–	±	–
14	HLA-DR$^+$, %	11.09	±	0.49	–	±	–	–	±	–
15	T-act., %	5.80	±	0.49	–	±	–	–	±	–
16	NPA, %	**69.7**	±	**0.7$^+$↓°°°↑**	**71.64**	±	**0.42°°°↑√√√↑**	61.4	±	0.77
17	CD19$^+$, %	**6.863**	±	**0.322$^+$↑**	**5.88**	±	**0.22°°°↓**	**6.95**	±	**0.20°↓**
18	CD19$^+$, abs	0.130	±	0.007	0.13	±	0.008	0.13	±	0.01
19	IgM, g/l	**1.57**	±	**0.04^{+++}↑°°↑**	**1.35**	±	**0.03√√√↑**	1.18	±	0.02
20	IgG, g/l	**14.55**	±	**0.29$^+$↑°°°↑**	**13.53**	±	**0.22°°°↑√√√↑**	**10.66**	±	**0.18°°°↓**
21	IgA, g/l	**1.97**	±	**0.06^{+++}↓**	**2.44**	±	**0.08**	2.38	±	0.07
22	IgE, ME	255.8	±	30.4↑	210.9	±	24.3↑	–	±	–

Reliability of differences:

°) – parameters of the immune status of liquidators from the control: ° – $p < 0.05$; °° – $p < 0.01$; °°° – $p < 0.001$;

$^+$) – parameters of the immune status of liquidators examined in 2001 from parameters obtained in 1999: $^+$ – $p < 0.05$; $^{++}$ – $p < 0.01$; $^{+++}$ – $p < 0.001$;

√) – parameters of the immune status of liquidators examined in 1999 from parameters obtained in 1995: √ – $p < 0.05$; √√ – $p < 0.01$; √√√ – $p < 0.001$.

For a group of patients examined in 2003 (Table 2, Figure 1), parameters of the cell immunity showed dynamic changes in indices of cytotoxic cells (↑) and activation markers (↑), decreasing indices of humoral immunity and increasing general IgE. The cell unit parameters indicated increasing concentration of lymphocytes, relative and absolute values of CTL and NK-cells, and insufficiency of immunoregulation due to CD8$^+$ increase, early activation marker increase for CD25$^+$ lymphocytes compared with the indices obtained in

2002, though the average values for $CD25^+$ are low. The immune status of liquidators, examined in 2003, indicated a considerable increase of $CD95^+$ and $HLA\text{-}DR^+$ molecule expression, as well as the average values of activated T-lymphocytes. B-lymphocytes are insufficient still, but in relation to parameters obtained in 2002 the average values are reliably higher. As compared with parameters obtained by the previous examination, the average values of serum immunoglobulins of all three classes were found reliably lower than in the control, the level of serum IgG. Hence, as compared with parameters determined by monitoring in the period of 1999 – 2003, disimmunoglobulinemia is observed that confirm existence of symptoms of the immune system aging in the liquidators.

Table 2. Average values of the immune status parameters of Chernobyl accident consequence liquidators from NW Region of Leningrad Oblast according to monitoring data obtained in 2002 – 2003

No	Immune status indices	Average values, $M \pm m$								
		2003 (n = 119)			**2002** (n = 108)			**Control values**		
1	Leukocytes	6.15	±	0.14	6.36	±	0.17	6.40	±	0,09
2	Lymphocytes, %	**33.20**	±	**0.76*↑**	30.23	±	0.88	31.8	±	0,5
3	Lymphocytes, abs	2.02	±	0.06	1.89	±	0.07	2.04	±	0,04
4	$CD3^+$, %	69.26	±	0.96	**69.24**	±	**0.92•↑**	69.1	±	0,17
5	$CD3^+$, abs	1.40	±	0.05	1.31	±	0.05	1.41	±	0,09
6	$CD4^+$, %	40.47	±	0.76	**42.28**	±	**0.93••↑**	42.0	±	1,3
7	$CD4^+$, abs	0.82	±	0.03	0.80	±	0.04	0.84	±	0,06
8	$CD8^+$, %	**34.03**	±	**0.76***↑°°↑**	**28.00**	±	**0.71•••↓**	29.3	±	1,20
9	CD8, abs	**0.68**	±	**0.03***↑°°↑**	**0.53**	±	**0.02••↓**	0.51	±	0,04
10	$CD4^+/CD8^+$	**1.29**	±	**0.05***↓°↓**	**1.66**	±	**0.07•••↑**	1.50	±	0,08
11	$CD16^+$, %	**20.22**	±	**0.89*↑°°°↑**	**17.32**	±	**0.72°°↓**	15.0	±	0,18
12	$CD16^+$, abs	**0.41**	±	**0.02**↑°°°↑**	0.323	±	0.017	0.31	±	0,01
13	$CD25^+$, %	**1.94**	±	**0.12**↑**	**1.29**	±	**0.12•••↓**	<10		
14	$CD95^+$, %	**46.43**	±	**0.86***↑**	25.80	±	1.51	20.0–40.0		
15	$HLA\text{-}DR^+$, %	**19.95**	±	**0.60***↑**	**13.51**	±	**0.54••↑**	8.0 – 15.0		
16	T-act., %	**13.84**	±	**0.63***↑**	**8.57**	±	**0.50••↑**	<10		
17	NPA, %	61.9	±	1.3	**59.7**	±	**1.1°↓•••↓**	63.3	±	1,1
18	$CD19^+$, %	**6.242**	±	**0.244*↑°°°↓**	**5.15**	±	**0.30°°°↓••↓**	7.5	±	0,15
19	$CD19^+$, abs	**0.126**	±	**0.007*↑**	**0.101**	±	**0.009°°↓•↓**	0.15	±	0,01
20	IgM, g/l	**1.29**	±	**0.05**↓**	**1.55**	±	**0.06°°↑**	1.28	±	0,06
21	IgG, g/l	**10.77**	±	**0.17***↓°°°↓**	**12.10**	±	**0.28•••↓**	11.96	±	0,18
22	IgA, g/l	2.43	±	0.09	**2.63**	±	**0.11°↑•••↑**	2.12	±	0,14
23	IgE, ME	185.6	±	33.3	**160.4**	±	**26.0•↓**	<= 120		

Reliability of differences:

°) – parameters of the immune status of liquidators from the control: ° – $p < 0.05$; °° – $p < 0.01$; °°° – $p < 0.001$;

•) – parameters of the immune status of liquidators examined in 2002 from parameters obtained in 2001: • – $p < 0.05$; •• – $p < 0.01$; ••• – $p < 0.001$;

*) – parameters of the immune status of liquidators examined in 2003 from parameters obtained in 2002: * – $p < 0.05$; ** – $p < 0.01$; *** – $p < 0.001$.

For the sake of comparison, groups of persons from extreme risk divisions "ERD" and citizens of the same city were examined. In contrast with the ERD cohort, the liquidators possessed reliably higher absolute values of lymphocytes, T- and B-lymphocytes, whereas in comparison with the group of citizens they possessed reliably higher absolute values of NK-cells with a tendency to increase $CD16^+$ relative value, $CD95^+$ expression is intensified, the indices of delayed activation of $HLA\text{-}DR^+$ cells and activated T-lymphocytes are increased. According to data on the humoral unit of the immune status reliable difference in parameters of liquidators and citizens are determined by parameters of serum IgM (↓) and IgA (↑), and general IgE (↑). Contrary to citizens, ERD group showed reliably decreased parameters of B-lymphocytes and serum IgM and increased activated T-lymphocytes.

All examined groups possessed extremely low average values of activated lymphocytes, specifically in liquidators and ERD (below 3.5%) that shows decreasing of the functional activity of the cell unit by the level of reception to IL-2. Increased activation markers represent the valuable difference in the immune status between Chernobyl accident liquidators and Chernobyl town citizens, which indicates the stressed functioning of the immune system among liquidators.

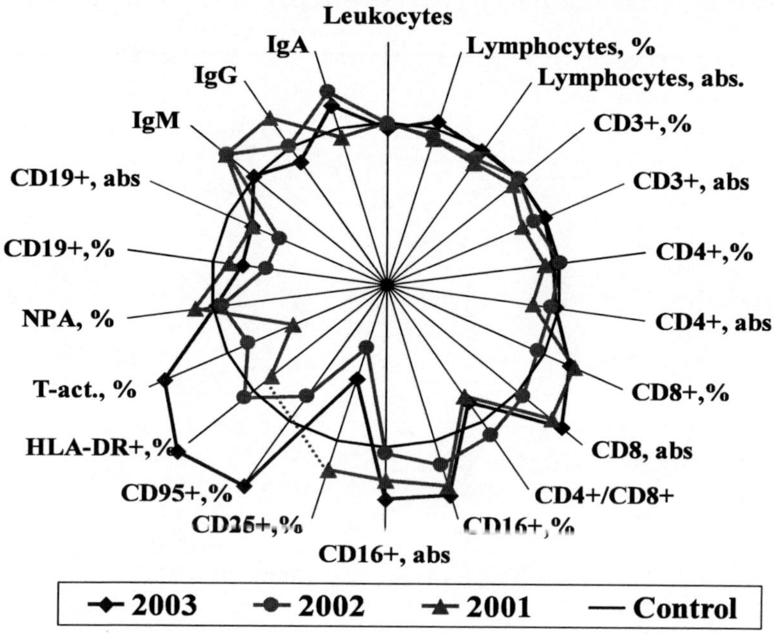

Figure 1. Dynamics of the immune status parameter variation for the Chernobyl accident consequence liquidators from NW Region according to monitoring data obtained in 2001 – 2003

Generally, according to the long-term observations (1992 – 2003), the immune status dynamics indicated the following changes (Figure 2):

- IMIN presence according to T-lymphocyte indices, determined in 1992, 1995, 1999 and 2001 with a tendency to decrease in the period of 1995 – 2001; T-helper indices in 1992, 1995 and 2001, and insufficiency of immunoregulation in 2001 and 2003;
- the presence of spontaneous NK-cell variations with the values increasing in 1993, 1995, 2001 and 2003, and considerable decrease of $CD16^+$ in 1994 and 1999;

- the presence of CTL – $CD8^+$ variations in the period of 1999 – 2003 by dynamics identical to that of natural killers – $CD16^+$, in this period and prelIMINary CTL insufficiency (1995 – 1999);
- decreasing early activation markers in 2002 – 2003 compared with 2001 and extremely low $CD25^+$ values in 2002 – 2003;
- gradual increase of delayed activation markers HLA-DR in 2001 – 2003 reaching the maximum at the upper reference border in 2003;
- the presence of disimmunoglobulinemia during all periods of observation with a tendency to increase serum IgM concentration in 1995 – 2001 and stable parameter in 2001 – 2002 and further decrease to control values; a tendency to increase serum IgG in 1995 – 2001 and further decrease in 2003 to the level of 1995; a tendency to increase serum IgA in 1995 – 2002 with a decrease in 2001 and equal values in 1999 and 2003;
- an increase of general IgE showing maximum of the average values in 2001 and the highest quantitative distribution in 1999 – 2001 (46.55 – 41.07%, respectively).

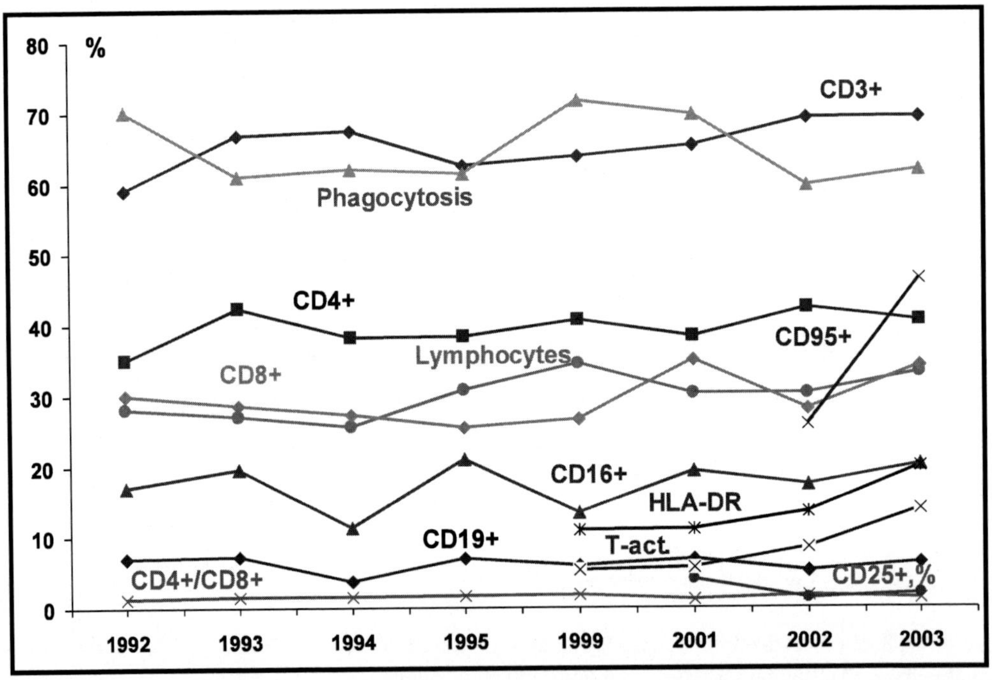

Figure 2. Dynamics of the immune status parameters of the Chernobyl accident liquidators from NW Region of Leningrad Oblast with respect to the year of examination

According to frequency analysis results dynamics of changing the immune status parameters among liquidators of the Chernobyl accident from NW Region in the period of 1995 – 2003 (Table 3) showed high occurrence T-cell unit insufficiency in 1995: T-lymphocytes in 1/3 examined patients, T-helpers and CTL in 1/4, B-lymphocytes in 1/3, increased NK-cell parameters in more than 40% of examined liquidators. Basing on the percentage of patients having normal parameters, NK-cells demonstrated the highest occurrence of deviating parameters of the immune status. Normal parameters of NT-cells were determined for 51.6% of examined patients only, for T-helpers with CD4$^+$ for 58.6%, then rising for T- and B-lymphocytes – 65.1% and 66.9%, respectively.

Table 3. Percentage of patients showing changes (↓↑) in indices of the immune status among liquidators of the Chernobyl accident consequences from NW Region in 1995 – 2003

Immune status parameters	Group (%) of patients with changes:									
	decreased (↓)					increased (↑)				
	1995 *n* =287	1999 *n* =177	2001 *n* =112	2002 *n* =108	2003 *n* =119	1995 *n*=287	1999 *n*=177	2001 *n* = 112	2002 *n* = 108	2003 *n* = 119
Leukocytes	9.8	7.01	3.60	4.72	1.82	7.0	7.64	9.91	8.49	3.64
Lymphocytes, %	3.5	8.11	9.91	7.55	1.82	2.1	16.22	6.31	4.72	1.82
Lymph., abs.	15.7	14.19	9.01	12.26	6.36	5.2	11.49	3.60	4.72	4.55
CD3$^+$, %	**32.1**	**27.85**	**22.22**	7.55	12.61	2.8	9.49	6.48	11.32	15.97
CD4$^+$, %	**24.7**	**14.47**	**20.18**	10.38	14.29	16.7	**20.75**	10.09	**24.53**	16.81
CD8$^+$, %	**24.0**	**20.13**	4.55	12.26	0.84	8.0	11.95	**36.36**	13.21	**36.97**
CD4$^+$/CD8$^+$	16.0	16.35	**37.61**	14.15	**29.41**	12.9	13.21	5.51	14.15	2.52
CD16$^+$, %	2.1	**27.67**	8.18	5.66	3.36	**46.3**	**18.87**	**41.82**	**35.85**	**44.54**
CD25$^+$, %	–	–	–	–	–	–	–	3.67	0	0
CD95$^+$, %	–	–	–	**42.45**	0	–	–	–	**18.87**	**70.59**
HLA-DR$^+$%	–	–	**27.27**	12.26	2.52	–	3.95	4.55	**30.19**	**47.06**
T-act., %	–	–	–	–	–	–	**10.17**	**13.39**	**33.33**	**63.03**
NPA, %	6.6	0	0	6.67	6.03	7.0	6.92	6.25	1.91	6.03
CD19$^+$, %	**32.75**	**42.77**	**30.91**	**55.24**	**37.82**	0.35	0.63	0.91	0	0
IgM	–	0	0	0	0.84	–	0	0	9.43	2.52
IgG	–	0	0	0	0	–	**19.5**	11.61	3.92	0
IgA	–	0.63	0	0	0	–	16.35	0	**22.64**	**15.13**
IgE, ME	–	**20.90**	**28.57**	**28.70**	**29.06**	–	**46.55**	**41.07**	**25.71**	**23.93**

In 1999, a tendency to decrease of percentage of patients with T-cell unit insufficiency and its considerable increase with reduction of NK-cells and B-lymphocytes were observed. A decrease of percentage of patients with normal values of B-lymphocytes was observed: from 66.9 to 56.6%, respectively. As compared with 1995, a considerable imbalance in the distribution of NK-cells with significant change of the ratio between increased and decreased CD16$^+$ parameters. The percentage of patients with increased level of general IgE was extremely high (46.55%); hence, the increase of IgE was detected both in the presence and the absence of allergic diseases. About 1/5 of examined patients had relatively low parameters of general IgE ($\leq$ 45 ME) that indicated disorder of the cell cooperation in the immune response, as well as decreasing tolerance to alien antigens (R.V. Petrov, 1995). The increase of general IgE may demonstrate an endotoxic state due to the effect of the risk factors of the radiation catastrophe and, as well indicate the change of allergic reactivity in remote periods after irradiation. Generally, in 1999 the percentage of patients showing deviations in the immune status parameters was extremely high.

Imbalance in the cell immunity system was also observed in the group of liquidators examined in 2001. However, this imbalance was of a type different from that determined in 1999: a significant increase of the part of patients with increased cytotoxic cell concentration (NK-cells – 41.82% and T-killers – 36.4%) and a decrease of the part of patients with normal parameters were observed. Moreover, T-cell subpopulation distribution was disturbed in more that 40% of liquidators and for 37.6% immunoregulation insufficiency was determined. In 30% insufficiency of B-lymphocytes was preserved. As in 1999, the imbalance in the humoral immunity system was characterized by both the presence of quantitative B-cell IMIN and disimmunoglobulinemia by Ig parameters of four classes determined.

The immune status of liquidators from NW Region, examined in 2002, was characterized by much higher percentage of patients with B-lymphocyte insufficiency (55.24%). In all observation periods (1995 – 2003) the quantitative insufficiency of B-lymphocytes with the function preservation was determined in more than 1/3 – 1/2 part of examined liquidators, for which in 1999 and 2002 the highest parameters of B-cell IMIN was observed. Percentage of patients with increased NK-cell concentration was high (35.85%), accompanied by considerably decreased level of T-killers – 13.2%. A tendency to decrease of the part of patients with T-helper normal level due to increasing number of people with increased $CD4^+$. Many patients (42.5%) possessing weakened $CD95^+$ expression, whereas for 1/5 part of patients examined in 2002 this parameters was found increased (mediated apoptosis).

Table 4. Dynamics of immune status parameters for the liquidators from NW Region at the main IPS according to monitoring performed in 2001 – 2002

Immune status parameters	Infectious syndrome			Allergic diseases			Infectious-allergic syndrome			Autoimmune diseases		
	2001	2002	2003	2001	2002	2003	2001	2002	2003	2001	2002	2003
Leukocytes												
Lymphocytes, %	*↓	↑	+⇑	↑	*↓+⇓		↓		↑			*↑
Lymphocytes, abs.												
$CD3^+$, %								*↑		*↓		
$CD3^+$, abs.												
$CD4^+$, %												
$CD4^+$, abs.												
$CD8^+$, %	↑	+⇓	*↑•⇑	*↑		↑	*↑	*↓+⇓	*↑•⇑	*↑	+⇓	
$CD8^+$, abs.			*↑		+⇓		*↑	+⇓	*↑•⇑			
$CD4^+/CD8^+$			*↓•⇓				*↓	+⇑	•↓	*↓	+⇑	
$CD16^+$, %			*↑			*↑•⇑			*↑	*↑	+⇓	
$CD16^+$, abs.			*↑			*↑•⇓			•⇑	*↑		
$CD25^+$, %		+⇓			+⇓			+⇓	•⇑		+⇓	
$CD95^+$, %	–		•⇑	–		•⇑	–		•⇑	–		•⇑
HLA-DR^+%		+⇑				•⇑			•⇑			•⇑
T-act., %		+⇑	↑			↑			•⇑			•⇑
NPA, %	*↑	+⇓		*↑	+⇓		*↑	*↓+⇓				
$CD19^+$, %			*↓					*↓+⇓	•⇑		*↓+⇓	*↓
$CD19^+$, abs.					+⇓			+⇓	•⇑			*↓
IgM, g/l	*↑	+⇓		*↑	*↑	•⇓	*↑					
IgG, g/l	*↑		*↓•⇓	*↑	+⇓		*↑	+⇓	*↓			•⇓*↓
IgA, g/l					+⇑	*↑				*↓	*↑+⇑	•⇓
IgE, ME	↑			↑↑	+⇓	+⇓	↑		↑			

The reliability of differences of:

*↓↑ – parameters of the immune status of liquidators from control values: * – $p < 0.05 – 0.001$;

+ ⇓⇑ – parameters of the immune status of liquidators according to data of examination performed in 2002 from those of 2001: + – $p < 0.05 – 0.001$;

• – – parameters of the immune status of liquidators according to data of examination performed in 2003 from those of 2002: • – $p < 0.05 – 0.001$;

↓↑ – tendencies to changes (unreliable)

The analysis of 2003 detected the following quantitative distribution of patients with changed parameters (↑↓). Increased levels were observed for the following: CTL – 37%; NK-cells – 44.54%; activation markers HLA-DR⁺ and CD95⁺ – 70%; activated T-lymphocytes – 60%. As compared with parameters, obtained in 2002, the above-mentioned are very high (Table 3). Moreover, 1/3 part of the liquidators examined in 2003 had insufficient immunoregulation, which is 2 times more frequent, too. Over 1/3 had B-lymphocyte insufficiency. The increased general IgE level was detected for 23.93% of patients that equaled the level of 2002, whereas 29.06% (34:117) had this index low (50 Me or lower) that indicated disordered cell cooperation. As compared with ERD and citizens of the city, liquidators demonstrated 3 times higher frequency of elevated general IgE level. All these data confirm the immune status assessment results indicating a strain of the immunity for the liquidators of the Chernobyl accident. Of special interest is circularity in increasing the number of patients having increased concentration of cytotoxic cells (2001, 2003), increased T-helper concentration (1999, 2002), insufficient immunoregulation (2001, 2003). The greatest group of examined liquidators had high concentration of activation markers.

Table 5. Dynamics of immune status parameters for the liquidators from NW Region at the main ChD according to monitoring performed in 2001 – 2002

Immune status parameters	Circulatory system diseases			Digestive system diseases			Bone and joint diseases			Nervous system diseases		
	2001	2002	2003	2001	2002	2003	2001	2002	2003	2001	2002	2003
Leukocytes												
Lymphocytes, %												
Lymphocytes, abs.	*↓			*↓			*↓	*↓				
$CD3^+$, %	*↓	+⇑		*↓	+⇑		*↓			↓		
$CD3^+$, abs.	*↓	+⇑					*↓					
$CD4^+$, %	*↓	+⇑			+⇑	•⇓						
$CD4^+$, abs.	*↓	+⇑			+⇑							
$CD8^+$, %	*↑	+⇓	*↑ •⇑	*↑	+⇓	*↑ •⇑	*↑	+⇓	*↑ •⇑	*↑	+⇓	↑
$CD8^+$, abs.	*↑	+⇓	*↑ •⇑	*↑		*↑ •⇑	*↑	+⇓	*↑ •⇑	*↑		*↑
$CD4^+/CD8^+$	*↓	+⇑	*↓ • ⇓	*↓	*↑ +⇑	• ⇓	*↓	+⇑	*↓ • ⇓		+⇑	
$CD16^+$, %	*↑	*↑+ ⇓	*↑ •⇑	*↑	+⇓	*↑ •⇑	*↑	*↑	*↑	*↑		*↑
$CD16^+$, abs.	*↑		*↑ •⇑			*↑ •⇑			*↑			
$CD25^+$, %		+⇓	• ⇑		+⇓	•⇑		+⇓	•⇑		+⇓	•⇑
$CD95^+$, %	–		• ⇑	–		•⇑	–		•⇑	–		•⇑
HLA-DR⁺%		+⇑	• ⇑			•⇑		+⇑	•⇑			•⇑
T-act., %		+⇑	• ⇑			•⇑		+⇑	•⇑			•⇑
NPA, %	*↑	*↓+ ⇓		*↑	*↓+ ⇓		*↑	*↓+ ⇓		*↑	*↓+ ⇓	
$CD19^+$, %		*↓+ ⇓	*↓ • ⇑		*↓+ ⇓	*↓ •⇑		*↓+ ⇓	*↓ •⇑	*↓	+⇓	*↓
$CD19^+$, abs.		*↓	*↓		*↓			*↓+ ⇓	*↓ • ⇓	*↓	*↓	
IgM, g/l	*↑	*↑	• ⇓	*↑	*↑	• ⇓	*↑	*↑	•⇓	*↑	*↑	
IgG, g/l	*↑	+⇓	*↓ • ⇓	*↑	+⇓	*↓ • ⇓	*↑	+⇓	*↓ • ⇓	*↑	+⇓	*↓
IgA, g/l		*↑+⇑			*↑ +⇑			*↑ +⇑			+⇑	
IgE, ME		+⇓										

The reliability of differences of:

*↓↑ – parameters of the immune status of liquidators from control values: * – $p < 0.05 – 0.001$;

+ ⇓⇑ – parameters of the immune status of liquidators according to data of examination performed in 2002 from those of 2001: + – $p < 0.05 – 0.001$;

• – – parameters of the immune status of liquidators according to data of examination performed in 2003 from those of 2002: • – $p < 0.05 – 0.001$;

↓↑ – tendencies to changes (unreliable)

The comparison of the immune status dynamics in the period of 1995 – 2003 indicated an imbalance in the main components of the immune system of liquidators from NW Region that evidenced about "distortion" of the immunogram (Figure 1). Most frequently, deviations are observed for cytotoxic cells CD16⁺, CD8⁺, T-helpers, B-lymphocytes and general IgE, and in 2002 – 2003 for CD95⁺. *In 2002 – 2003, a significant increase of activation markers and extension of the group of patients with generally elevated parameters were observed.*

As induced by changes in the immune status of liquidators, high frequency of clinical symptoms of the immune system dysfunction and somatic diseases and the high risk of proliferative disease occurrence in remote time, we have analyzed dynamics of the immune status parameters at basic immunopathological syndromes (Table 4), main classed of chronic diseases (Table 5) and some pathological processes (Table 6).

Table 6. Dynamics of immune status parameters for the liquidators from Moscow and NW Regions according to monitoring performed in 2001 – 2003

No.	Immune status parameters	Year of examination								
		2001			2002			2003		
		NW Region	Moscow Region		NW Region	Moscow Region		NW Region	Moscow Region	
1	**Leukocytes**		***⇓			*⇓				
2	**Lymphocytes, %**		*** ⇓	••↓				* ⇑	*** ⇓	•••↓
3	**Lymphocytes, abs.**	*⇓		•↓		*⇓			*** ⇓	•••↓
4	**CD3⁺, %**	**⇓		•↑						
5	**CD3⁺, abs.**								** ⇓	•••↓
6	**CD4⁺, %**	*⇓	* ⇓						* ⇑	•••↑
7	**CD4⁺, abs.**									
8	**CD8⁺, %**	**⇑	N	•••↓	N	**⇑	•••↑	** ⇑	N	•••↓
9	**CD8⁺, abs.**	**⇑	N	•••↓	N		•↑	** ⇑	N	•••↓
10	**CD4⁺/CD8⁺**	**⇓	N	•↑	N	*⇓	•••↓	* ⇓	N	•••↑
11	**CD16⁺, %**	***⇑	*** ⇓	•••↓	** ⇑	*⇓	••↓	*** ⇑		••↓
12	**CD16⁺, abs.**	*⇑	*** ⇓	•••↓		**⇓	••↓	*** ⇑	* ⇓	•••↓
13	**CD25⁺, %**			•↑			•••↑			•••↑
14	**CD95⁺, %**	–	–	–			•↓	⇑		•••↓
15	**HLA-DR⁺, %**						•↑	⇑		••↓
16	**T-act., %**						•↓	⇑	*** ⇑	•••↓
17	**Phagocytosis, %**	***⇑		•••↓	*⇓		•••↑			•↑
18	**CD19⁺, %**				*⇓	**⇑	•••↑	*** ⇓	***⇑	•••↑
19	**CD19⁺, abs.**		*⇓				•••↑			•↑
20	**IgM, g/l**	**⇑			**⇑				* ⇑	•↑
21	**IgG, g/l**	***⇑	**⇑	••↓				*** ⇓	*** ⇓	
22	**IgA, g/l**			••↑	*⇑	*⇑			* ⇑	
23	**IgE, ME**						•↑			

* ⇓⇑ – changes of the parameter with respect to control values;

•) – reliability of differences in parameter for liquidators from Moscow and MO from corresponding parameter for liquidators for NW Region: •) – $p < 0.05$; ••) – $p < 0.01$; •••) – $p < 0.001$

At a definite variability of immunological parameters for each immunopathological syndrome separately, in 2001 for all IPS the following tendencies were general: increasing CD8⁺ (%), NPA, IgM, IgE and IgG (except for autoimmune pathology); low level of immunoregulation at combined pathology of infectious-allergic syndrome and autoimmune diseases, especially at infectious-allergic syndrome induced by considerable disorder in T-cell subpopulation distribution ($CD4^+\downarrow$ $CD8^+\uparrow$, Table 4). At all IPS general IgE level was elevated, being the most valuable in case of allergic and autoimmune diseases. In 2002, changes in the immunological parameters generally tended to decreasing levels of CTL-$CD8^+$(%), early activation $CD25^+$ lymphocyte markers, NPA, serum IgG at the combined pathology of infectious-allergic syndrome and AD, and increasing serum IgA at AD and autoimmune pathology. The general tendency in the immune status variation at all IPS for the patients, examined 2003, was a decrease of serum IgG (except for AD) and elevation of activation markers: $CD95^+$, activation T-lymphocytes, HLA-DR^+ (except for the infectious syndrome), CTL-$CD8^+$(%, except for autoimmune diseases) and general IgE (except for AD, which average parameters remained normal, which is untypical of this pathology). The combined infectious-allergic pathology showed the highest number of changes in dynamics of the processes.

Much higher similarity of tendencies in the change of immunological parameters is observed for the main classes of diseases: diseases of the circulatory system, the digestive system, bones and joints and, to a lower extent, diseases of the nervous system. In 2001, for the main classes of diseases predominant among liquidators, general tendencies in changes of the immune status parameters were reduction of:

- diminution of absolute values of lymphocytes;
- T-lymphocyte concentration;
- Immunoregulation index (except for nervous system diseases);

and increasing concentration of cytotoxic cells: CTL and natural killers *and* NPA, serum IgM, IgG, general IgE, which is most significant in case of nervous systems diseases (Table 5).

In 2002, dynamics of changes in immunological indices indicated a decrease of:

- CTL;
- NPA;
- Early activation marker $CD25^+$;
- B-lymphocytes;
- Serum IgG,
- and an increase of:
- Immunoregulation index;
- HLA-DR^+;
- Activation T-lymphocytes;
- Serum IgA.

In 2003, similarity in dynamics in relation to 2002 was observed in increasing all activation markers and by other parameters as follows: $CD8^+(\uparrow)$, $CD16^+(\uparrow)$, $CD4^+/CD8^+(\downarrow)$, $CD19^+(\uparrow)$, IgM($\downarrow$), IgG($\downarrow$). High similarity in dynamics of deviation in the case of three main classed of diseases (circulatory system, digestive system, bones and joints) is explained by their high occurrence in all mentioned groups. For example, CSD occurred in 85.58% of examined liquidators, digestive system diseases – in 66.35%, and musculoskeletal system –

in 53.85%. For some nosological CSD forms, such as atherosclerosis and HD, IHD, and CVS:DEP, dynamics of the immune status are generally characterized by the regularities, previously described for the whole class of CSD. This is explained by high frequency of this pathology among diseases of the circulatory system. Similar regularities are observed for other nosological forms and classes of diseases (for example, GIT → digestive system, deforming osteochondrosis → bone and joint diseases). Therefore, it is indicated that the liquidators possessing some disorders in the immune status, observed at some nosological forms of diseases, demonstrate dynamical changing of immunological parameters, independent of the type of pathology.

At functional dysfunctions in the circulatory system (VD and NCD) no decrease of T-cell component is observed, similar to CSD accompanied by atherosclerosis. On the contrary, a tendency to increase $CD3^+$ and $CD4^+$ is observed. The increase of $CD8^+$ is not so high; T-cell subpopulations show relatively normal distribution, and the immunoregulation index equals control values. Similar to organic diseases of the circulatory system, dynamics of $CD16^+$ variation demonstrated two maxima (2001 and 2003) and a minimum (2002). Accompanying VD and NCD determined in patients, examined in 2003, an increase of activation markers compared with the data of the prior examination. In 2003, dynamics of changes in activation T-lymphocytes significantly (2.6 times) exceeded their average values compared with the data of 2002. The changes in NPA parameters demonstrated maxima of average values in 2001 and 2003 and reliable minimum in 2002. At VD and NCD, the indices of the humoral unit of the immune status had no significant differences from the case of organic CSD.

Table 7. The abundance of liquidators (%) from NW Region with decreased (↓) and elevated (↑) indices of the immune status in case of CSD (2003)

Immune status indices	Diseases, abundance of patients with the values:							
	decreased (↓)				increased (↑)			
	AD, HD	IHD	CVD: DEP	VD, NCD	AD, HD	IHD	CVD: DEP	VD, NCD
$CD3^+$, %	13.64	14.29	9.68	6.67	13.64	14.29	12.90	26.67
$CD4^+$, %	15.15	7.14	14.52	0	15.15	21.43	14.52	26.67
$CD8^+$, %	1.52	0	0	0	**45.45**	**40.48**	**51.61**	**40.0**
$CD4^+/CD8^+$	**34.85**	**28.57**	**37.10**	**13.33**	3.03	4.76	1.61	6.67
$CD16^+$, %	1.52	4.76	3.23	0	**46.97**	**47.62**	**35.48**	**40.0**
$CD95^+$	0	0	0	0	**68.18**	**71.43**	**75.81**	**80.0**
HLA-DR^+, %	3.03	2.38	3.23	6.67	**46.97**	**42.86**	**48.39**	**46.67**
T-act., %	–	–	–	–	**65.15**	**57.14**	**62.90**	**66.67**
Phagocytosis, %	7.58	14.29	6.56	0	4.55	2.38	1.64	6.67
$CD19^+$, %	**39.39**	**35.71**	**33.87**	**53.33**	0	0	0	0
IgA	0	0	0	0	**15.15**	**14.29**	**16.13**	6.67
IgE, ME	–	–	–	–	**24.24**	**26.19**	**24.59**	**40.0**

For more than 1/3 part of patients examined in 2003 the frequency analysis (Table 7) indicated T-cell insufficiency by the immunoregulation index and B-lymphocyte insufficiency, which were higher in the case of VD and NCD, high percentage of patients with increased CTL (40 – 51.6%) and NK-cells (36 – 48%), and extremely high occurrence of activation antigens and activation T-lymphocytes. In the case of CSD accompanied by atherosclerosis, compared with functional diseases, the group of patients with increased serum IgA is more abundant. This group also show frequent occurrence of increased general IgE. At all other ChD of the circulatory system the group of patients possesses a uniform distribution of increased general IgE.

Table 8. Average indices of the immune status for the liquidators having benign tumors (NW Region, 2003)

No.	Immune status parameters	Average values, $M \pm m$ 2001 (n = 15)	2002 (n = 18)	2003 (n = 13)
1	Leukocytes	6.53 ± 0.53	**5.75 ± 0.29°↓**	6.75 ± 0.44
2	Lymphocytes, %	29.40 ± 2.29	32.0 ± 3.01	**37.27 ± 1.71°°↑**
3	Lymphocytes, abs	1.83 ± 0.13	1.79 ± 0.17	**2.54 ± 0.23*↑°↑**
4	$CD3^+$, %	63.20 ± 3.20	65.77 ± 2.53	63.78 ± 4.02
5	$CD3^+$, abs	**1.14 ± 0.09°↓**	1.16 ± 0.11	1.62 ± 0.21
6	$CD4^+$, %	40.87 ± 2.31	39.98 ± 2.22	39.29 ± 2.65
7	$CD4^+$, abs	0.74 ± 0.06	0.74 ± 0.09	0.95 ± 0.11
8	$CD8^+$, %	**34.36 ± 1.92°↑**	**27.27 ± 2.19⁺↓**	**35.91 ± 2.17*↑°↑**
9	CD8, abs	0.62 ± 0.05	**0.47 ± 0.06**	**0.93 ± 0.11**↑°°↑**
10	$CD4^+/CD8^+$	1.31 ± 0.19	1.75 ± 0.25	**1.18 ± 0.12*↓°↓**
11	$CD16^+$, %	21.85 ± 2.48	**20.63 ± 2.31°↑**	**25.31 ± 4.17°↑**
12	$CD16^+$, abs	0.41 ± 0.06	0.381 ± 0.057	**0.66 ± 0.12*↑°↑**
13	$CD25^+$, %	4.26 ± 0.66	**1.36 ± 0.32⁺⁺↓**	2.10 ± 0.40
14	$CD95^+$, %	– ± –	21.59 ± 3.33	**44.30 ± 3.08***↑**
15	$HLA\text{-}DR^+$, %	12.15 ± 1.07	14.03 ± 1.15	**20.71 ± 1.59**↑**
16	T-act., %	6.88 ± 0.98	9.06 ± 0.98	**15.82 ± 1.57**↑**
17	NPA, %	**71.6 ± 1.4°°°↑**	**61.7 ± 2.4⁺⁺↓**	**54.1 ± 4.1°↓**
18	$CD19^+$, %	6.719 ± 0.929	**5.28 ± 0.85°↓**	6.308 ± 0.595
19	$CD19^+$, abs	0.123 ± 0.018	0.108 ± 0.027	0.155 ± 0.018
20	IgM, g/l	1.35 ± 0.10	1.44 ± 0.12	1.19 ± 0.06
21	IgG, g/l	**13.73 ± 0.51°°↑**	**11.85 ± 0.57⁺↓**	**11.41 ± 0.39⁺↓**
22	IgA, g/l	2.15 ± 0.19	2.27 ± 0.26	2.37 ± 0.22
23	IgE, ME	208.3 ± 74.2	129.6 ± 43.4	151.7 ± 73.0

Reliability of differences in the immune status parameters for:

°) – liquidators of the Chernobyl accident from control values: ° – $p < 0.05$; °° – $p < 0.01$; °°° – $p < 0.001$;

*) – liquidators examined in 2003 from parameters, obtained in 2002: * – $p < 0.05$; ** – $p < 0.01$; *** – $p < 0.001$;

⁺) – liquidators examined in 2002 from parameters, obtained in 2001: ⁺ – $p < 0.05$; ⁺⁺ – $p < 0.01$; ⁺⁺⁺ – $p < 0.001$.

High percentage of patients with increased activation markers, cytotoxic cell parameters, insufficient immunoregulation and B-lymphocytes, cyclic variations of CTL and $CD4^+/CD8^+$, and increased NK-cell parameters, observed in 2003, initiated analysis of the variation dynamics of the immune status parameters among liquidators having diseases of the proliferative type – benign tumors. The immune status of liquidators possessing benign proliferative processes also presented lymphocytosis: the average values lymphocyte concentration approached the upper border of the physiological norm and reliably exceeded control values and corresponding indices at other ChD and IMIN clinical symptoms (Table5). As compared with other IMIN clinical symptoms, T-cell indices ($CD3^+$, $CD4^+$) tended to decrease with irregular distribution of T-cell subpopulations and reliable decrease of $CD4^+/CD8^+$(↓). In all three years of monitoring average values of NK-cells were above the upper border of physiological norms with highly expressed tendency to increase in 2003, which reliably and significantly exceeded control values. Parameters of NPA are reliably decreased, which reliably decreased with years compared with the values of 2001. Cyclic variations of CTL and immunoregulation parameter were determined, also observed with other IMIN and ChD symptoms. Lymphocyte activation markers are increased: $CD95^+$ expression, delayed HLA-DR^+ activation marker and activation T-lymphocytes at low values of early $CD25^+$ activation, which was observed before for the case of chronic pathology. Simultaneous increase of all three activation markers (CD95, HLA-DR and act. T-lymphocytes) for the liquidators was observed for 38.66% (46:119). At the same time, $CD95^+$ and HLA-DR^+ are increased in 41.18% (49:119), and $CD95^+$ and activated T-lymphocytes – in 49.58% (59:119). Therefore, activation processes are intensified due to apoptotic and delayed activation of cells. Apparently, similar to cytotoxic cells, activation marker parameters may also display spontaneous variations between detected maxima and minima. Hence, further examinations may give both a decrease of activation by markers increased in 2003 and a tendency to increase. Since these parameters were included in the later monitoring period for estimating the immune status of liquidators, the true dynamics is stipulated by current observation. Compared with other pathological processes and ChD, there are no considerable differences by the humoral immunity parameters.

Thus, in the presence of benign proliferative states two directions for changing immunological parameters occur:

1. the totality of the change undetectable in case of other IMIN and ChD clinical symptoms:
   - lymphocytosis;
   - T-cell component parameter decrease of the immune status: $CD3^+$, $CD4^+$;
   - disordered distribution of T-cell subpopulations;
   - immunoregulation insufficiency;
   - NK-cell concentration increase;
   - NPA decrease.
2. Immunological changes, also observed in case of other pathological states according to data of monitoring performed in 2003:
   - Cyclic variations of CTL and IRI;
   - Increasing activation markers of lymphocytes: increased expression of $CD95^+$, HLA-DR^+ and activated T-lymphocytes;
   - Low values of early $CD25^+$ activation marker.

Apparently, the presence of both signs in the immune status allows relation of examined patients to the high risk group of proliferative disease development. According to earlier monitoring results, obtained for liquidators from NW Region of Leningrad Oblast, the presence of spontaneous variations of cytotoxic cells, preferably natural killers, as the risk factor for possible development of proliferative states (Figure 5). Data of monitoring in 2004, performed for patients with the above-mentioned changes in the immune status observed in 2003, several cases of proliferative disease development were detected, including malignant ones (gastric adenocarcinoma; prostate cancer, with prior long-term benign hyperplasia; recrudescent urinary bladder cancer, B-cellular lymphoma, etc.). As compared with the character of immunological parameter deviation in the previous period (1999 – 2002), we suggest uniting the changes observed into the immunological characteristic of proliferative syndrome, shown below.

IMMUNOLOGICAL CHARACTERISTIC OF PROLIFERATIVE SYNDROME
• Lymphocytosis within 2 – 3 years
• High (мах) values / $CD3^+$ insufficiency
• Spontaneous variations (↓↑↓↑) of $CD8^+$ and $CD16^+$ cytotoxic cells
• $CD4^+/CD8^+$(↓↓) immunoregulation insufficiency
• insufficiency/significant increase of NK-cells during 2-3 years, negative dynamics tended to increase
• Significant increase of activation marker parameters:
• $CD95^+$ expression from (↓↓) to (↑↑↑),
• delayed HLA-DR (↑) activation marker,
• activation T-lymphocytes
• NPA decrease
• B-lymphocyte deficiency

Patients with such changes should be considered as subject to development of proliferative states. They have to pass deep clinical and instrumental examination 2 – 3 times a year, especially in the presence of a “target organ”, i.e. in the case of any chronic disease, which may transform to a tumor process. This is the high risk group of radiation impact consequence development.

To compare detected changes, let us present dynamics of the immune status parameters for the liquidators having increased fatigability syndrome (IFS) (Table 9) principally different from the proliferative syndrome characteristic. No lymphocytosis is detected; no tendency to decrease T-cell component parameters of the immune status and disturbance of T-cell subpopulation distribution are observed. The immunoregulation index falls within the range of normal values. Which is more valuable, in the case of IFS, no cyclic variations of cytotoxic cell values are observed: CTL parameters, highly elevated in patients, examined in 2001, are reliably decreased to the norm in 2002 and remained unchanged in 2003. Parameters of NK-cells show tendency to increase compared with the control and have identical values in all three periods of monitoring. The affinity is observed in dynamics of the activation markers, similar to other ChD. The level of B-lymphocytes is reliably decreased. Serum IgG parameters display a tendency to decrease (reliably in 2003), whereas for IgA a tendency is to increase. Hence, it may be noted that in the presence of the increased fatigability syndrome at the end of the fourth five-year period after the accident no T-cell component insufficiency is observed, a decrease of T-helpers, detected in 2001, and immunoregulation index are then leveled, and at the background of expressed disimmunoglobulinemia an insufficiency by humoral immunity parameters (B-lymphocytes and IgG level, which was not observed in 2001) is determined. Insufficiency of the humoral component was also observed in much earlier periods of works in the Chernobyl accident zone. The dynamics determined testifies about a valuable imbalance in all components of the immune status, at the same time different from dynamics observed for proliferative states.

Table 9. Average values of the immune status parameters among liquidators having the increased fatigability syndrome (NW Region, 2003)

No.	Immune status parameters	Average values, $M \pm m$								
		2001 (n = 17)			2002 (n = 18)			2003 (n = 18)		
1	Leukocytes	6.15	±	0.41	**5.51**	±	**0.28**$^{\circ\circ}$↓	**6.71**	±	**0.52***↑
2	Lymphocytes, %	32.29	±	2.57	35.44	±	2.47	32.24	±	1.54
3	Lymphocytes, abs	1.90	±	0.13	1.95	±	0.17	2.12	±	0.17
4	$CD3^{+}$, %	69.14	±	2.83	71.19	±	1.66	69.72	±	2.12
5	$CD3^{+}$, abs	1.31	±	0.10	1.39	±	0.13	1.48	±	0.13
6	$CD4^{+}$, %	39.33	±	2.28	44.34	±	2.11	43.66	±	2.09
7	$CD4^{+}$, abs	0.76	±	0.07	0.87	±	0.09	0.94	±	0.11
8	$CD8^{+}$, %	**37.15**	±	**1.89** $^{\circ\circ}$↑	**29.44**	±	**1.78**$^{++}$↓	30.88	±	1.93
9	CD8, abs	**0.72**	±	**0.06**$^{\circ}$↑	0.57	±	0.06	**0.64**	±	**0.05**$^{\circ}$↑
10	$CD4^{+}/CD8^{+}$	**1.12**	±	**0.10** $^{\circ}$↓	**1.65**	±	**0.16**$^{+}$↑	1.53	±	0.14
11	$CD16^{+}$, %	18.77	±	2.30	18.25	±	2.01	18.50	±	1.86
12	$CD16^{+}$, abs	0.37	±	0.05	0.350	±	0.052	0.40	±	0.05
13	$CD25^{+}$, %	5.19	±	0.79	**1.24**	±	**0.20**$^{+++}$↓	1.91	±	0.36
14	$CD95^{+}$, %	–	±	–	29.13	±	3.37	**45.09**	±	**2.11*****↑
15	$HLA\text{-}DR^{+}$, %	13.40	±	1.20	13.06	±	1.31	**17.52**	±	**1.36***↑
16	T-act., %	7.99	±	1.43	7.90	±	1.22	**11.75**	±	**1.27***↑
17	NPA, %	**69.2**	±	**1.9**$^{\circ}$↑	**57.6**	±	**2.1**$^{+++}$↓$^{\circ}$↓	64.9	±	3.6
18	$CD19^{+}$, %	7.118	±	0.770	**5.16**	±	**0.54**$^{\circ\circ\circ}$↓	**5.767**	±	**0.642** $^{\circ}$↓
19	$CD19^{+}$, abs	0.136	±	0.017	**0.104**	±	**0.016**$^{\circ}$↓	0.126	±	0.019
20	IgM, g/l	1.47	±	0.11	1.49	±	0.12	1.31	±	0.13
21	IgG, g/l	**14.26**	±	**0.47**$^{\circ\circ\circ}$↑	**11.38**	±	**0.65**$^{++}$↓	**10.27**	±	**0.35**$^{\circ\circ\circ}$↓
22	IgA, g/l	1.77	±	0.16	2.05	±	0.12	2.33	±	0.22
23	IgE, ME	162.1	±	39.4	138.6	±	43.7	211.1	±	94.7

Reliability of differences in the immune status among liquidators:
$^{\circ}$) – from control values: $^{\circ}$ – $p < 0.05$; $^{\circ\circ}$ – $p < 0.01$; $^{\circ\circ\circ}$ – $p < 0.001$;
*) – examined in 2003 compared with the data obtained in 2002: * – $p < 0.05$; ** – $p < 0.01$; *** – $p < 0.001$;
$^{+}$) – examined in 2002 compared with the data obtained in 2001: $^{+}$ – $p < 0.05$; $^{++}$ – $p < 0.01$; $^{+++}$ – $p < 0.001$.

Dynamics of the Immune Status Parameters for the Liquidators from Moscow and MO According to the Monitoring Data Obtained in 2001 – 2003

The comparison of the immune status parameters for the liquidators from NW Region of Leningrad Oblast and Moscow region, obtained during three years of monitoring, indicated differently directed changes in the majority of immunological parameters (Table 6). CTL variation periods are different: the liquidators from NW Region showed CTL increase in 2001 and 2003 and reduction to normal values in 2002, whereas the liquidators from The Moscow region possessed elevated CTL in 2002 and its reduction to the norm in 2001 and 2003. Similarly, the periods of the immunoregulation index changes showing the opposite dynamics compared to CTL were also different for these groups of liquidators. The indices of activation markers are also changed in different directions: the liquidators from The Moscow region had considerably higher markers of the early activation of lymphocytes, whereas the group from NW Region had them at a low level. Parameters of other activation markers are the following: in 2002 – 2003, $CD95^+$ expression, delayed $HLA\text{-}DR^+$ **and act.** T-lymphocyte activation (except for $HLA\text{-}DR^+$ in 2002) parameters are higher for the liquidators from NW Region compared with the group from Moscow region. In 2001, the liquidators from both regions showed equal average values of $HLA\text{-}DR^+$ and act. T-lymphocyte parameters and they always possessed disimmunoglobulinemia.

The comparison performed for the earlier monitoring periods indicated the contrary to the group from NW Region, for the group from The Moscow region no spontaneous variations in NK-cell parameters were observed. Since 1999, the same situation was observed for CTL, which prior to this period showed clear tendency to decrease (Figures 2 and 3).

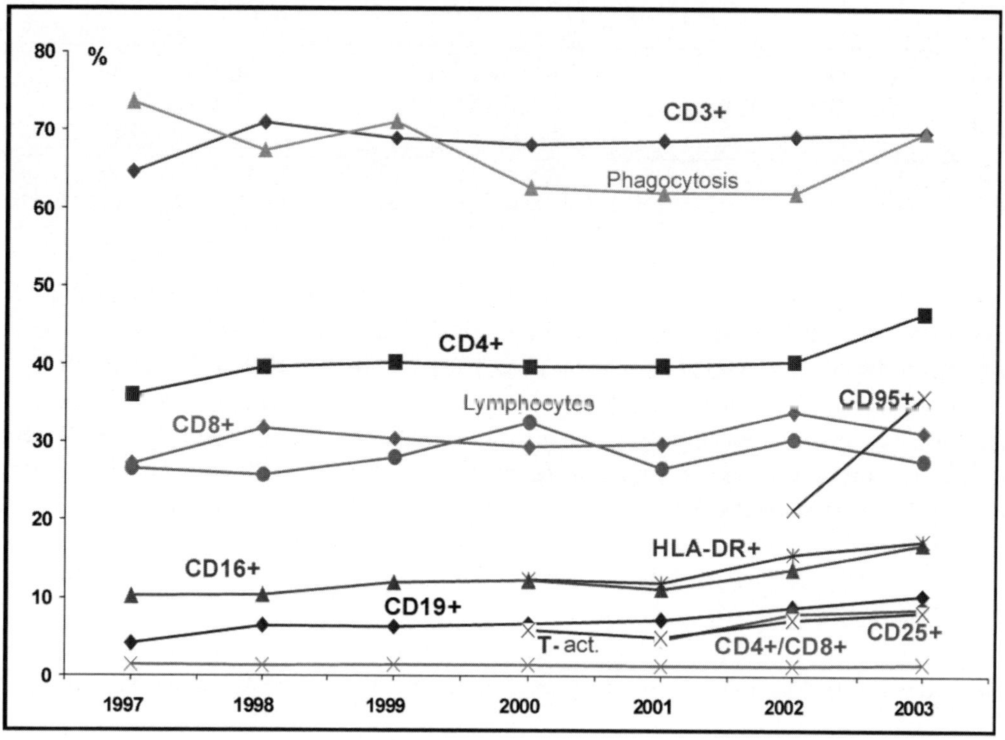

Figure 3. Dynamics of the immune status parameters for the liquidators form The Moscow region according to monitoring data obtained in the period of 1997 – 2003

For the liquidators from Moscow region, CTL – $CD8^+$ parameters demonstrated wave-like variations having maxima 1998 and 2002 and decreased to normal average values in 2000 – 2001 and 2003. In the period of 1997 – 2001, NK-cell parameters for the liquidators from The Moscow region were decreased (Figure 3), whereas in the period of 2001 – 2003 they uniformly tended to increase, anyway, being reliably lower compared with the liquidators from NW Region.

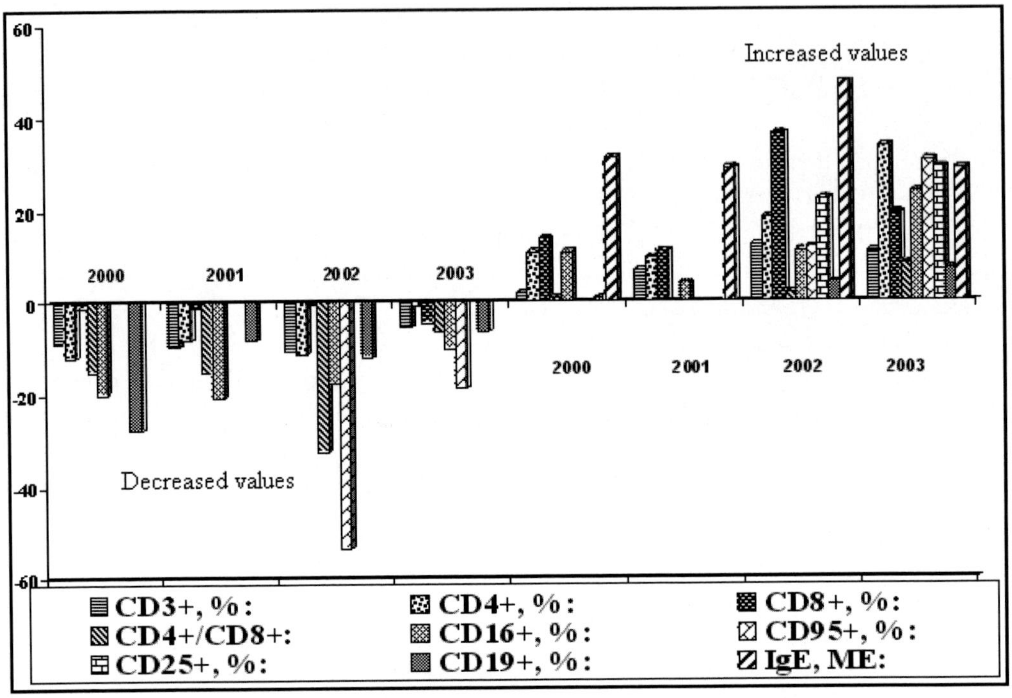

Figure 4. Percentage of patients with changing parameters of the immune status among examined groups of liquidators from The Moscow region according to monitoring data obtained in 2000 – 2003

At the background of natural killer low level, compared with other groups of liquidators, the patients from The Moscow region showed considerably higher frequency of IMIN clinical symptoms during the whole period of observation. Another difference observed in later periods concerned the change of the activation markers: low early activation markers for the liquidators from NW Region and considerably higher average values of $CD25^+$ for the liquidators from The Moscow region at the background of elevated parameters for more than 1/4 part of the examined patients. The difference in the character and deviation frequency is also observed for other markers: CD95, HLA-DR, act. T-lymphocytes with higher increase of $CD95^+$ and $HLA\text{-}DR^+$ for the NW Region liquidators and act. T-lymphocytes for the Moscow region liquidators (Figure 4).

Dynamics of the quantitative distribution of the groups of patients with changed parameters of the immune status among liquidators from The Moscow region indicated increasing group of examined patients with elevated T-helpers, NK-cells, $CD25^+$ and $CD95^+$ expression, and decreasing abundance of patients showing reduced $CD95^+$ expression and increased CTL. Also, differently directed deviations are observed for $CD95^+$ (2002 (↓), 2003 (↓↑)) and $CD4^+/CD8^+$ (2002 (↓), 2003 (↓↑)) parameters possessing lower deviation frequency (Figure 4).

Therefore, the main dynamic directions in the period of 2002 – 2003 are the following:

Reliable decrease (↓)	Reliable elevation (↑)
Lymphocytes (%, abs.); CTL – $CD8^+$; Serum IgG.	T-helpers (%); NK-cells; Immunoregulation index; Activation antigen HLA-DR^+; Act. T-lymphocytes; $CD95^+$ expression; NPA; Total IgE.

The results of monitoring performed in 2002 – 2003 indicated a considerable imbalance in the humoral and cell immunity stat for the liquidators from Moscow region. The comparison of the immune status phenotype for the liquidators from NW and Moscow regions, monitored during these three years, demonstrated its deficiency by many parameters of the cell immunity for the first group of patients: lymphocytes, T-lymphocytes, T-helpers, and immunoregulation index, whereas parameters of cytotoxic cells ($CD8^+$, $CD16^+$ and NPA) were elevated (Figure 1). The humoral component of the immune status demonstrated B-lymphocyte insufficiency, moderately decreasing serum IgA and elevating IgM and IgG. The cell activity parameters possessed low values of early and delayed activation markers ($CD25^+$, HLA-DR^+) and activated T-lymphocytes. In 2002, dynamics of the immune status indicated increasing parameters of the T-cell component: $CD3^+$, $CD4^+$ with dissociation of some (N) and absolute (↓) values, CTL decrease, $CD4^+/CD8^+$ elevation, NK-cell decrease but not approaching the control level, and NPA. Dynamics of the humoral component indicated development of much higher B-lymphocyte insufficiency and differently directed changes of serum IgG (↓ to N) and IgA (↑). The change of activation markers indicated low $CD95^+$ with a tendency to increase for HLA-DR^+ and $CD25^+$ expression. The main characteristic of the immune status dynamics according to 2003 is much more higher, compared with 2001, elevation of cytotoxic cell and activation marker indices: $CD95^+$, $CD25^+$, HLA-DR^+ and equality to the control of the cell component parameters reduced in 2001. Monitoring of the humoral component indicated remained B-lymphocyte insufficiency and disimmuno-globulinemia with reduction of IgM to normal level, IgG insufficiency and elevated IgA level.

The phenotype of the immune status for liquidators from Moscow region, monitored in 2001, was also characterized by insufficient cell immunity with dissociation of relative and absolute T-cell component and B-lymphocyte values, insufficiency of immunoregulation and natural killers, and moderate elevation of serum IgM, IgG and IgA, and the early lymphocyte activation marker $CD25^+$ (Figure 5). Dynamics of the immune status in 2002 for the liquidators from The Moscow region is characterized by lower insufficiency of the cell immunity with preserved dissociation of relative and absolute T-cell component values, uniform elevation of the humoral component by all monitored parameters: B-lymphocytes and the levels of serum immunoglobulins of all three classes; by data of cell activity monitoring – $CD25^+$ and HLA-DR^+ expression is elevated and $CD95^+$ expression is decreased. In 2003, the cell component line for the liquidators from the Moscow region generally simulates the phenotype line of 2001 possessing higher dissociation of relative and absolute T-cell component values ($CD3^+$, $CD4^+$, $CD8^+$ and $CD16^+$) and insufficiency by absolute values , increasing immunoregulation index, NPA, NK-cells, and elevation of cellular activity markers and B-lymphocytes, also accompanied by dissociation of the values: elevated level of relative values, normal level of absolute values, and disimmunoglobulinemia with decreasing IgG and elevating IgM and IgA. The specific feature of dynamics monitored in 2003 is the inverse character of changes in cytotoxic cells in relation to parameters, obtained in 2002: $CD8^+$ (↑) – 2002 and (N→↓) – 2003; $CD16^+$ (↓) – 2002 and (↑-%,↓-abs.) – 2003.

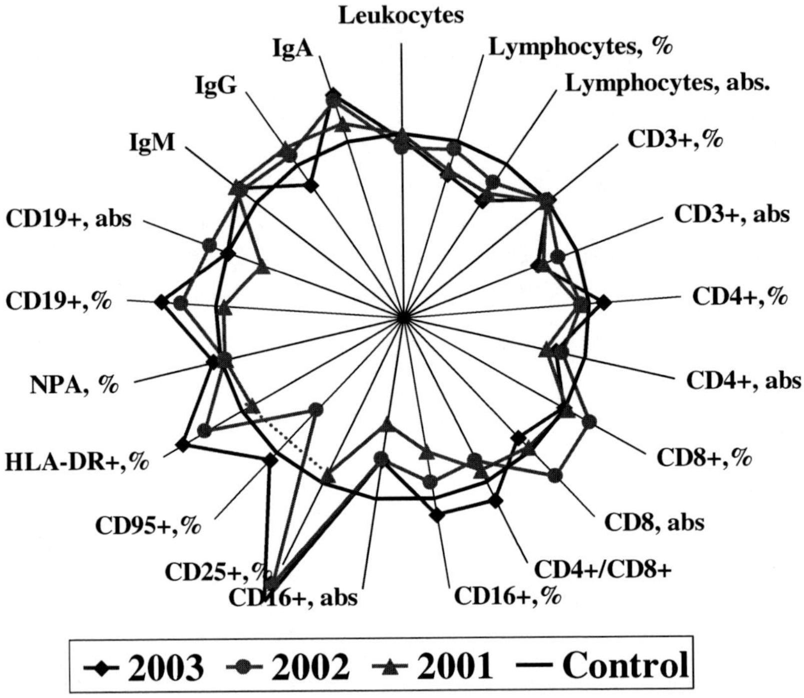

Figure 5. The immune status of the liquidators from the Moscow region

Analysis of the Immune Status Parameter Dynamics for the Liquidators from Krasnoyarsky krai According to Monitoring Data for the Period of 2001 – 2003

According to monitoring performed in 2001, the phenotype of the immune status for the liquidators from Krasnoyarsky krai is moderately activated by the majority of the cell immunity parameters: lymphocytes, T-helpers, CTL, natural killers with higher elevated cytotoxic cells $CD8^+$ and $CD16^+$ (Figure 6). The humoral component of the immune status had a moderate insufficiency of B-lymphocytes (%) at normal absolute values, elevated serum IgG tending to increase serum IgM and IgA. The level of $CD95^+$ and $CD25^+$ expression corresponded to control values; HLA-DR^+ and act. T-lymphocyte parameters were elevated. Similar to other groups of liquidators, the immune status of patients from Krasnoyarsky krai demonstrated accelerated elevation of total IgE level. The immune status phenotype of the liquidators from Krasnoyarsky krai, examined in 2002 and 2003, demonstrated similar values of the cell immunity indices for both years and differed from the data of 2001 by decreasing leukocytes, dissociation in parameters of lymphocytes, T-lymphocytes, T-helpers, NK-cells, NPA, B-lymphocyte insufficiency, similar $CD25^+$ and $HLA\text{-}DR^+$ values, and elevated $CD95^+$ expression in 2003. The current differences were determined by the level of serum immunoglobulins: 2002 – IgG (↑↑) and IgA (↑); 2003– IgG (N) and IgA (↑↑). The level of IgM was elevated in 2002 and 2003, which was higher in 2003.

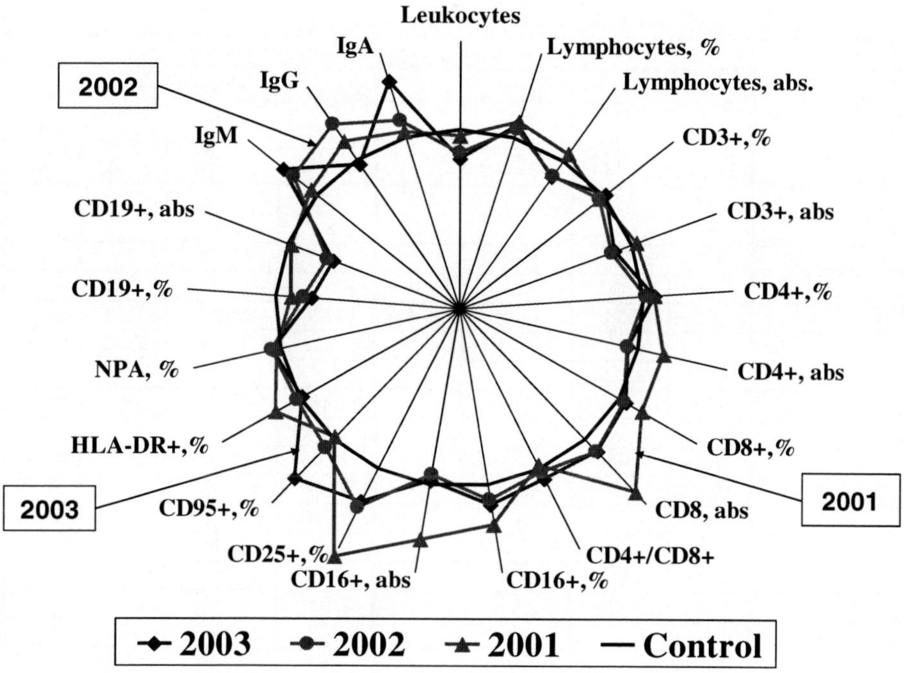

Figure 6. The immune status of the liquidators from Krasnoyarsky krai

As compared with dynamics of the immune status of the liquidators from NW and Moscow regions, the liquidators from Krasnoyarsky krai have not shown so considerable change in the activation markers, which was significant in the immune status of the liquidators from NW Region.

According to four-time examination, dynamics of the quantitative distribution of patients with changes in the immune status parameters in examined groups of liquidators from Krasnoyarsky krai indicated higher frequency of B-cell IMIN, especially in autumn 2002, elevation of $CD95^+$ expression, observed during two latest examinations, variations in NK-cell parameter deviation, mostly tented to elevation of parameters, especially in 2001 and 2003, high frequency of elevated CTL in 2001, at the primary examination in this remote period after the accident and expected tendency to decrease the parameter (Figure 6, Table 10).

More than 1/3 of patients examined in 2003 had elevated T-helpers; examination in autumn 2002 indicated high level of total IgE for 40% of the liquidators. Also, the IgE level was elevated in other periods, which was observed for 1/4–1/3 of monitored liquidators. About 1/4–1/3 of monitored liquidators from Krasnoyarsky krai also showed elevated activated T-lymphocyte parameters. Thus, among the cell activity parameters $CD95^+$ and activated T-lymphocyte elevation are most frequently deviating ones. The comparison of the immune status phenotype of the liquidators from three monitoring regions in the period of 2001 – 2003 indicates a single general effect – the elevation of total IgE both in the presence and in the absence of allergic diseases.

Table 10. Percentage of patients with changed (↓↑) parameters of the immune status in the liquidators from Krasnoyarsky krai, obtained in 2001 – 2003

Immune status parameters	Percentage of patients showing parameters:							
	decreased (↓)				elevated (↑)			
	2001, $N = 231$	**2002,** spring, $n = 201$	**2002,** autumn $n = 80$	**2003** $n = 77$	**2001** $N = 231$	**2002,** spring, $n = 201$	**2002,** autumn $n = 80$	**2003** $n = 77$
Leukocytes	2.28	5.97	7.5	10.39	3.42	3.48	3.75	1.3
Lymphocytes, %	3.04	3.48	1.25	0	10.27	6.47	3.75	11.69
Lymphocytes, abs.	6.46	9.45	13.75	18.18	8.75	3.48	2.5	2.6
$CD3^+$, %	14.45	13.85	20	9.09	9.89	7.69	6.25	12.99
$CD4^+$, %	8.37	8.21	13.75	11.69	**27.0**	**29.74**	**23.75**	**32.47**
$CD8^+$, %	1.9	6.67	8.75	6.49	**34.22**	**20.51**	**21.25**	**19.48**
$CD4^+/CD8^+$	**22.05**	**17.95**	**17.5**	**12.99**	3.8	8.21	7.5	5.19
$CD16^+$, %	9.51	11.79	8.75	9.09	**32.7**	**27.18**	**27.5**	**31.17**
$CD25^+$	–	–	–	–	7.22	2.49	0	1.3
$CD95^+$	**31.94**	**28.86**	2.5	5.19	**25.48**	**37.81**	**53.75**	**50.65**
HLA-DR^+, %	**10.27**	**18.41**	**18.75**	**19.48**	**17.11**	**9.45**	**10.0**	**11.69**
T-act., %	–	–	–	–	**29.28**	**17.91**	**21.25**	**20.78**
Phagocytosis, %	0.76	0	0	0	0.38	0	1.25	1.33
$CD19^+$, %	**33.84**	**28.87**	**60.0**	**50.65**	2.28	0.52	0	1.3
IgM	0.38	0	0	0	0	1.99	13.75	9.09
IgG	0.76	0	0	0	7.22	15.42	1.25	2.6
IgA	0.38	0	0	0	4.94	**12.94**	**18.75**	**25.97**
IgE, ME	–	–	–	–	**23.57**	**33.83**	**47.5**	**24.0**

Dynamics of the Immune Status Parameters for the Liquidators from the Moscow Region in Case of Proliferative Diseases According to Monitoring Data Obtained in 2001 – 2003

Analysis of the immune status dynamics in the presence of IMIN clinical symptoms, the main classes of the most high-frequency ChD, some nosological ChD forms, and polymorbidity syndrome among the liquidators from the Moscow region indicated similar tendencies, observed from the monitoring data on the liquidators from other regions:

- the data obtained that in case of organic CSD accompanied by atherosclerosis greater number of directions in the changes of immunological parameters (compared with VD and NCD) were confirmed numerously. This testifies about higher imbalance in the immune system operation with CSD, related to atherosclerosis development;
- in the case of gastric diseases and deforming osteochondrosis frequently occurring in the remote period, dynamics of the immune status coincides with the main directions of variations in the immunological parameters in case of chronic pathology, polymorbidity syndrome and CSD;

- in the case of hepatosis or adipose degeneration of liver in the liquidators (2003) demonstrated high T-helper parameters, which exceeded the upper border of the physiological variation range, and high values of the immunoregulation index, no increase of NK-cell parameters typical of the majority of ChD nosological forms, relatively decreased level of serum IgG (as in the presence of IMIN clinical symptoms of the infectious-allergic syndrome, thyroid gland enlargement, VD and NCD), as well as high level of total IgE, including data obtained in 2003, at the general tendency to decrease of the parameter compared with prior circumstances;
- quantitative distribution of changed values in the presence of arterial hypertension but without developed atherosclerosis differed from that, observed for the case of CSD accompanied by atherosclerosis. For the case of arterial hypertension without atherosclerosis, insufficient natural killers were observed that was not detected in other CSD groups, and the highest occurrence of depressed $CD95^+$ expression, elevated T-lymphocytes, T-helpers and activation markers, except for $CD95^+$, and elevated total IgE. For the cases of HD with atherosclerosis and HD without it, deviation algorithms for the immunological parameters may be both similar and different ($CD3^+$, $CD16^+$, $CD25^+$, IgE) at high frequency of deviations according to similar algorithms. These data were obtained for the Moscow region exclusively;
- for the liquidators from the Moscow region, dynamic tendencies in the immune status observed are similar to these for the liquidators from NW Region: at low variability of immunological deviations observed in cases of some nosological forms of diseases the monitoring detects changes of the immunological parameters independent of the pathology character.

For the sake of comparing the immunological characteristic of the proliferative syndrome deduced from long-term observations of the liquidators from NW Region, let us present data on the immune status dynamics in benign proliferative states among the liquidators from the Moscow region. In the presence of benign tumors, dynamics of the immune status (Table 11) also in many ways reproduces basic regularities of changes in the immunological parameters at chromic pathologies and radically differs from that for the liquidators from NW Region.

The main differences are the following:

- in contrast with the liquidators from NW region, no lymphocytosis is observed;
- no decreases of T-cell component parameters of the immune status is observed: $CD3^+$, $CD4^+$; T-lymphocyte parameters equal control values, and T-helpers show a tendency to increase;
- no disturbance of T-cell subpopulation distribution is observed, the immunoregulation index corresponds to the norm;
- considerable differences in NK-cell parameters with expressed exceeding over control values for the liquidators from NW Region are observed;
- for the liquidators from the Moscow region NPA parameters equal control values, whereas for the liquidators from NW Region in 2003 they are decreased;

Table 11. Dynamics of the immune status parameters for the liquidators from the Moscow region and NW Region in the presence of benign tumors (2001 – 2003)

No.	Immune status parameters	Year of examination								
		2001			2002			2003		
		NW Region	Moscow region		NW Region	Moscow region		NW Region	Moscow region	
1	Leukocytes				*⇓	**⇓			*⇓	•↓
2	Lymphocytes, %		* ⇓	↑				** ⇑	*⇓	•••↓
3	Lymphocytes, abs.					***⇓		* ⇑#↑	***⇓	•••↓
4	$CD3^+$, %									↑
5	$CD3^+$, abs.	*⇓	* ⇓			**⇓			** ⇓	•↓
6	$CD4^+$, %			↑					#↑$^{+}$↑	↑
7	$CD4^+$, abs.		* ⇓			**⇓			*⇓	•↓
8	$CD8^+$, %	*⇑		•↓	$^{+}$↓	*⇑$^{+}$↑	•↑	* ⇑#↑		↓
9	$CD8^+$, abs.							** ⇑##↑		•••↓↓
10	$CD4^+/CD8^+$							* ⇓#↓	#↑	•↑
11	$CD16^+$, %			•••↓	** ⇑		↓	* ⇑	$^{+}$↑	↓↓
12	$CD16^+$, abs.		*** ⇓	•••↓		*⇓	•••↓	* ⇑#↑	**⇓	••↓↓
13	$CD25^+$, %				$^{++}$↓	$^{+++}$↑	•••↑		$^{+++}$↑	•••↑
14	$CD95^+$, %	–	–	–	⇓	⇓⇓		###↑	###↑	•↑
15	HLA-DR^+, %					$^{+++}$↑		##↑	$^{+++}$↑	
16	T-act., %							##↑	⇑	•••↓
17	NPA, %	***⇑		↑	$^{++}$↓			* ⇓		•↑
18	$CD19^+$, %		*⇓		*⇓	$^{+}$↑	•↑		***⇑$^{+++}$↑	•••↑
19	$CD19^+$, abs.		** ⇓							
20	IgM, g/l									
21	IgG, g/l	**⇑			$^{+}$↓		↑	$^{+}$↓	**⇓##↓$^{+}$↓	
22	IgA, g/l									
23	IgE, ME						↑			

Reliability of differences:

* ⇓⇑) – a reliable difference of a parameter from corresponding control value;

$^{+}$) – immune status parameters obtained in 2002 from these obtained in 2001: $^{+}$ – $p < 0.05$; $^{++}$ – $p < 0.01$; $^{+++}$ – $p < 0.001$;

#) – immune status parameters obtained in 2003 from these obtained in 2002: # – $p < 0.05$; ## – $p < 0.01$; ### – $p < 0.001$;

•) – a parameter for the liquidators from the Moscow region from that for the liquidators from NW Region: • – < 0.05; •• – $p < 0.01$; ••• – $p < 0.001$;

↑↓ – tendencies in the variation of the immune status parameters.

CTL variation periods are different: for the liquidators from the Moscow region it is increased in 2002, whereas for the liquidators from NW region in 2001 and 2003, and vice versa, the reduction periods for the liquidators from Moscow were observed in 2001 and 2003, and for the liquidators from NW Region – in 2002;

For the liquidators from the Moscow region, parameters of the early activation marker $CD25^+$ tend to increase; for the liquidators from NW Region, the average values of all activation markers considerably exceeded these values for the liquidators from the Moscow region.

Thus, the liquidators from the Moscow region have no the totality of changes in the immunological parameters observed for the liquidators from NW region, typical of benign proliferative states. Generally, dynamics of changes in the immunological parameters for the liquidators from the Moscow region coincide with dynamics for other pathological states.

CONCLUSION

For the liquidators from NW Region examined in the period of 1995 – 2003, dynamics of the immunological status parameters indicated annually detectable imbalance in the conditions of the main components of the immune system that strictly proves highly "corrupted" state of the immunogram. The following cytotoxic cells demonstrate the most frequent deviations: CD16$^+$, CD8$^+$, T-helpers, B-lymphocytes, total IgE, and in 2002-2003 – CD95$^+$. In 2002-2003, considerable elevation of the activation markers and the frequency of patients with increased values are also observed.

In the remote period after the Chernobyl accident, all three groups of liquidators monitored demonstrated phenotypes of the immune status with changes typical of each region. At higher frequencies of IMIN clinical symptoms and distribution of chronic somatic diseases among the liquidators from Moscow and Moscow Oblast, the most considerable changes in parameters are detected for the liquidators from NW Region. For all three groups, quantitative and immunoregulation imbalance with increasing parameters of the cellular activity is detected: CD95$^+$, HLA-DR$^+$, T-activated lymphocytes, variations in cytotoxic cells most considerable for the liquidators from NW Region. The liquidators from Krasnoyarsky Krai manifested not so considerable change in the activation markers. Disimmunoglobulinemia with differently directed variations of serum Ig with B-lymphocyte insufficiency is strictly determined.

It is found that in the presence of benign proliferative states among the liquidators from NW Region, two directions in the immune status variation are formed. The totality of changes undetectable at other IMIN and ChD clinical symptoms are the following:

- lymphocytosis;
- decreasing parameters of T-cell component of the immune status, which are CD3$^+$ and CD4$^+$;
- disordered distribution of T-cell subpopulations;
- immunoregulation insufficiency;
- NK-cell elevation;
- NPA decrease and the presence of immunological variations, which were also observed in other pathological states: circular CTL and immunoregulatory index variations;
- increasing lymphocyte activation markers: elevating CD95$^+$, HLA-DR$^+$ and activated T-lymphocyte expression;
- low marker of CD25$^+$ early activation.

Basing on long-term monitoring of the Chernobyl accident liquidators, the immunological characteristic of the proliferative syndrome was detected. It was expressed in the presence of lymphocytosis during 2 – 3 years, maximum high T-lymphocyte values and their insufficiency, immunoregulation insufficiency $CD4^{+}/CD8^{+}$ (↓↓), the presence of spontaneous variations (↓↑↓↑) of cytotoxic cells $CD8^{+}$, $CD16^{+}$ with insufficiency or considerable elevation of NK-cell parameters within 2 – 3 year period, and negative dynamics towards elevation. A considerable intensifying elevation of the activation markers is observed: expression of $CD95^{+}$ from (↓↓) to (↑↑↑), $HLA\text{-}DR^{+}$ delayed activation marker (↑), and activated T-lymphocytes, negative NPA dynamics (↓), and deficiency of B-lymphocytes.

At a definite variability of immunological deviations observed for the main classes and nosological forms of the most frequent chronic pathology, the monitoring performed among liquidators determined similarity in dynamics of the immune status variation independent of the character of particularly analyzed somatic pathology, which is explained by polyorganic damage and the presence of polymorbidity syndrome.

ACKNOWLEDGMENT

To I. V. Oradovskaya "The health state of the Chernobyl accident liquidators: immunologist's opinion 20 years after"

REFERENCES

[1] Petrov R.V., Khaitov R.M., Pinegin B.V. *et al.*, *Immunologia*, 1992, No. 6, pp. 51 – 62. (Rus)

[2] Petrov R.V., Khaitov R.M., Oradovskaya I.V., and Pinegin B.V., *Immunologia*, 1992, No. 4, pp. 43 – 53. (Rus)

[3] Pinegin B.V., Eremina O.F., and Tamarchenko I.S., "The study of the immune status of flying school cadets using a complex of standard and unified level I tests", In Coll.: *Methodology of Organization and Results of Extended Immunological Examinations*, Thes. All-Union Conference, Moscow – Angarsk, 1987, pp. 243 – 249. (Rus)

[4] Simonova A.V., Liozner A.L., Sidorenko I.V. *et al.*, *Immunologia*, 1989, No. 6, pp. 72 – 75. (Rus)

[5] Urakova E.B., Zamaldinova O.A., Simonova A.V *et al.*, *Immunologia*, 1989, No. 1, pp. 61 – 63. (Rus)

In: 20 Years after the Chernobyl Accident
Editors: E. Burlakova and V. Naidich, pp. 171-180
ISBN 1-60021-249-2

Chapter 10

To the 20th Anniversary of the Chernobyl Accident. Medical Consequences in Armenia

N. M. Hovhannisyan, Sh. M. Petrosyan, M. I. Mirijanyan, K. V. Asryan, A. S. Poghosyan, A. N. Hovhannesyan, A. K. Abrahamyan, G. R. Harutyunyan, A. G. Karapetyan, E. G. Gevorkyan, Zh. H. Petrosyan and N. K. Harutyunyan

Sector of Radiation Medicine and Burns[1], Centre of Traumatology, Orthopedics, Burns and Radiology, Ministry of Health, Republic of Armenia

The Chernobyl Accident emitted a large quantity of radioactive substances to the atmosphere, induced radiation impact to a great number of people and radionuclide contamination of wide territories in Ukraine and Russia, and it also touched Armenia. To say nothing of radioecological problems, which were also significant, Armenia obtained a cohort of people (about 3,000 persons) who participated in liquidation of the accident consequences and then formed a large "risk group" among the population (hereinafter: Liquidators).

The aim of this investigation is to analyze health parameters and the structure of morbidity in Armenian citizens who had immediate participation in liquidation of the Chernobyl NPP accident consequences based on the results of standard medical and hospital examinations, to reveal a probable dependence between the morbidity rate for the most widespread classes of diseases and the radiation dose perceived, as well as to determine the role of other factors influencing health parameters. The interest is conditioned not only by accident factors (both radiation and non-radiation), which are general for all countries, but by the factors specific for Armenians as well (geographical location, nationality, close-together inhabitancy on a small territory, socioeconomic conditions, psychoneurological factors, etc.).

[1] WHO Collaborating Center

Materials and Methods

For revealing epidemiological features of health status in Chernobyl accident liquidators, the morbidity rate, prevalence of diseases, invalidism and mortality rate were studied. At the Sector of Radiation Medicine and Burns (SRM&B) 2,210 liquidators are registered for follow-up and dispensary monitoring. Basing on the specificities of labor in different periods of the accident consequences liquidation, all liquidators under examination were subdivided into three groups with respect to the year of work in the accident zone: 51.2% in 1986, 33.8% in 1987 and 14.8% in 1988 – 1989.

The cohort under investigation was subdivided into four groups with respect to external radiation dose (ERD) perceived: group I – liquidators with the ERD below 10 cGy (32.9%); group II – from 11 to 20 cGy (13.7%); group III – above 20 cGy (5.6%); group IV – people who could not confirm perceived ERD (48.0%).

At present, over a half of the cohort examined is older than 51 – 51.8%, while 26.0% are at the age of 41 – 50 and 22.2% are 31 – 40 in age.

The liquidators were examined in accordance with a three-stage system, developed by the authors. It includes ambulatory/dispensary, hospital and rehabilitation stages. The ambulatory monitoring includes annual medical examination and, if required, hospitalization of liquidators for full clinical and laboratory examination, including ECG with physical ergometric tests, ultrasound examination (US), electroencephalography (EEG), endoscopic examinations, Doppler analysis of hemodynamics in peripheral and cerebral blood vessels, biochemical and immunological statuses, cytogenetic examinations, thyroid hormones study, etc. If required, this complex of examinations and procedures can be supplemented by radionuclide scintigraphy, computer tomography (CT) and other complex modern methods.

Epidemiological investigations are based on the analysis of long-term monitoring of the health status and annual enhanced examinations of liquidators, covered by the Republican Register system, in which personal information including medical and dosimetric data of each liquidator is stored. The databank includes information about ERD, work place or site, activity duration and type, as well as morbidity rate.

Data analysed were compared with similar parameters of the control group of 91 persons, who were examined at the SRM&B due to assignment for work at Chernobyl, but took no part in emergency response. They were also subject to dynamic observation.

Results and Discussion

Analysis of the health parameters of liquidators, recorded to the Register, is an extremely complicated task. Firstly, currently monitored cohort has not been examined before the accident. Secondly, changes in socioeconomic conditions proceeding in the country had a considerable effect on the formation of the health parameters. The basic reliability criterion for changes in the health parameters of victims is their interactive assessment during post-accidental years with respect to the radiation rate. While analysing data on health status in liquidators, the regularities observed in formation of some parameters are assumed statistical, considering that the results obtained might form the basis for deeper cause-effect dependencies.

As found in the investigation, health parameters of liquidators deteriorated during the entire period of monitoring, which generally coincides with data of scientific publications[1, 3, 4, 11, 14, 15, etc.]. The comparative analysis allowed to reveal that, on the one hand, there was registered a considerable increase of pathologies, not typical for radiation damages, but reflecting morphofunctional disorders of polyetiological origin in protective and regulatory systems of the human organism. On the other hand, an increase of a number of diseases, which appeared to be radiation-induced, was also detected.

As indicated by analysis of the morbidity rate among liquidators, the following classes of diseases have tendency to increase: diseases of nervous and cardiovascular systems, gastrointestinal tract, and respiratory organs (Table 1). At present, in accordance with dispensary monitoring data pathologies of cardiovascular system (CVS) are also at the leading position: 82.0% compared with 13.7% in 1987. Other classes are ranked as follows[2]: the nervous system – 68.0% (32.0%); the respiratory organs – 76.0% (15.0%); the digestive tract – 55.0% (16.3%); the urinogenital system – 21.0% (2.5%); the musculoskeletal system – 49% (21.9%).

For proper assessment of the accident consequences of high significance is mortality rate in liquidators. This parameter increased gradually and by now reached 7.8% (172 persons for the whole period after the accident). The reasons for mortality among liquidators can be ranked as follows: diseases of CVS – 27.9%, neoplasms –17.5%, traumas (car accidents), digestive tract diseases; urinogenital system diseases (chronic renal insufficiency); respiratory tract diseases; killed at the war in Nagorny Karabakh – 8 persons.

Table 1. Dynamics of morbidity rate among liquidators according to long-term monitoring (%)

Morbidity rate (by organs and systems)	Year of monitoring				
	1987	1990	1995	2000	2005
Nervous system	32.0	60.4	78.0	84.6	68.0
Cardiovascular system	13.7	10.8	48.1	81.8	82.0
Respiratory organs	15.0	26.1	39.0	57.0	76.0
Digestive apparatus	16.3	30.8	27.4	59.4	55.0
Musculoskeletal system	21.9	43.8	30.1	23.2	49.0
Urinogenital system	2.5	4.1	6.3	10.1	21.0
Neoplasms	0.1	0.2	1.3	2.5	3.0
Total number	560	413	361	560	242

Thus, the mortality reasons are not clearly related to activity on liquidation of the Chernobyl accident consequences and all cases fall within the range of common statistics for mentioned pathologies.

For the period after the Chernobyl accident, epidemiological interactive studies of malignant neoplasm development in liquidators revealed no considerable increase of cancer disease cases. In contrast with the previous apprehensions, no increase in hemoblastosis morbidity rate was observed. To-date, malignant neoplasms were detected in 1.36% of liquidators (30 persons). Among these cases, 9 persons had gastric carcinoma.

[2] Values in brackets give the abundance of this class of diseases for liquidators in 1987.

In initial years after the accident, the diseases of nervous system were the most widespread (32.0% in 1987). At present, their rate is 68.0%. During several initial years, an elevation of in-between neuropsychic disorders and vegetative dystonia rates was observed. These dysfunctions are precursors to cerebrovascular insufficiency (Table 1). It is known that vasoneurotic syndrome is a consequence of the regulatory function disturbances, induced by direct damaging of a vessel wall and cerebral neurons. Further on, in addition to intracerebral vessels, the process involves great vessels of the cerebrum, specifically of vertebrobasilar basin. Obviously, this is promoted by early development of cervical spine osteochondrosis. The whole complex of disorders brings forth development of more expressed encephalopathy. Thus, vicious circle appears, which then defines the progress in cerebrovascular process [1, 5, 11 – 20].

In 1987, organic pathology represented by cerebral blood circulation disturbances and encephalopathy equalled 14% (Figure 1). In the following years, its part was gradually elevated to 53.2% in 2005; parallel decrease of the functional changes from 86% in 1987 to 46.8% in 2005 was observed. It should be noted that, generally, the number of organic diseases is elevated due to patients with disturbances of cerebral blood circulation and discirculatory encephalopathies of vascular genesis. Hence, generally, functional diseases are presented by vasoneurotic and asthenoneurotic syndromes, and various types of neurasthenia.

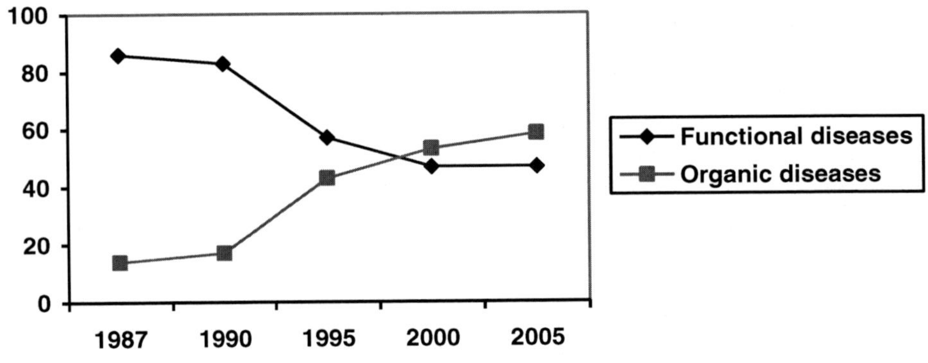

Figure 1. Development of functional and organic diseases of the nervous system in liquidators (%)

In the majority of patients with organic pathology (92%), asthenic and depressive syndromes were observed with manifestation of such symptoms as weakness, somnolence, unstable mood, mental capacity decrease, memory dysfunctions, and sexual activity decrease.

It is significant that even small doses of alcohol caused highly expressed alcoholic intoxication in liquidators. EEG parameters of liquidators indicate changes shaped as desynchronizing of cerebral cortex bioactivity and disturbances of its interaction with subcortical structures. In the majority of cases, the use of Doppler method allowed to detect decreasing cerebral blood flow in their vertebrobasilar basin accompanied by cervical spine osteochondrosis and gradual elevation of atherosclerotic effects in cerebral vessels.

Thus, in the initial years after the accident central nervous system (CNS) disorders were functional for the liquidators. However, during several years these changes were intensified by chronic stress, alcoholic and nicotinic intoxications and, with age, by atherosclerosis; all these symptoms gradually caused organic damages of CNS – encephalopathy, i.e. functional CNS disorders are converted to organic pathology. It should be noted that the highest expression of these changes was observed in liquidators of the accident who worked at Chernobyl NPP in 1986 (Figure 2).

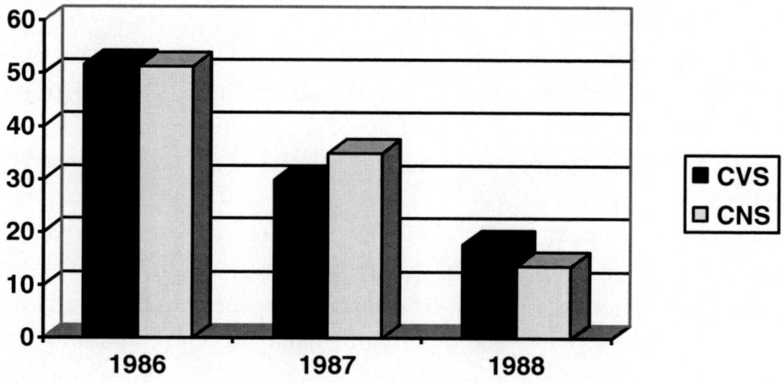

Figure 2. The dependence of morbidity rate for nervous and cardiovascular systems on the year of presence in the Chernobyl accident zone (%)

For the purpose of studying the features of cerebrum functional state (CFS) in liquidators, our Centre has created a computerized system for complex CFS assessment. This system is based on original methods of psychological and psychophysiological testing (levels of anxiety, asthenia, hypochondria, psychic depression, visual-motor reactions, productivity of a short-term verbal ability of memory, and voluntary attention processes) and mathematical analysis of cardiac rhythm variability (CRV), aimed at detection of regulatory abilities of CNS vegetative section (statistic, histogram, autocorrelation and spectral indices of CRV). The results of these tests indicated some progressing negative shifts in psychological status of liquidators of the Chernobyl accident consequences: reactivity of neurodynamics, CNS excitation tonicity, concentration and fatigability of attention, short-term memory productivity, etc. These parameters are much lower in liquidators than in control group. Moreover, liquidators demonstrate a definite stress in the operation of vegetative nervous system and increased involvement of subcortical nervous centers, which is untypical of the normal state (data obtained by the method of CRV mathematical analysis).

The highest rate in morbidity is shown by CVS diseases, which were studied using a complex of clinical instrumental and laboratory methods. In the period of 1987 – 2005, a considerable dynamic increase of CVS diseases was observed for the liquidators (from 13.7 to 82.0%). In the initial years of investigation, they were determined as functional disorders then transformed to an organic pathology (ischemic heart disease, arterial hypertension). Compared with the whole population, in this group organic diseases of CVS were observed in relatively young persons (48.2% persons aged 31 – 50). The gross part of liquidators having CVS diseases were involved in operations in the accident zone in 1986 (51.8%, on average), 35.2% in 1987, and 13% in 1988. These data are reliable enough to confirm the dependence of CVS pathology on perceived dose of external radiation (Figure 2). The number of patients with electrocardiographic (ECG) changes increased with time. The studies in dynamics indicated deeper disorders of automatic functions, excitation, and conduction. Myocardial perfusion was found also dynamically decreased. The liquidators also possessed disorders of equilibrium between symptomatic and parasymptomatic sections of the vegetative nervous system. The interactive veloergometry (VEM) studies indicated transformation of normokinetic type of hemodynamics to hyperkinetic and suppression of contractile function of the heart, metabolism intensity, cardiac efficiency, and coronary reserve. Initially, the group of liquidators displayed disorders of lipid exchange (stable increase of total lipid levels, cholesterol, β-lipoproteins, background lipid peroxidation, coagulation–anticoagulation (hypercoagulation phenomena) and immunological systems (immune deficiency of cellular type, phagocytal function disturbance, immunoglobulin level increase), which became more intense with time.

Thus, the accident after-effects, in particular, ionizing radiation exposure accompanied by other factors induced development of gradually progressing cardiovascular diseases in liquidators. Primary functional shifts transformed to the organic pathology, which was reliably confirmed by ECG, VEM and other clinical and laboratory parameters.

According to data of long-term interactive monitoring, the morbidity structure in liquidators shows high significance of chronic non-specific diseases of lungs (CNDL) at the background of uncomplicated pulmonary anamnesis. Interactive analysis of morbidity rate in liquidators indicated a significant increase of CNDL frequency (from 15% in 1987 to 76% in 2005). The course of lung diseases in liquidators is progressive, accompanied by corresponding pathology of nervous and other systems and with time they are interrupted by gradually shorter remissions, and may hardly be treated. The study of the respiratory function (RF) helped in determining a reliable decrease of spirographic indices during all years of monitoring compared with the control group and progress in ventilation disturbances with predominance of mixed changes. The dependence of liquidators' morbidity, as well as ventilation, immunological and biochemical disorders on the year of participation in liquidation of the Chernobyl accident consequences was shown. This testifies to the important role of radiation in the pathology development. Correlations between different parameters were revealed. As the pathological process in lungs of liquidators with CNDL is intensified, interdependent immunological and biochemical shifts testifying both suppression of immunological reactivity and intensification of oxidative processes proceed, simultaneously.

Thus, systematic clinical and paraclinical investigations promote for corrections and rehabilitation measures at different stages of irradiation consequences. The results obtained stipulate selection of the optimal treatment methods (by antioxidants, in particular) for pathogenetic correction of deep and stable disorders in homeostasis, increase of adaptive and compensatory abilities of the human organism.

Besides the above-described changes, the performed clinical and laboratory analysis has detected increasing frequency of gastrointestinal tract (GIT) diseases in liquidators from 16.3% in 1987 to 55.0% in 2005. Taking into account analysis of etiological factors of GIT diseases in liquidators, the role of non-radiation factors should also be noted: imbalanced nutrition, stress, pernicious habits, etc. It should be mentioned that long-existing functional disorders may transfer into organic diseases. Increasing severity of the course of diseases, namely, hyperacid gastritis and ulcerous disease of stomach should be noted. These diseases can hardly be treated and possess severe manifestations and shorter remissions that allows for taking this group of patients for the risk group requiring special attention. Many years after the accident the liquidators demonstrate morbidity rate of chronic hepatitis and cirrhosis increasing from 2.8% in 1987 to 19.9% in 2005 due to some reasons, including intensification of medical activity, the use of ultrasound and radionuclide examination methods, and expansion of a set of functional diagnostic procedures. One more reason increasing morbidity rate is extended alcohol taking and tobacco smoking among the liquidators.

Clinical and laboratory studies, implemented in our Centre in the period of 1987 – 2005 indicated no significant changes in hemocytogenetic red bud in liquidators during initial 4 – 5 years after the accident. Compared with the control group before 1991, concentration of hemoglobin and erythrocytes was preserved within the range of the physiological norm. Obviously, starting from the 6th year of monitoring the reliable decrease of erythrocyte quantity should also reflect hemoglobin concentration decreasing it. In the following years of monitoring before 2000, the reliable decrease of the erythrocyte quantity was followed by a decrease of hemoglobin quantity. Since 2000 until now, concentrations of erythrocytes and hemoglobin approach the physiological norm, however, some delay is observed. It is our opinion that a significant corresponding decrease in erythrocytes and hemoglobin concentration in the period of 1991 – 2000 cannot be fully considered a result of remote consequences of low-dose irradiation of the organism. In this period, overlapping of various non-radiation factors is possible, among which socioeconomic conditions, expansion of the list of chronic diseases, aging of Armenian liquidators, etc. should be mentioned. The study of the morphological composition of leukocytes indicated an increase of abundance of band form neutrophils and basophils in all examined groups.

For the whole period of observation blood tolerance to heparin and prothrombin were found decreased. Fibrinogen concentration was wave-like changed. Recently, statistically reliable increase of fibrinogen concentration was observed in investigations. All these changes indicate intensification of hypercoagulation processes in blood samples of liquidators. For all investigation periods, stable changes in the lipid exchange and lipid peroxidation (LPO) were observed in all groups of examined liquidators, independently of the dose and pathology. A considerable increase of background blood plasma LPO, enzymatic and non-enzymatic LPO in erythrocyte membranes was observed.

In recent 20 years, the liquidators manifested disturbances of the immune system, both quantitative and functional (predominant deficiency of the immunity T-unit, elimination of suppressor mechanisms, increasing concentration of immunoglobulins of the main classes and moderate suppression of immunophagocyte and complimentary activities).

The problem of cytogenetic changes is still urgent. During recent 20 years, cytogenetic indices of blood lymphocytes manifested clear tendency to increase. In 1996, the percentage of chromosomal aberrations (ChA) more than 3 times exceeded the level of 1987. At the same time, we failed to detect dicentrics, specific for irradiation. No differences between indices I and II mitoses was found that indicated their elimination, if they have ever been present. A wave-like occurrence of polyploids (tetraploids) was determined, the highest level of which, 6.72 ± 0.43, was detected in 1998, compared with 1.46 ± 0.14 in 1991. Since 1994, multiaberrant cells have been also observed. The studies performed in the period of 2002 – 2005 indicate a considerable increase of ChA frequency in liquidators compared with the control: (12.08 ± 0.38) and (2.75 ± 0.45)%, respectively.

Of special scientific and practical interest are spermatogenesis changes in liquidators. The study of gonad function indicated functional changes in the initial years after the accident; in particular, the activity of spermatozoids and ejaculate volume were decreased. This testified to hormonal disorders manifested as decreasing androgenous saturation of the organism. However, in the following years more significant changes were revealed, manifested as teratospermia (increasing amount of teratological spermatozoid forms), which was observed, however, at the background of normal concentration of spermatozoids in ejaculate. It is apparent that results obtained may testify to the beginning of mutagenic effects of the accident.

Thus, 10 – 20 years after the radiation impact, the chromosome apparatus of peripheral blood lymphocytes in the liquidators is still suppressed and even deteriorates. Obviously, instability of genome is stipulated by both radiation and effects of non-radiation factors (environmental, socioeconomic, neuropsychological, etc.). Therefore, it is necessary to take into consideration changes of clastogenic factor (CF), which is commonly known [26 – 28] as a cellular product, formed due to activation of free-radical LPO processes. Even 10 – 15 years after the accident, 2 – 5 times elevation of blood plasma CF level was found in liquidators compared with the control.

The application of regressive curves allowed to particularize regularities determined in the studies on the nervous, cardiovascular and respiratory systems and gastrointestinal tract, and provided for forecasting variations in health parameters with acceptable accuracy in the coming years. The dispersion two-factor analysis helped in detection of a part of radiation and non-radiation factors in the total effect on increasing morbidity rate during almost 20 years and in dynamics by the following classes: NS, CVS, CNDL, and GIT. In relation to NS, CNDL and GIT classes, one may speak about both early and remote post-accidental effect of ionizing radiation on the morbidity rate. As to diseases of the cardiovascular system, the dose dependence is clearly observed in the early post-accidental period, whereas later on age factor becomes the predominant one. With the use of system analysis methods applied in this investigation, in the post-accidental period there was demonstrated the summation of both radiation and non-radiation effects/factors (such as the age and tobacco smoking). Mortality rate analysis demonstrated that either there is no dose dependence or, otherwise, it is masked by more significant factors affecting life duration.

Application of cluster method allowed determining the time of presence under the radiation impact (and, definitely, the external radiation dose perceived) for 83% of liquidators, who immediately after return from the accident zone were examined for a complex of peripheral blood parameters; this latter facilitated differentiation/ranking liquidators exposed to ionizing radiation in relatively narrow range of dose loads. The two-factor dispersion analysis allowed detecting clear dependence of immunological changes on the time of participation in liquidation of Chernobyl accident consequences that testifies to a significant role of radiation factor during the whole period of investigation.

Thus, apparently, steady increase of Chernobyl post-accidental morbidity is the integral result of polyfactor effect of the catastrophe and stable tendency to morbidity increase, among which radiogenic and psychogenic ones are most valuable and durable. Basing on dynamic monitoring, one may state that the accident factors, in particular, ionizing radiation accompanied by other factors (radiophobia, general stress situation, deterioration of social conditions, rearrangement of socio-political system, etc.), induce development of somatic diseases in liquidators. These diseases are in gradual progress and initial functional shifts are transformed into organic pathology. This completely coincides with the data of instrumental studies (ECG, VEM, spirography, etc.) and clinical and laboratory parameters – LPO, immunology and biochemistry. It should also be noted that these changes were more expressed in liquidators of 1986; this indicates their definite dose dependence.

The entire variety of environmental factors causes dose and time («chrono») dependant effect on the immune mechanisms and casually distorting their physiological proceeding forms an unfavourable background for development of various pathological processes in the organism. It is of special importance that the development of pathologies in liquidators is complicated by age changes, heavy social conditions, pernicious habits (tobacco smoking, alcohol drinking). The whole complex of mentioned factors intensifies the pathology and is accompanied by immunological insufficiency and blood biochemical disturbances.

The increasing level of lipid peroxidation, chromosomal aberrations, clastogenic factor, spermatogenesis disorders, etc. require additional monitoring and creation of new prophylactic and treatment approaches. In this context, further development of antioxidants problem seems the most promising. Our long-term investigations in this direction, carried out in cooperation with the scientists from France, Japan, Belarus, Russia, and Ukraine, commonly known antioxidants (α-tocopherol, ascorutin, aevit, etc.) and new ones («Tanacan», France; «AOB», Japan; «Loshtak», Armenia; «Lofegal», Belarus) possess highly expressed antioxidant and anticlastogenic properties. Therefore, the significance of their application for prophylaxis and treatment of radiation damages can hardly be overestimated.

REFERENCES

[1] Aleksandrovsky Yu.A. *et al.*, *Zh. Nevropatologii i Psikhiatrii im. Korsakova*, 1989, vol. 89(5), pp. 111 – 117. (Rus)

[2] Baevsky R.M., 'Scientific theoretical bases of the use of heart rhythm analysis for assessment of the stress state of regulatory systems of the organism', *Proc. Intern. Symp. "Computer Electrocardiography at the Edge of the Centuries"*, Moscow, 1999, Electronic edition, Website: www.ECG.ru. (Rus)

[3] Burlakova E.B., 'Urgent questions of radiation medicine', *Med. Radiologia*, 1996, vol. 23, pp. 72 – 77. (Rus)

[4] Voloshin P.V., *Ukr. Vestnik Psikhonevrologii*, 1996, No. 1, pp. 3 – 8. (Rus)

[5] Gus'kova A.K., *Proc. Conference "Urgent and Forecasted Disorders of Psychic Health After Chernobyl Nuclear Catastrophe"*, Kiev, 1995, pp. 22 – 23. (Rus)

[6] Gus'kova A.K., *Med. Radiologia i Radiats. Bezopasnost'*, 1997, No. 5, pp. 12 – 17.(Rus)

[7] Gus'kova A.K., *Vrach*, 1999, No. 6, pp. 35 – 37. (Rus)

[8] Gus'kova A.K., *Kazansky Med. Zh.*, 1994, vol. LXXV(5), pp. 341 – 344. (Rus)

[9] Gus'kova A.K. and Barabanova A.V., *Thes. Rep. Symposium on the Effect of Low Dose Radiation*, Minsk, 1968, p. 9. (Rus)

[10] Gus'kova A.K., *Zh. Nevropatologii i Psikhiatrii*, 1989, vol. 89(2), pp. 138 – 142. (Rus)

[11] Ivanov V.K. and Tsyb A.F., *Medical Radiological Consequences of the Chernobyl Accident for the Population of Russia: Assessment of Radiation Risks*, Moscow, Meditsina, 2002. (Rus)

[12] Il'in L.A., *Med. Radiologia i Radiats. Bezopasnost'*, 1998, No. 1, pp. 8 – 17. (Rus)

[13] Krasnikova L.I., *Proc. 2nd Intern. Conference on "Remote Medical Consequences of the Chernobyl Catastrophe"*, Kiev, 1998, pp. 85 – 86. (Rus)

[14] Kulikova T.A., *Proc. Intern. Conference on "Radiobiological Effects"*, Kiev, 2005, pp. 86 – 87. (Rus)

[15] Loganovsky K.N., *Sots. Klin. Psikhiatria*, 1999, vol. 1, pp. 5 – 17. (Rus)

[16] Morozov A.M. and Kryizhanovskaya L.A., *Clinics, Dynamics and Treatment of the Border Psychic Dysfunctions in Liquidators of the Chernobyl Accident*, Chernobylinterinform, 1998, p. 330. (Rus)

[17] Nyagu A.I. *et al.*, *Mezhdunarodnyi Zh. Radiats. Med.*, 1999, vol. 2(2), pp. 3 – 25. (Rus)

[18] Sushkevich G.N. and Tsyb A.F., 'International consequences of the Chernobyl accident', *IFECA IHO Scientific Report*, Geneva, 1995.

[19] Torubarov F.S., Nikolaeva M.K., and Dakhno D.V., *Med. Radiol.*, 1991, No. 3, pp. 54 – 56. (Rus)

[20] Kharchenko V.P. *et al.*, *Proc. Intern. Conference on "Urgent and Forecasted Disorders in Psychic Health After the Chernobyl Nuclear Catastrophe"*, Ukraine, Kiev, 1995, pp. 72 – 73. (Rus)

[21] Kharchenko V.P. *et al.*, *Med. Radiol.*, 2001, vol. 3(1-2), pp. 307 – 311. (Rus)

[22] Khayutin V.M. and Lukoshkova E.V., 'Heart rate variations: spectral analysis', *Vestnik Aritmologii*, 2002, vol. 26, pp. 10 – 21. (Rus)

[23] 'Chernobyl ten years after', *IAEA Assessment Report*, OECD, Paris, 1995.

[24] Yarmonenko S.P. and Konoplyannikov A.G., *Clinical Radiobiology*, Moscow, 1992, pp. 92 – 114. (Rus)

[25] Altimaras J., 'Understanding autonomic sympatovagal balance from short-term heart rate variations. Are we analyzing noise?', *Comparative Biochemistry and Physiology*, Part A, 1999, vol. 124, pp. 447 – 460.

[26] Emerit I., Levy A., Cernjavsky L., Arutyunyan R., Oganesian N., Pogosian A., Mejlumian H., Sarkisian T., Gulkandanian M., Quastel M., Goldsmith J., Riklis E., Kordysh R., Poliak S., and Merklin L., 'Transferable clastogenic activity in plasma from persons exposed as salvage personnel of the Chernobyl reactor', *J. Cancer Res. Clin. Oncol.*, 1994, vol. 120, pp. 558 – 561.

[27] Emerit I., Oganesian N., Sarkisian T., Arutyunyan R., Pogossian A., Asrian K., Levy A., and Cernjavsky L., 'Clastogenic factors in the plasma of Chernobyl accident recovery workers: anticlastogenic effect of Gingko biloba extract', *Radiation Res.*, 1995, vol. 144, pp. 198 – 205.

[28] Emerit I., Oganesian N., Arutyunyan R., Pogossian A., Sarkisian T., Cernjavsky L., Levy A., and Feingold J., Oxidative Stress-Related Clastogenic Factors in Plasma from Chernobyl Liquidators: Protective Effects of Antioxidant Plat Phenols, Vitamins and Oligoelements. Elsevier, Paris, 1997, pp. 239 – 246.

In: 20 Years after the Chernobyl Accident
Editors: E. Burlakova and V. Naidich, pp. 181-194
ISBN 1-60021-249-2

Chapter 11

GENETIC MONITORING OF NATURAL DROSOPHILA POPULATIONS IN RADIATION CONTAMINATED REGIONS OF BELARUS

I. B. Mosse*[3]*, M. E. Mikhailova, I. V. Glushkova, S. I. Kasinskaya, I. P. Anoshenko, T. V. Aksyutik, N. I. Tikhanovich and N. A. Kamyish
Institute of Genetics and Cytology, NAS of Belarus, Minsk, Belarus

ABSTRACT

Genetic monitoring of natural Drosophila melanogaster populations inhabiting in the regions of Belarus with different background radiation (Vetka and Svetilovichi villages), radionuclide-contaminated after the Chernobyl accident, compared with populations from the Berezinsky biosphere reserve (the control area) has been performed. The dominant and recessive lethal mutation levels and genetic structure of the populations were analyzed for frequencies of F- and S-alleles of Adh (alcohol dehydrogenase), Gpdh (glycerinophosphate dehydrogenase) and Sod (superoxide dismutase) loci. Populations inhabiting in the regions with high background radiation exhibited higher frequency of lethal mutations and higher heterozygosity than these in the control area. Moreover, higher frequency of polymorphous Sod locus S-allele was detected in these populations. Apparently, SOD S- alleles are more adaptively valuable under conditions of high background radiation, because as is known, superoxide dismutase is an effective radioprotector at all levels: molecular, cellular and organism. Since 1998 adaptation of populations to stress impacts has been analyzed. Nonspecific adaptation of natural Drosophila melanogaster populations from Vetka and Svetilovochi villages of Gomel Oblast was revealed. They are higher adapted than the control population from the Berezinsky biosphere reserve to both ionizing radiation effect and chemical mutagen

[3] Main addressee: 27 Akademicheskaya street, 220072 Minsk, Belarus; Tel: (375-17) 284-19-42; Fax: (375-17) 284-19-17; E-mail: I.Mosse@igc.bas-net.by (Irma Borisovna Mosse)

EMS. After laboratory cultivation without irradiation within 6 – 8 generations adaptation to radiation in the population from radiocontaminated regions remained. Maintenance of the control natural drosophila population samples under laboratory conditions is an environmental stress that leads to formation of nonspecific adaptation to unfavorable factors, including ionizing radiation, within 6 – 8 generations. It should be taken into account that the population adaptation is formed via elimination of sensitive genotypes at various stages of ontogenesis.

Keywords: Drosophila melanogaster, natural populations, genetic structure, allozyme loci, Adh, Gpdh, Sod, F- and S-alleles, radiation, chemical mutagen EMS, stress, lethal mutations, nonspecific adaptation

The Chernobyl accident has attached particular urgency to studying response regularities of individual organisms and entire populations to an increased background radiation in their habitats. Affected by low-dose chronic radiation, natural fauna and flora populations demonstrated significant changes in vital processes and genetic systems [1 – 3].

Despite the importance of the population investigations, multiple regularities of genetic processes in populations of living organisms are still insufficiently disclosed. This specially concerns the question of long-term mutagenic effects on the populations.

In the years passed since the Chernobyl accident, various genetic changes, including those which could cause adaptation, have proceeded in natural populations affected by permanent radiation impact.

Besides a high genetic load, the long-term radiation also raises the frequency of mutations favorable for the organism and intensifies other genetic processes increasing radiation resistance of populations. Such genetic (evolutionary) adaptation provides preservation of populations under extreme conditions by eliminating sensitive individuals.

The adaptive response of the population to radiation impact is not a simple mechanical selection of more resistant genotypes. This process is accompanied by transformation of genetic structure of the populations, which is caused by formation of optimal genotypes [4]. To date radioadaptation formation was shown in natural populations of plants [2], animals [5 – 8] and microorganisms [1].

The most effective adaptation is observed in populations with rapid succession of generations and high fertility that is particularly typical of insects. Investigations of drosophila allows for analyzing its genetic variability degree, the variety of mutation spectrum, the level of heterozygosity, etc. These data are of importance for general biology and allow prediction of genetic situations, which may also be observed in other higher organisms.

Adaptation of long-term irradiated experimental drosophila populations to mutagenic action of radiation is shown [5 – 8]. For example, radiosensitivity of chronically irradiated laboratory populations of *Drosophila melanogaster* was determined by the mutation induction at additional acute radiation impact [8]. In this case, the populations irradiated during 55 generations turned out higher resistant as compared with the control. Nonspecific adaptation pattern of irradiated populations was demonstrated in the laboratory drosophila populations [8].

Therefore, this work is targeted at the study of the question if natural drosophila populations inhabiting in the regions impacted by Chernobyl accident become adapted to long-term radiation and, if so, whether this adaptation is specific or not. To ascertain this question we have analyzed the mutation process levels, radiosensitivity and the genetic structure of natural *Drosophila melanogaster* populations inhabiting radiation-contaminated areas of Belarus and the Berezinsky biosphere reserve.

MATERIALS AND METHODS

Natural drosophila populations in Vetka and Svetilovichi settlements of Gomel' region (24 Ci/km^2) and the Berezinsky biosphere reserve (for the control) were sampled. The genetic structure of the populations was analyzed by the frequency of F- and S-allele loci of alcohol dehydrogenase (Adh), glycerophosphate dehydrogenase (Gpdh) and superoxide dismutase (Sod) using electrophoresis in 12.5% starch gel [9]. On the enzyme manifestation the gel was fixed in a mixture containing acetic acid, ethanol and water in 1:5:5 ratio.

For estimating genetic radioresistance, drosophila males from the studied natural drosophila populations were exposed to single acute γ-radiation of 30 or 40 Gy dose in a γ-irradiation device LMB-γ-IM with ^{137}Cs radiation source and exposure dose rate equal 0.1 Gy/s.

Sigma multifunctional alkylating agent ethyl methane sulfonate (EMS) in 25 mmol/l concentration was taken as a chemical mutagen. The mutagen with 1% saccharose solution applied on paper filters was fed to adult individuals (males) for 24 h.

The frequency of recessive sex-linked lethal mutations (RSLLM) was estimated by standard technique using the tester Basc line [10]. The frequency of dominant lethal mutations (DLM), comprising two types – embryonic lethal mutations (ELM) and postembryonic lethal mutations (PELM), were determined by standard techniques [10]. ELM was estimated as a ratio between the numbers of undeveloped and laid eggs. PELM was determined as the ratio between the numbers of developed eggs except for imago and developed eggs. DLM (total lethal cases) was determined as the ratio between the numbers of laid eggs except for imago and total number of laid eggs.

To determine duration of preserving an adaptation state of the population after radiation removal, samples of natural populations from radiation-contaminated and control regions were placed to a thermostat and exposed to standard laboratory conditions free from radiation in test tubes with a standard nutritional medium.

Six and eight generations after radiation effect removal and cultivation in the laboratory, males were exposed to acute γ-radiation of 30 Gy (in γ-irradiation device LBM-γ-IM with ^{137}Cs radiation source and exposure dose rate equal 0.1 Gy/s). The frequency of dominant lethal mutations (DLM) was determined using standard techniques [10].

RESULTS AND DISCUSSION

Variability of Adh, Gpdh and Sod Loci in Natural Drosophila Populations Inhabiting in Radiation-Contaminated Regions of Belarus

The gene-enzymatic system of alcohol dehydrogenase represents one of the most suitable models for studying the effect of particular genes on adaptation of the organisms owing to relatively easy identifiability of an enzyme, high polymorphism, and the role played by Adh in the vital activity of drosophila. According to its electrophoretic mobility, Adh is represented by two classes of allozymes – Adh-F (fast) and Adh-S (slow), which, as is generally known, differ in some functional properties [11]. It has been found that allozyme F exhibits higher activity and lower stability, whereas allozyme S demonstrates opposite properties [12].

Hence, in natural populations genotype frequencies of Adh locus correlate significantly with geographic and climatic parameters, latitude and temperature, in particular [12].

Table 1 presents data, obtained by the authors of the current work in the period of 1998 – 2000, on the genetic structure of natural drosophila populations for the ratio of FF, FS and SS genotypes frequencies and Adh locus F-allele frequency.

As indicated by these results, the frequency of Adh-F gene in populations sampled from the Berezinsky biosphere reserve equaled 0.89, with the frequencies of FF- and FS-homozygotes being equal 0.81 and 0.17, respectively.

Table 1. The ratio of frequencies of genotypes and Adh locus F-allele in Drosophila melanogaster natural populations

Population habitat	Number of species analyzed, n	Frequency of genotypes			F-allele frequency
		FF	FS	SS	
Berezinsky reserve	189	0.81	0.17	0.02	0.89 ± 0.02
Vetka settlement, Gomel' region	102	0.58	0.30	0.12	0.73 ± 0.04*
Svetilovichi settlement, Gomel' region	100	0.50	0.30	0.20	0.65 ± 0.05*

* $p < 0.001$.

At the same time, in both populations from radiation-contaminated regions, the frequency of heterozygotes equaled 0.30, and the frequencies of FF-homozygotes and SS-homozygotes in Vetka and Svetilovichi settlements equaled 0.58 and 0.50, and 0.12 and 0.20, respectively.

Thus, the ratio of genotype frequencies for Adh gene in populations from radiation regions differed from the Berezinsky reserve populations ($p < 0.001$). In these populations, the frequencies of heterozygotes and SS-homozygotes were slightly increased that obviously provided higher plasticity for the populations. Similar data were obtained for *Centaurea scabiosa* L. populations exposed to chronic radiation in the East Ural radiation trail [2]. The authors emphasize that such populations exhibit higher heterozygosity and greater number of genotype combinations. As is generally known, heterogeneity in the populations is the basis for their adaptation to varying environmental conditions. The attainment of long-term periods of stable equilibrium for genotype frequencies is characteristic for the natural population. However, anthropogenic impacts in any forms of their manifestation change this natural equilibrium and elimination of less adapted forms and formation of combinations of genotypes, whose relative fitness is higher under given conditions of existence, take place.

Of course, during generation succession more radioresistant organisms stand a better chance of progeny reproduction. However, when analyzing natural populations, besides the radiation factor, one should take into account the whole set of environmental conditions: temperature, humidity, amount of precipitations, food resources, which may directly or indirectly affect the genetic structure of the populations. A mediated mechanism of Adh polymorphism maintenance is possible owing to linking to cosmopolitan chromosomal inversion In(2L)t [13]. As shown by the laboratory tests [14-15], Adh-S and Gpdh-F alleles located in the 2nd chromosome are related to In(2L)t inversion. The advantageous combination of Adh-F and Gpdh-S allozymes for balancing the reduced and oxidized NAD coenzyme form – $NADH/NAD^+$, is also known. Extra-mitochondrial NADH molecules themselves are incapable of permeating inside mitochondria through the membranes. However, electrons they emit may be involved in the mitochondrial electron transfer chain by means of glycerophosphate shuttle mechanism [16].

Thus, one may suggest an existence of functional interrelation between Adh and Gpdh loci. Therefore, we have analyzed frequencies of genotypes and Gpdh locus alleles in natural drosophila populations. Data in Table 2 indicate the presence of two Gpdh locus alleles – fast (F) and slow (S) – in natural drosophila populations studied.

Table 2. The ratio of genotype and Gpdh locus F-allele frequencies in Drosophila melanogaster natural populations

Population habitat	Number of analyzed species, *n*	Frequencies of genotypes			F-allele frequency $p^F \pm S_p$
		FF	FS	SS	
Berezinsky reserve	158	0.48	0.39	0.13	0.68 ± 0.04
Vetka settlement, Gomel' Oblast	86	0.45	0.50	0.05	0.70 ± 0.05
Svetilovichi settlement, Gomel' Oblast	80	0.46	0.48	0.06	0.70 ± 0.05

In the Berezinsky reserve population, the frequencies of FF, FS and SS equaled 0.48, 0.39 and 0.13, respectively, whereas the F-allele frequency equaled 0.68.

The F-allele frequency in the samples from Vetka and Svetilovichi settlements was the same as in the Berezinsky reserve population, however, they were not identical by the genotype frequency. These populations show heterozygote frequencies increasing to 0.50 and the SS-homozygote levels decreasing to 0.05 – 0.06.

Thus, the increase of heterozygote frequency for glycerophosphate dehydrogenase gene is typical of both populations inhabiting in the settlements with elevated background radiation. However, similar to locus Adh, we may not relate changes in genotype frequencies exclusively to elevated background radiation. Most likely, as combined with the radiation contamination of the environment, the whole complex of environmental conditions affect the ratio of genotype frequencies for the studied glycerophosphate dehydrogenase gene. As is also shown [17, 18], of polymorphism by superoxide dismutase (Sod) locus in populations of various taxons under ionizing radiation impacts possesses the adaptive character. It is known that superoxide dismutase is highly effective radioprotector at the molecular, cellular and organism levels. As proved in the experiments, performed on *Drosophila melanogaster*, the organism–synthesized SOD implements a protective function. The protection level correlated with the specific activity of isoenzyme, which is noticeably higher for the S-form of SOD. Moreover, it has been shown that chronic irradiation of laboratory drosophila populations induces an increase in the frequency of SOD S-allelic forms exhibiting the increased specific activity. In irradiated populations, highly active S-form displaced low-active F-form [19].

Table 3. The ratio of frequencies of genotypes and S-allele of Sod locus in Drosophila melanogaster natural populations

Population habitat	Number of analyzed species, *n*	Frequencies of genotypes			F-allele frequency $p^F \pm S_p$
		FF	FS	SS	
Berezinsky reserve	104	0.78	0.20	0.02	0.12 ± 0.04
Vetka setlement, Gomel' region	77	0.10	0.60	0.30	0.60 ± 0.06*
Svetilovichi setlement, Gomel' region	80	0.11	0.50	0.39	0.64 ± 0.05*

* $p < 0.001$.

We have also analyzed the genetic structure of samples taken from natural populations, which inhabit the regions with increased background radiation and from Berezinsky biosphere reserve, for Sod locus. Table 3 shows the analysis results for the frequency ratio of genotypes and alleles of SOD-locus.

As is observed, the frequency of Sod locus S-allele in the populations from Vetka and Svetilovichi settlements equaled 0.60 and 0.64, respectively. The frequency of heterozygotes equaled 0.60 and 0.50, respectively. The frequency of S-allele in the irradiated population, monitored for over 13 generations, increased to the 60% level.

As a consequence, the frequency of S allozyme in these populations was much higher compared with the control (Berezinsky reserve population), where its level was only 0.12 ($p <$ 0.001). Apparently, so high frequency of Sod locus S-allele in the samples taken from the populations of Vetka and Svetilovichi settlements is related to elevated background radiation in these regions.

The authors' data correlate well with the results by other investigators [20], who have studied the frequency dynamics of F- and S-alleles in irradiated and non-irradiated *Drosophila melanogaster* populations, both alleles showing1:1 initial frequency.

An increase in the frequency of S-allele to 65% and a decrease in F-allele of Sod polymorphic locus to 35% within 12 generations exposed continuous ionizing radiation impact were demonstrated in chronically irradiated *Centaurea scabiosa* L. populations [19] inhabiting during a long time in the East Ural radioactive trail.

Therefore, basing on the authors' data and data from the literature, it may be concluded that SOD S-allozyme plays an important functional role in adaptation of the populations. Owing to high protective role, S-allozyme occurs most frequently in populations living under conditions of increased background radiation and other extreme environmental conditions.

For estimating the level of genetic variability of populations we have calculated their heterozygosity (H) for three loci under study. Table 4 shows the mean heterozygosity of the Berezinsky reserve population equal 0.25. For the sampls of Vetka and Svetilovichi settlement populations, the average heterosygotity was much higher and made up 0.47 and 0.43, respectively.

Table 4. Heterozygosity of drosophila natural populations for Adh, Gpdh and Sod loci

Locus	Drosophila populations from:					
	Berezinsky reserve		Vetka settlement		Svetilovichi settlement	
	real	expected	Real	expected	real	Expected
Adh	0.17	0.20	0.30	0.40	0.30	0.46
Gpdh	0.39	0.44	0.50	0.42	0.48	0.42
Sod	0.20	0.22	0.60	0.48	0.50	0.46
Mean hetero-zygosity	0.25	0.29	0.47	0.43	0.43	0.45

Most likely, transformation of the genetic structure of the populations, exposed radiation contamination conditions, leads to formation of optimal genotypes with higher resistance to radiation impact and thereby providing survival of populations under new ecological conditions.

Genetic Response of Drosophila Natural Populations to Additional Radiation Impact

In 1998, radiosensitivity of the individuals from Berezinsky biosphere reserve and Vetka settlement natural populations were compared. For this purpose a method of additional acute γ-irradiation by 40 Gy dose was used.

After additional radiation impact, the ELM, PELM and DLM frequencies were significantly higher in males from the control Berezinsky reserve population than in Vetka settlement population (Table 5). This indicates adaptation of it to radiation.

Table 5. The frequency of dominant lethal mutations in additionally irradiated individuales from natural drosophila populations

Dominant lethal mutation parameters	Drosophila populations from:	
	Berezinsky reserve	Vetka settlement
Number of laid eggs	2831	3192
Number of undeveloped eggs	1786	1365
Number of imago	438	1349
ELM, %	63.09 ± 0.91	42.76 ± 0.88*
PELM, %	58.09 ± 0.93	26.16 ± 0.78**
DLM, %	84.53 ± 0.68	57.74 ± 0.87**

* $p < 0.05$ as compared with Berezinsky reserve population;
** $p < 0.001$ as compared with Berezinsky reserve population.

To confirm the results obtained in 1999, the levels of ELM, PELM and DLM were studied in natural populations from Svetilovichi settlement and Berezinsky reserve. These parameters were significantly higher in the species from Svetilovichi settlement rather than in the Berezinsky reserve population (Table 6), which indicates higher level of the mutation process in the population from the radiation-contaminated region compared with the control. However, under additional single irradiation by 40 Gy dose, an increase in PELM and DLM frequencies is more significant for Berezinsky reserve population than for Svetilovichi one. For example, additional acute irradiation of the species from the Berezinsky reserve population resulted in 8.6-fold elevations of ELM and 3.8-fold elevations of PELM and DLM, while the Svetilovichi population demonstrated the following elevations of these parameters: 5.4-fold for ELM, and 2.0-fold for PELM and DLM (Table 6). Thus, it is indicated that the Svetilovichi settlement population was higher adapted to the radiation impact than the control.

Table 6. The effect of additional single irradiation with 40 Gy dose on the frequency of dominant lethal mutations in natural populations of drosophila

Parameters of dominant lethal mutations	Drosophila populations from:			
	Berezinsky reserve		Svetilovichi settlement	
	Before exposure	After irradiation with 40 Gy	Before exposure	After irradiation with 40 Gy
Number of laid eggs	3245	5877	4626	5384
Number of undeveloped eggs	125	1946	269	1674
Number of imago	2573	1231	2978	1374
ELM, %	3.85 ± 0.34	33.11 ± 0.61	5.81 ± 0.34*	31.09 ± 0.63
PELM, %	17.53 ± 0.67	68.68 ± 0.6	31.65 ± 0.68*	62.96 ± 0.66**
DLM, %	20.71 ± 0.71	79.05 ± 0.53	35.62 ± 0.70*	74.48 ± 0.59**

* $p < 0.001$ as compared with the mutation frequency in Berezinsky reserve population before the exposure;
** $p < 0.001$ as compared with the Berezinsky reserve population after irradiation with 40 Gy

Simultaneously with the study of additional radiation impact by the DLM test, this question was also investigated by RSLLM test. In 1998 after additional single irradiation of natural population from Vetka settlement by 40 Gy dose, the RSLLM frequency was significantly lower (Table 7) than in the control population from the Berezinsky reserve after the same radiation exposure.

Table 7. The frequency of recessive sex-linked lethal mutations (RSLLM) in the species of natural drosophila populations exposed to additional 40 Gy radiation

Parameters of recessive sex-linked lethal mutations	Drosophila population:	
	Berezinsky reserve	Vetka settlement
Number of analyzed chromosomes	831	1429
Number of lethal mutations	105	95
RSLLM, %	12.64 ± 1.15	6.65 ± 0.66*

* $p < 0.001$ as compared with the Berezinsky reserve population

Similar data were obtained in 2000, after additional 30 Gy dose irradiation the RSLLM frequency in the Berezinsky reserve population was much higher than in the Vetka settlement population.

Thus, the results obtained indicate an intensified mutagenic process for drosophila populations inhabiting in the areas with increased background radiation. Moreover, adaptation of the populations from radionuclide-contaminated regions to mutagenic action of radiation was revealed for both dominant and recessive lethal mutations. This is in agreement with the data obtained for laboratory drosophila populations [13], according to which the samples of populations irradiated during 48 generations indicated higher resistance to additional acute irradiation compared with the control.

Comparison of Sensitivity of Drosophila Natural Populations to EMS Impact

It was also of interest to find out if the increased resistance of natural populations from radionuclide-contaminated regions is specific or not. For this purpose, chemical mutagen EMS was chosen as the test factor.

In 1999, males sampled from Berezinsky reserve and Svetilovichi settlement populations were treated by 25 mmol/l EMS. The data on DLM frequencies are presented in Table 8.

Males from the Berezinsky reserve population exposed to EMS treatment showed a significant increase in ELM, PELM and DLM frequencies, whereas the mutagen effect on males from the Svetilovichi population did not cause an increase in PELM and DLM frequencies. The analogous EMS effect was observed for the natural Vetka settlement population. After EMS treatment of males from the populations studied, a significant increase in PELM and DLM parameters was revealed in the Berezinsky reserve population compared with the population inhabiting in Vetka settlement with increased background radiation.

The data obtained indicates the elevated resistance of populations exposed to long-term chronic irradiation, not only to additional acute γ-radiation, but also to chemical mutagen EMS impact, i.e. to nonspecific character of their adaptation. These results are in agreement with the data from the literature, obtained on drosophila laboratory populations [6, 8] and other biological objects [19]. Here it is shown that nonspecific resistance is a generality manifesting itself under the long-term exposure to various environmental and anthropogenic factors. This should be taken into account at the assessment of genetic effects of the environment contamination.

Table 8. The influence of EMS on the frequency of dominant lethal mutations in natural populations of drosophila

Parameters of dom-inant lethal mutations	Drosophila populations from:			
	Berezinsky reserve		Svetilovichi settlement	
	Before exposure	After EMS treatment	Before exposure	After EMS treatment
Number of laid eggs	3245	4777	4626	3779
Number of undeveloped eggs	125	324	269	341
Number of imago	2573	3318	2978	2399
ELM, %	3.85 ± 0.34	6.78 ± 0.36*	5.81 ± 0.34*	9.02 ± 0.47**
PELM, %	17.53 ± 0.67	25.49 ± 0.63*	31.65 ± 0.68*	30.22 ± 0.75
DLM, %	20.71 ± 0.71	30.54 ± 0.67*	35.62 ± 0.70*	36.52 ± 0.78

* $p < 0.001$ as compared with the mutation frequency before exposure in the Berezinsky reserve population

** $p < 0.001$ as compared with the mutation frequency before exposure in the Svetilovichi settlement population

Adaptation Processes in Natural Drosophila Populations from Radiation-Contaminated Regions of Belarus After Radiation Effect Removal

In 2000, we decided to check if the adaptation of populations is preserved after removal of radiation influence. The samples from natural populations from Vetka settlement and the Berezinsky reserve were placed under the laboratory conditions (in the thermostat without irradiation). The adaptation to irradiation was studied after their cultivation under such conditions during 8 generations. As for the Vetka settlement population, the adaptation was preserved. For example, no change in the DLM frequency in F_8 was observed before and after additional irradiation of this population: (39.66 ± 1.23) and (39.51 ± 1.18), respectively (Figure 1). It proved to be surprising that in the 8th generation the Berezinsky reserve population was found more resistant to additional acute irradiation rather than in F_1 (Figure 1). In both populations, DLM frequencies were found practically equal. This indicates that the maintenance of natural populations under laboratory conditions (place limitation, overpopulation, alteration of light, food and temperature regimes, etc.) may cause a stress and formation of nonspecific adaptation of the population to unfavorable factors, including ionizing radiation.

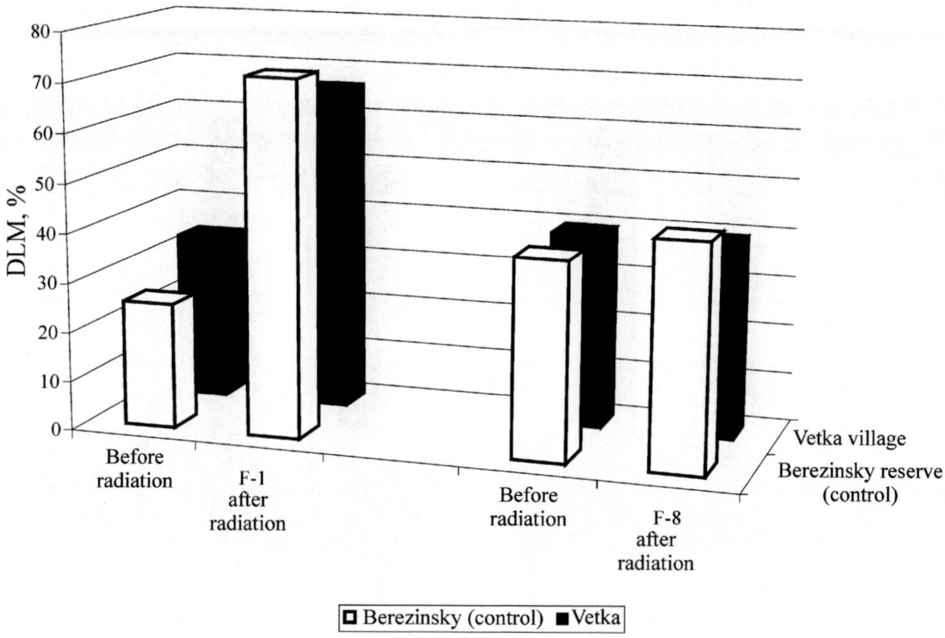

Figure 1. DLM frequencies in natural populations of drosophila before and after of 30 Gy irradiation in the 8-th generation after radiation impact removal (measured in 2000)

In 2001, to determine how fast the control population has become resistant to radiation, the current test was repeated and, hence, the analysis was performed for the 6th generation after sample cultivation in a thermostat. As caught, the flies from the Vetka settlement population were found more adapted to acute irradiation compared with the Berezinsky reserve population (Figure 2). DLM frequencies in the natural Vetka settlement population before and after additional radiation impact equaled (33.0 ± 0.66) before the additional acute irradiation and (58.59 ± 0.99) after the acute irradiation, whereas in the Berezinsky reserve population these parameters equaled (41.98 ± 1.08) before and (78.96 ± 1.57) after the additional radiation ($p < 0.001$). This indicates adaptation to the radiation impact for Vetka populations under study. Six generations after radiation impact removal, a significant increase in DLM frequency was observed in both populations under study. However, the Berezinsky reserve population showed 1.8-fold DLM frequency increase (from (35.42 ± 1.02) to (62.39 ± 1.14) at $p < 0.001$), whereas for the Vetka settlement population a 1.2-fold increase only was determined (from (54.10 ± 0.95) to (66.18 ± 1.11) at $p < 0.001$) (Figure 2).

As a consequence, both tests implemented in 2000 – 2001 indicated that radiation adaptation of natural drosophila populations from radionuclide-contaminated regions was preserved at least within 8 generations after the radiation removal. Hence, the account must be taken of two moments.

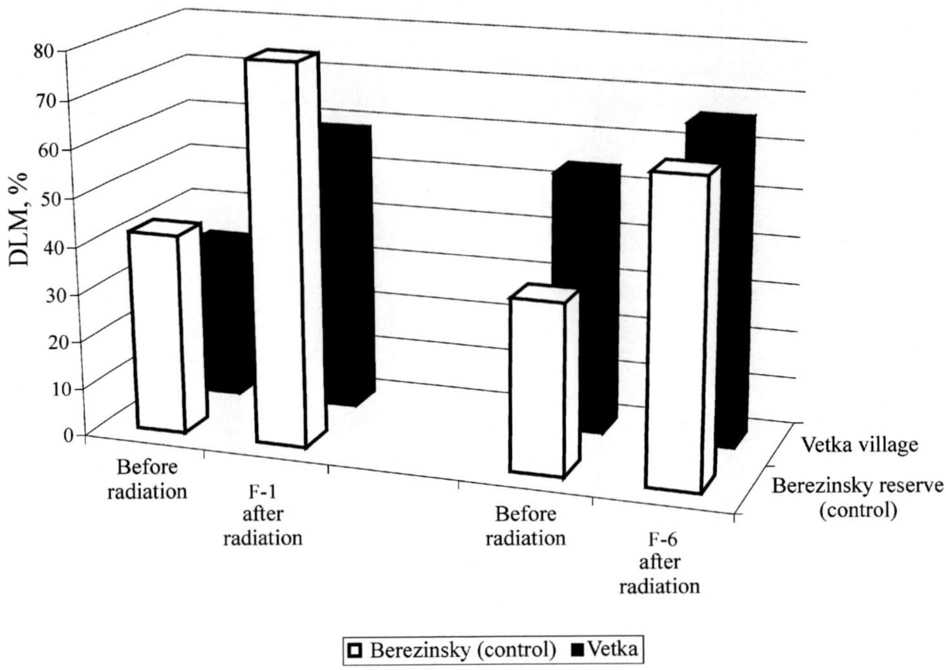

Figure 2. DLM frequencies in natural drosophila populations before and after 30-Gy irradiation in the 6-th generation after radiation impact removal (measured in 2001)

Firstly, the increase in spontaneous mutation frequency in the natural sampls from the control populations, maintained under laboratory conditions (without radiation impact) during 6 – 8 generations (Figures 1 and 2). Secondly, cultivation under laboratory condition during 6 – 8 generations has induced an adaptation in the control population. The fact is that during 6 generations of maintaining under experimental conditions the Berezinsky reserve population has just approached the Vetka settlement population in its adaptation to radiation (Figure 2), whereas after 8 generations of cultivation in the laboratory the resistance of both populations to ionizing radiation became practically equal (Figure 1). As a consequence, maintaining under the laboratory conditions may represent a stress similar to natural existence in conditions of low dose radiation. Cultivation under experimental conditions may cause formation of nonspecific adaptation of the populations to unfavorable factors, including the radiation impact. These facts should be taken into account in the study of mutation frequency dynamics during elimination of radionuclides from the animals, caught in radiation-contaminated regions and placed to a laboratory or vivarium conditions.

CONCLUSION

We have analyzed genetic structure of natural drosophila populations from different regions of Belarus for allele frequencies of Adh, Gpdh and Sod loci. Higher heterozygosity values as well as higher S-allele frequency of superoxide dismutase locus exhibiting as is generally known, radioprotective properties were noted in the populations inhabiting in the regions with increased background radiation.

It was revealed that natural populations of *Drosophila melanogaster* from Vetka and Svetilovichi settlement of Gomel region had higher mutation frequencies as compared with the control population from the Berezinsky biosphere reserve. At the same time, nonspecific adaptation of populations from radiation-contaminated regions was established: the populations demonstrated higher resistance to mutagenic effects of not only additional high-dose radiation, but also of chemical mutagen EMS compared with the control population. It is of particular interest that the transfer of flies from the control natural population to the laboratory conditions is a stress which induces nonspecific adaptation to unfavorable environmental factors, including ionizing radiation.

REFERENCES

[1] Dubinin N.P., Shevchenko V.A., Alekseenko A.Ya. *et al.*, *Uspekhi Sovrem. Genetiki*, 1972, No. 4, pp. 170 – 205. (Rus)

[2] Kal'chenko V.A., Kalabushkin B.A., and Rubanovich A.V., *Genetika*, 1991, vol. 27(6), pp. 676 – 683. (Rus)

[3] Makeeva E.N., Klimets E.P., Mosse I.B. *et al.*, *Vestsi AN Belarusi, Ser. Biyal. Navuk*, 1996, vol. 116(1), pp. 97 – 101. (Bel)

[4] Glushkova I.V., Mosse I.B., Anoshenko I.P. *et al.*, *Radiats. Biol. Radioekol.*, 2002, vol. 42(2), pp. 124 – 129. (Rus)

[5] Cordeiro A.R., Marques E.K., and Veiga-Neto A.I., *Mutat. Res.*, 1973, vol. 19(3), pp. 325 – 329.

[6] Nothel H., *Mutat. Res.*, 1976, vol. 36(2), pp. 245 – 248.

[7] Nothel H. and Strese B., *Mutat. Res.*, 1982, vol. 103(1), pp. 87 – 90.

[8] Mosse I.B., *Radiation and Heredity: Genetic Aspects of Radiation Safety*, Minsk, Izd. Universitetskoe, 1990, 208 p. (Rus)

[9] Ayala F.J., Powell J.R., Tracey M.J. *et al.*, *Genetics*, 1972, vol. 70, pp. 113 – 139.

[10] Tikhomirova M.M., *Genetic Analysis*, Leningrad, Izd. LGU, 1990, 280 p. (Rus)

[11] Chambers G.K., Wilks A.V., and Gibson J.B., *Biochem. Genet.*, 1984, vol. 22(1–2), pp. 153 – 168.

[12] Heinstra P.W.H., Sharloo W., and Thoring G.E.W., *Genetics*, 1987, vol. 117(1), pp. 75 – 84.

[13] Bubliy O.A. and Imasheva A.G., *Genetika*, 1997, vol. 33(7), pp. 924 – 933. (Rus)

[14] Knibb W.R., Oakeshott J.G., and Gibson J.B., *Genetics*, 1981, vol. 98, pp. 833 – 847.

[15] VanDelden W. and Kamping A., *Evolution*, 1989, vol. 43, pp. 775 – 793.

[16] Sactor B., *Adv. Insect. Physiol.*, 1970, vol. 7, pp. 267 – 338.

[17] Oberley L.W., Leindgren A.L., Baker S.A., and Stevens R.H., *Radiat. Res.*, 1976, vol. 68(2), pp. 320 – 328.

[18] Petkau A., Kelly K., Chelack W.S., and Barefoot C., *Biochem. Biophys. Res. Commun.*, 1976, vol. 70(2), pp. 452 – 458.

[19] Kal'chenko V.A., Rubanovich A.V., and Shevchenko V.A., *Genetika*, 1996, vol. 32(11), pp. 1509 – 1512. (Rus)

[20] Peng Tong Xu, Moya A., and Ayala F., *Proc. Natl. Acad. Sci. U.S.A.*, 1986, vol. 83, pp. 684 – 687.

In: 20 Years after the Chernobyl Accident
Editors: E. Burlakova and V. Naidich, pp. 195-229 ISBN 1-60021-249-2

Chapter 12

RESULTS OF INVESTIGATIONS PERFORMED IN THE CHERNOBYL ACCIDENT ZONE AND PRACTICAL RESULTS OF THE ACCIDENT CONSEQUENCES LIQUIDATION IN THE AGRICULTURAL SECTOR OF THE RUSSIAN FEDERATION

***R. M. Alexakhin*[1], *N. I. Sanzharova*[1*], *S. V. Fesenko*[1], *S. A. Geras'kin*[1], *S. I. Spiridonov*[1], *A. V. Panov*[1], *N. N. Isamov, Jr.*[1], *A. V. Nizhebovsky*[2], *P. V. Prudnikov*[3] *and A. D. Pasternak*[4]**

[1] Russian Institute of Agricultural Radiology and Agroecology, Obninsk, Kaluga Oblast, Russia
[2] Federal Agency on Agriculture, Moscow, Russia
[3] Bryansk Agrochemradiology Center, Bryansk, Russia
[4] Bryansk Radiological Veterinarian Laboratory, Bryansk, Russia

ABSTRACT

Radioecological conditions in the agriculture after the Chernobyl accident are analyzed. The consequences of radiation impact on agricultural plants and animals are estimated. The regularities and dynamics of aerial and subsoil pathways of radionuclide migration in agricultural ecosystems are described. Parameters of radionuclide sorption in soils and dynamics of their biological mobility are determined. Radionuclide migration models for agricultural ecosystems are suggested. Protective measures applied to agriculture are classified, designed to decrease the contamination level of output

* The main addressee: 109 km Kievskoe Ave., 249032 Obninsk, Kaluga Oblast, Russia; Russian Institute of Agricultural Radiology and Agroecology; Tel: (7-48439) 6-72-05; Fax: (7-48439) 6-80-66; E-mail: riar@obninsk.org, natsan2004@mail.ru

products. The strategies of the agriculture management in different periods after the accident are scientifically justified. The scale and efficiency of protective measures applied to the agricultural sector of the Russian Federation are estimated.

Keywords: the Chernobyl accident, agricultural ecosystems, radiation impact, migration, protective measures, rehabilitation strategies

INTRODUCTION

The Chernobyl accident is the largest catastrophe in the history of the nuclear engineering which caused a large-scale contamination of agricultural lands. The complexity of the situation, caused by the accident, was induced by heterogeneity of radioactive fallout, differences in radionuclide composition of emissions, variability of natural-climatic conditions in the contamination zone, and huge socioeconomic consequences.

The Chernobyl accident caused extremely severe consequences for the agriculture. On the significant territories, the density of radioactive contamination was so high that made impossible consumption of output agricultural products. Radioactive depositions were mostly composed of biologically mobile ^{131}I, ^{90}Sr and $^{134,137}Cs$, which intensively migrate via agricultural chains. Thickly settled regions with developed agriculture were also contaminated. The feature of the main contamination zone is represented by specific soil conditions: Polesye territories demonstrate predominance of low fertility soils with low sorption capacity. The accident happened in the period, when the animal stock was moved to pastures and obtained maximum exposure to radiation.

Despite the obviousness of the Chernobyl accident consequences for the agriculture, only in 2001, 15 years after at the International Conference in Kiev this accident was primarily accepted as the "rural accident". Firstly, at all stages of the after-accident period the internal irradiation due to consumption of radionuclide-contaminated agricultural products gave about 50% of total radiation dose burden to people. Secondly, the main contaminated territory is represented by agricultural lands. Thirdly, among the population exposed to additional radiation impact caused by the accident, the country folk are predominant having higher radiation doses absorbed compared with the city dwellers.

The problem of agricultural production management and support of population with the products meeting the sanitary-hygienic standards was found most difficult to decide after the accident, because it touched upon socioeconomic questions concerning keeping and restoration of habitual living conditions of the country folk on the extremely large territory. High contamination level of agricultural lands induced a necessity to apply protective measures in all branches of the agriculture and the use of specific technologies for processing of the agricultural raw products.

The feature of the Chernobyl accident is dynamism of the radiation conditions, which required principally new approach to arrangement of the radiation monitoring system and adoption of protective measures. Optimization of the countermeasures and development of special strategies of their application in different periods of the accident that would provide for prudent management of material, human and financial resources became one of the new tasks. This task can be resolved owing to development of the decision support systems for the agricultural production management in case of radiation contamination, based on designing dynamic models of radionuclide migration in agricultural ecosystems.

The result of realizing programs on the accident consequence liquidation in the agriculture became an abrupt decrease of radionuclide concentration in the main kinds of products (milk, beef, grain) which provided a significant reduction of radiation burdens to the population. Already 2 – 3 years after the accident agricultural production was terminated in all contaminated regions (except for Bryansk Oblast) with maximum permissible concentrations (MPC) of radionuclides exceeding the accident level. In large areas temporarily terminated agricultural production was started up or restrictions applied were lifted. Practical implementation of protective systems reduced negative consequences of the accident for the population and economy of the contaminated areas.

The experience in liquidating the Chernobyl accident elucidated principally new features in the system of the radiation safety maintenance for the population related to the problem of introducing sanitary standards for radionuclide concentration in products on contaminated territories. These standards should be identical with these assumed for uncontaminated territories, e.g. for the prior-to-accident period. A transition from temporarily permissible levels to prior-to-accident standards increased nonstandard production. Therefore, prior-to-accident standards may only be introduced gradually with respect to contamination levels of the agricultural lands, soils features and economic conditions in the contamination zone.

Thus, liquidation of the accident consequences required solution of the fundamental scientific questions concerning the study of radionuclide behavior in agricultural ecosystems; development and adoption of the product radiation control and radiation condition monitoring; renovation and development of principally new methods and ways for rehabilitation of contaminated territories providing for stable development of the agriculture and safe residence of the population.

RADIATION CONDITIONS IN AGRICULTURE OF THE ACCIDENT REGION

The Chernobyl accident emitted a great amount of radionuclides to the environment, including fission fragments and non-spent reactor fuel. The emergency emission was spread in the time (April 26 – May 10), which produced a complicated configuration of radioactive trails on the landscape providing specific features of radionuclide fallout composition and physicochemical form.

Concerning the environmental problems, the most urgent factor was the presence of biological mobile radionuclides ^{131}I, ^{137}Cs and ^{90}Sr in the emission. The role of short- and medium-lived radionuclides (^{95}Zr, ^{106}Ru, ^{141}Ce, ^{144}Ce) was limited by the first year after the accident. Of special importance was fallout of short-lived ^{131}I ($T_{1/2}$ = 8,04 days), whose emergency emission equaled $2.7 \cdot 10^{17}$ Bq. This radionuclide mostly migrates by the following chain: radioactive fallout – stock forage – milk – human body (up to 80% of ^{131}I appeared in the body with milk).

In a considerable area of the Chernobyl trail ^{137}Cs ($T_{1/2}$ = 30.17 years) was found to be the main dose-forming radionuclide. In the initial period after the accident, the presence of shorter lived radionuclide ^{134}Cs ($T_{1/2}$ = 2.04 years) was rather urgent. Total emission of ^{137}Cs was estimated to be equal to $4\cdot10^{16}$ Bq (about 35% of emission was deposited on the Russian territory) and ^{134}Cs – $1.5\cdot10^{17}$ Bq. The emission of ^{90}Sr ($T_{1/2}$ = 28.1 years) equaled $2.3\cdot10^{17}$ Bq, whose role as the source of the radiation hazard was much lower compared with ^{137}Cs. The contribution of this radionuclide into total radiation dose of the population was relatively high in Ukraine and Belarus, whereas for the Russian Federation it was out of any practical interest.

Plutonium radionuclides mostly concentrated in the nearest to the reactor zone and being very inert in agricultural chains of transfer are insignificant as a radiological factor for the population living outside the 30-kilometer restriction zone.

Two forms of Chernobyl fallout are identified: fuel particles, mostly represented by nuclear fuel overcooling particles, construction materials, etc. and gas condensate fallout including fine aerosols. Nuclear fuel particles were mostly deposited in the nearest (30-km) zone around the Chernobyl NPP.

Table 1. Distribution of agricultural lands by ^{137}Cs contamination density (ha) [1]

Oblast	Density of contamination by ^{137}Cs, kBq/m^2				
	37-185	185-555	555-1480	>1480	Total
Bryansk	401400	186600	97600	17106	702706
Kaluga	111700	33100	700	-	145500
Orel	396400	22800	-	-	419200
Tula	653000	125700	-	-	778700

Summing up the accident consequences, 3.2% of the former USSR territory was contaminated by ^{137}Cs, whose concentration exceeded 37 kBq/m^2. The areas with the highest contamination levels were located within 200 – 300 km zone around the Chernobyl NPP; the maximum level (over 1500 кBq/m^2 by ^{137}Cs) was detected in the 30-km zone. The NNE (north-north-east) zone of contamination (200 km from the epicenter) including Gomel, Mogilev and Bryansk Oblasts was generally formed by radioactive fallout in the period of April 27 – 29, 1986. In some settlements of this zone the contamination density by ^{137}Cs exceeded 1500 kBq/m^2. The northwest (NW) zone 500 km distant from the epicenter included some regions of Kaluga, Tula and Orel Oblasts. It was formed by atmospheric fallouts in late April – early May, 1986. In this zone, contamination by ^{137}Cs was generally below 600 kBq/m^2. Outside the most highly contaminated areas, there were many regions with ^{137}Cs concentration within the range of 37 – 200 kBq/m^2. It is believed that 21 administrative subjects of the Russian Federation were contaminated by emissions of the Chernobyl accident. For the majority of these regions contamination by ^{137}Cs did not exceed 37 kBq/m^2. For the four regions (Bryansk, Kaluga, Orel and Tula Oblasts), the level of contamination by ^{137}Cs was detected between 37 and 1500 kBq/m^2.

THE STRUCTURE OF AGRICULTURAL SECTOR IN CONTAMINATED REGIONS OF RUSSIA

The regions of Russia exposed to the highest contamination represent the territories of intensive agricultural production. In 1986, four regions of Russia with the highest level of contamination produced over 2 million tons of grains, about 1.3 million tons of potatoes, over 750,000 tons of milk and 200,000 tons of beef.

Table 2. Agricultural production scope in the areas of the Russian Federation contaminated due to the Chernobyl accident, thousand tons (1986)

Oblast	Grain	Potatoes	Vegetable	Milk	Beef
Bryansk	296.6	691.4	46.9	57.6	14.4
Kaluga	38.5	50.5	0.04	49.6	8.2
Orel	370.4	177.5	17.8	450	106.7
Tula	1437	327	51	670	195
Total	2142.5	1246.4	115.74	1227.2	324.3

Contaminated regions of Bryansk, Orel and Tula Oblasts produced from 40 to 70% of grain, up to 60% of potatoes and from 30 to 60% of livestock products in the total production scope of these Oblasts.

In the accident zone, in the areas with ^{137}Cs concentration exceeding 37 kBq/m^2, more than 6,000 settlements with about 3 million citizens were located. In Bryansk Oblast, which was most highly contaminated, in the contamination zone about 700 settlements with 327,400 citizens were located.

RADIATION IMPACT ON AGRICULTURAL PLANTS AND ANIMALS

The highest radiation impact was observed for ecosystems in the 30-km restriction zone of the Chernobyl accident. The accident happened in the end of April, in the period of fast growth and reproductive organs formation, when phytocoenoses and many representatives of mesofauna demonstrate the highest radiosensitivity. The highest radiation impact was detected in the initial 10 – 20 days after the accident, when a significant contribution into the absorbed dose was made by short-lived radionuclides. The second phase included summer and early autumn of 1986, during which the dose rate decreased to 20 – 50%of the initial value [2]. In 1986 – 1988, in the 30-km zone of the Chernobyl accident radiation damage to plants and animals was observed, though it did not disturb the ecosystems, except for the pine forest nearby Chernobyl NPP (the “Red forest”).

In 1986, the main contribution into doses absorbed by plants was made by β-radiation, and external γ-radiation gave about 5 – 10% of the total dose [3, 4, 5]. In the 30-km zone the dose absorbed by herbs in the initial period equaled 1 – 100 Gy (1000 Gy as a maximum). The two third parts of this dose was formed during the first month.

In the vegetation period of 1986, radiobiological effects in herbs were not visually observed even in the 10-km zone. However, the harvest of winter rye was reduced to 50% and grains were partly sterilized at the level of 15 Gy dose absorbed during the first month [6, 7]. At the exposure dose of 7.5 mR/h on the 15th day after the accident, winter wheat demonstrated reduced number of grains in heads, sterility reached 25%, and the crop yield was 10 centner/ha [8]. In 1986, the quality of cock's-foot grass seeds gathered in the areas with different radioactive contamination level depended on the radiation impact intensity: a reliable tendency to decrease of germinating power and mass of 1000 grains with increasing γ-radiation intensity was observed [9]. Simultaneously, winter wheat grains (which absorbed 10 – 400 Gy dose during the first month) demonstrated satisfactory germinating power (67 – 95%), and morphological parameters of plantlets were normal [7].

Cytogenetic analysis of root meristem cells from winter rye and winter wheat plantlets harvested in 1986 indicated a dependence of aberrant cell yield on the radiation dose [4, 6]. A reliable exceeding of the control level of aberrations was detected at 3.1 Gy absorbed dose, mitotic activity suppression – at 1.3 Gy, germinating power – at 12 G, e.g. in 1986, according to the basic tests, the radiation damage to agricultural plants seemed to be similar to the effect induced by comparable doses of γ-radiation. The frequency of aberrant cells in intercalary meristem in the second and the third generations of winter rye and winter wheat at the most highly contaminated site reliably exceeded this parameter for the first generation.

In 1987, the increased level of mutations in phytocenoses apearing as various morphological deformations was observed. The morphoses observed in plants included fasciation and culm branching, doubling, changes in blossom trusses, color and size of lamina and flowers [7]. Morphoses were detected when the exposure dose rate of γ-radiation reached the level of 20 – 30 mR/h on 10 May, 1986. As the dose rate increased to 75 – 150 mR/h, intensification of vegetative propagation (heath) and gigantism were observed.

Radiation damage of farm livestock was caused by thyroid gland disorders due to radioactive iodine accumulation in it. 240 days after the accident, cows in Gomel Oblast demonstrated the ratio of absorbed doses from all radiation sources for the thyroid GIT mucosa and the whole body of 230:1.2:1 [10]. Partial atrophy or complete degradation of the thyroid gland was observed in evacuated cows 5 – 8 months after the accident and some animals died demonstrating clinical symptoms of myxedema [11].

Doses absorbed by thyroid glands of cattle from Khoiniksk and Bragin regions of Belarus equaled 44 – 124 and 25 – 314 Gy, respectively. These doses were generally formed within the first month after the accident. By the end of the first year after the accident, the radiation doses to the whole bodies of animals were below 0.2 Gy [12].

Some part of cattle (1987) and lambs (1988), evacuated from the 30-km zone (doses absorbed by the thyroid gland exceeded 200 Gy), demonstrated chronic radiation disease [12]. The level of thyroid gland radiation damage relates to the concentration of stable iodine in the forage. For example, thyroid glands of lambs with reduced level of iodine nutrition from Belarus absorbed greater amount of radioactive iodine, which promoted formation of 2 – 2.5-fold higher doses compared with the control [12].

The change in the concentration of thyroid hormones and adenylate cyclase activity, determined in productive cattle in the first year after the accident, were reversible. The results obtained testify about the presence of cAMP activation system in animals with reduced secretion of thyroid hormones and radioiodine damage to the thyroid gland [11].

MIGRATION OF RADIONUCLIDES IN AGRICULTURAL ECOSYSTEMS

By the moment of the Chernobyl accident the facts were well-known about increased intensity of long-lived ^{90}Sr and ^{137}Cs radionuclides migration in trophic agricultural chains in woodlands (A.N. Marei, A.A. Moiseev, R.M. Barkhudarov, V.A. Knizhnikov). A considerable volume of information about the behavior of radionuclides (mostly ^{90}Sr) in the accident area in South Urals and a large data massif of model tests (V.M. Klechkovsky, N.V. Timofeev-Resovsky, N.V. Kulikov, I.V. Gulyakin, F.A. Tikhomirov, R.M. Alexakhin, N.A. Korneyev, E.V. Yudintseva, E.A. Fedorov, B.S. Prister, N.P. Arkhipov, A.N. Sirotkin, etc.) were accumulated.

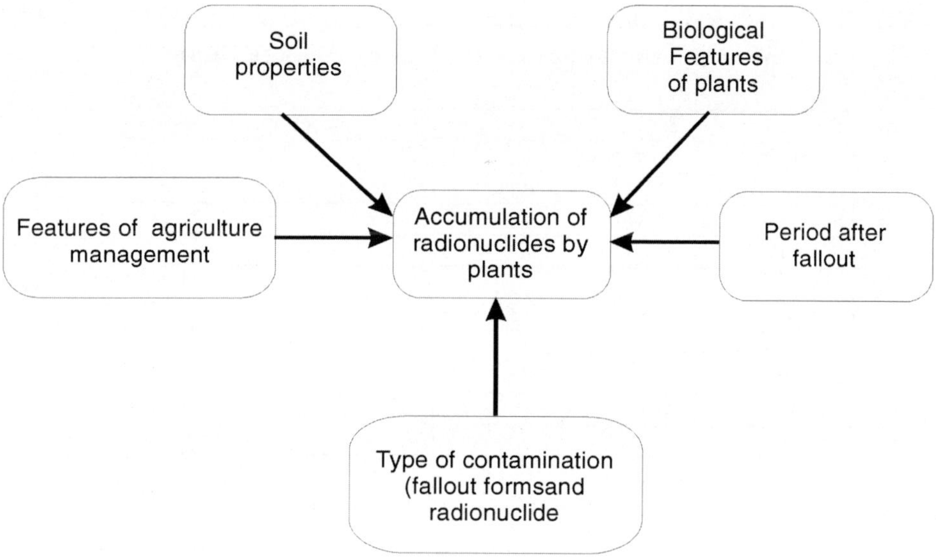

Figure 1. Basic parameters defining the radionuclide migration rate in the soil – agricultural plants system

The Chernobyl accident set up several new questions before the agricultural radioecology. They are:

- estimation of aerial contamination parameters for the soil part of agricultural ecosystems for early spring fallouts;
- study of the influence of fallout physicochemical forms on migration of radionuclides by agricultural chains;
- study of the dynamics of radionuclide biological mobility with respect to fallout forms;
- estimation of radionuclide migration rates in various soil-climatic zones.

Five groups of factors determining behavior of radionuclides were identified:

- radioactive fallout characteristics;
- soil properties;

- biological features of plants;
- period after fallout;
- specific features of agricultural practice.

AERIAL CONTAMINATION

Radionuclide fallout after the accident induced foliar contamination of the herbal part of plants having at that time highly developed herbal mass – winter crops, natural and sown perennial grasses [13].

Table 3. ^{137}Cs concentration in plants standardized by fallout density by June 10, 1986 (the northern radioactive trail zone), (Bq/kg)/kBq/m^2)

Crop	^{137}Cs concentration in plants
Natural grasses (dry meadows)	23-75
Natural grasses (floodplain meadows)	60-175
Sown perennial grasses	15-225
Winter rye	16-118
Barley	2-4
Oats	2-5
Corn	4-20
Greens	15-17
Lettice	12-14
Dock	10-12

The observations in the 50-kilometer zone of the Chernobyl accident indicated that accumulation of radionuclides in plants during the aerial contamination is dependent on the fallout characteristics, meteorological conditions and some parameters characterizing the interaction between radioactive mixture and plants. The accumulation of radionuclides by plants was determined by the scope of their primary absorption. The main part of fallout radionuclides was present in the soil and root mat – up to 60 – 80% [14]. The grass stand of natural meadows absorbed 30 – 45% of fallout γ-irradiating nuclides and up to 30% of ^{137}Cs. For the winter rye, variations in the primary absorption of radionuclides fell within 10 – 30% and 7 – 20%, respectively. The differences in the initial absorption for the same culture depended on both the conditions of seeds during the fallout period and the distance from Chernobyl NPP.

During the vegetation period in 1986, the concentration of radionuclides in plants was reduced due to radioactive decay, biomass increase and seeds cleaning by meteorological factors. The ^{131}I half-life period in the grass stand of pastures varied from 2-3 to 9.6 days [14]. For winter rye, the half-life period estimated for the totality of γ-irradiating nuclides, equaled 8 – 10 days; the second period took 28 – 35 days, for grass stands – 7 – 12 and 30 – 35 days, respectively. ^{137}Cs concentration decreased slower than the totality of γ-irradiating nuclides: by June, vegetative mass of winter rye contained 20 – 30% of primarily absorbed ^{137}Cs and by harvesting period – about 9 – 20%; for the grass stands these values were 15 – 40 and 5 – 10%, respectively.

DYNAMICS OF BIOLOGICAL AVAILABILITY OF RADIONUCLIDES IN SOILS

Dynamics of radionuclide behavior in the soil defines their absorption in all components of agricultural chains. With time, radionuclides captured by soil are fixed by the solid phase. In 1986, the concentration of exchangeable ^{137}Cs in soils was maximal and varied from 9.5 to 30.1% [15]. Two periods are identified, when the exchangeable ^{137}Cs concentration changed at different rates: the first period lasted since 1986 till 1988 and the second – since 1989 till 1995. The first and the second half-life periods of decrease of the exchangeable ^{137}Cs concentration in the soil lasted 2.4 – 3.5 and 4.6 – 12.6 years, respectively (Table 4).

For the zone with the fuel component valence in the fallout, half-life periods of the exchangeable ^{137}Cs concentration decrease in mineral soils is longer than for the aerosol fallout zone. Generally, 5 years after the accident ^{137}Cs forms present in the soils in the trails with different fallout characteristics are relatively stabilized [16].

The behavior of ^{90}Sr in soils had some features caused by the fallout properties. Due to ^{90}Sr leaching from fuel particles, during the initial 5 years after the accident the concentration of its available forms in soils increased, whereas since 1990 it was stabilized [17]. For the aerosol fallouts, these regularities are not so significant as for contamination by fuel particles.

Table 4. The half-life periods of exchangeable ^{137}Cs concentration decrease in soils, years

Half-life period	Soils			
	Automorphous		Hydromorphous	
	Soddy-podzolic, soddy		Soddy-gley, peated	
	Fuel component zone	Aerosol fallout zone	Fuel component zone	Aerosol fallout zone
Average	5.3	5.2	3.7	3.5
First	3.5	2.8	3.2	2.4
Second	4.5	9.5	8.9	12.6

Thus, 5 years after the accident a period of relative stabilization of radionuclide adsorption in the soil occurs. Quite high concentration of ^{137}Cs available forms (the totality of exchangeable and mobile forms) is typical of soddy-poszolic and soddy soils of light mechanical composition – from 14.5 to 23.3% [18]. Medium and highly loamy soddy-podzolic, grey forest soils and chernozems fix radionuclides more strongly. The concentration of ^{137}Cs available forms in them varies from 4.2 to 15.6%. For ^{90}Sr, no considerable differences for various types of soils were detected, because the exchangeable type of sorption by soils is typical of this radionuclide.

VERTICAL MIGRATION OF RADIONUCLIDES IN SOILS

Vertical migration of ^{90}Sr and ^{137}Cs in soils of natural meadows is a slow process: radionuclides are mostly concentrated in 0 – 10 cm layer. The fastest migration is typical of peat soils, 7 – 8 years after the accident ^{137}Cs was detected at the depth of 20 cm in them.

Maximal coefficients of ^{137}Cs quasi-diffusion were obtained for bog meadows: from 0.097 to 0.464 cm^2/year for the "slow" component and from 0.40 to 1.28 cm^2/year for the "fast" component [18]. The average parameters of quasi-diffusion and convective transfer of ^{137}Cs by dry meadows is 4 – 5 times lower than for peats, equaling 0.015 - 0.062 and 0.09 - 1.39 cm^2/year for "slow" and "fast" component, respectively. The main contribution is made by the "slow" component of migration. One of the integral parameters used in forecasting the radiological state is the half-cleaning period of the root area of the soils – the period during which the concentration of radionuclides in the root area two-fold decreases [20]. The longest ecological periods of the half-cleaning of the soils from ^{137}Cs were observed for dry meadows (T_{ec} = 55 – 143 years) and the shortest for bog meadows (T_{ec} = 15 – 21 years).

For all types of meadows, quantitative parameters of ^{90}Sr migration were higher compared with ^{137}Cs. For dry meadows, the quasi-diffusion "fast" component coefficients vary from 0.08 to 0.65, whereas for bog meadows – from 0.49 to 0.69 cm^2/year. Elimination of ^{90}Sr from the root area proceeds twice faster compared with ^{137}Cs. The half-cleaning periods for the root area from ^{90}Sr varied from 30 to 96 years for dry meadows and from 13 to 18 years for lowland meadows.

Table 5. Half-life periods of ^{137}Cs concentration decrease in the grass stands, years

Half-life period, year	Soils				
	Automorphous		Hydromorphous		
	Soddy-podzolic, soddy		Soddy-gley, peated; soddy-podzolic, gleyed		Peat bog
	Fuel component zone	Aerosol fallout zone	Fuel component zone	Aerosol fallout zone	Aerosol fallout zone
Average	3.8	5.0	10.1	4.1	7.9
First	6.5	3.0	1.3	5.3	1.3
Second	3.7	17.0	4.2	28.9	-

DYNAMICS OF RADIONUCLIDE BEHAVIOR IN THE SOIL – PLANT SYSTEM

Accumulation of radionuclides in plants was defined by the fallout form and properties of the soils. In 1986, ^{137}Cs concentration in plants in the aerosol fallout zone was about 1.5 times higher than in the areas with predominant fuel component [15]. In the second year after the accident ^{137}Cs concentration in plants was reduced 3.2 – 5.5 times, because the dominating way of ^{137}Cs delivery became the root consumption. Since 1987 till 1995 ^{137}Cs transfer coefficients were decreasing, however, at lower rate, by 1.4 – 6.8 times, on average. Since 1989 – 1990, variations of ^{137}Cs transfer coefficients to the plants were mostly caused by differences in the soil properties (Table 5) [16].

The specificity of ^{90}Sr behavior in soils, induced by fallout properties, has defined dynamics of its accumulation in plants. In the early period after the accident, the grass stand contamination was caused by aerial fallouts. In 1987, ^{90}Sr concentration in the grass stands was 4 times lower. During ^{90}Sr leaching from the fuel matrix, it appeared in a form accessible for plants. As a result, transfer coefficients of this radionuclide increased during initial 5 – 6 years after the accident; thereafter, biological availability was stabilized [17]. At present, ^{90}Sr transfer coefficients approach the values observed for Polesye regions during global fallouts.

Table 6. Half-life periods of ^{137}Cs concentration decrease in plants, years

Crop	Groups of soils	Half-life periods
Cereals	Sands	7 - 11.8
	Loamy	5.1 - 22.0
	Argillaceous	14.4 - 40.4
	Peaty	9.2 - 14.4
Potatoes, root crops	Sands	7.0 - 9.2
	Loamy	10
	Argillaceous	13.3
	Peaty	10
Sown perennial grasses	Sands	6.4
	Loamy	4.0
	Argillaceous	9.8
	Peaty	17.3
Natural grasses	Sands	5.8
	Loamy	18.0
	Argillaceous	13.3
	Peaty	17.3

The study of dynamics of ^{137}Cs accumulation in plants revealed two periods of radionuclide delivery decrease: the first period was 1987 – 1989, when an abrupt decrease of ^{137}Cs accumulation in plants, on average, by 3 – 5 times was observed; the second period was 1990 – 1995, when the rate of ^{137}Cs transition to plants was decreased [16, 20, 21]. Since 1996, no significant differences in concentration of the radionuclide in separate years were observed. Large data scope processing, accumulated in the database of Russian Institute of Agricultural Radiology and Agroecology, indicated further reduction of ^{137}Cs accumulation in plants. Usually, T_{ec} values obtained in the long-term periods after the accident, which exceeded 10 years and even approached 15 years, are close to values observed for global fallouts.

Migration of Radionuclides in the Ration – Farm Cattle – Agricultural Production System

For the first period after the accident fallout unique data on animal products contamination during aerial contamination of pastures were obtained. In the first period after the accident, ^{131}I concentration in milk decreased with its concentration in the pasture vegetation. By the middle of June this radionuclide was not identified in milk. Gradually decreasing concentration of ^{137}Cs in milk reflected dynamics of pasture grass stand contamination. In this period, the standardized coefficient of ^{137}Cs transfer to milk equaled 0.42 calculated per its concentration in grass of 1 Bq/kg [22].

The transfer of ^{137}Cs to animal products is defined by the following factors: the contamination level of pastures and feeding ration, species and age of animals, type and composition of ration, productivity of animals, and technique of their maintaining. Since fodder plants forming the ration of animals possess different ability to accumulate ^{137}Cs from the soil, the ration composition plays an important role in delivery of the radionuclide to the organism. Moreover, the ration composition for animals for the pasture and stall periods are significantly different. Seasonal dynamics in the radionuclide content in forage and products of cattle breeding were observed in the studies, carried out for both global fallouts and in the period after the Chernobyl accident (Figure 2).

The highest ^{137}Cs transfer coefficients from the day ration to 1 liter of milk were observed in the pasture period, which was caused by consumption of soil particles by animals with forage [23].

As the pasture productivity was high, the average consumption of soil particles by cattle equaled 0.5 kg/day. If the pasture was poor, the consumption of soil particles exceeded 2 kg/day, which gave about 20% of dry forage mass. Another reason for increasing delivery of radionuclides to milk on poor pastures is consumption of the lower level of the grass stand by cattle, where ^{137}Cs concentration is 2 – 3 times higher than in the upper layer [24].

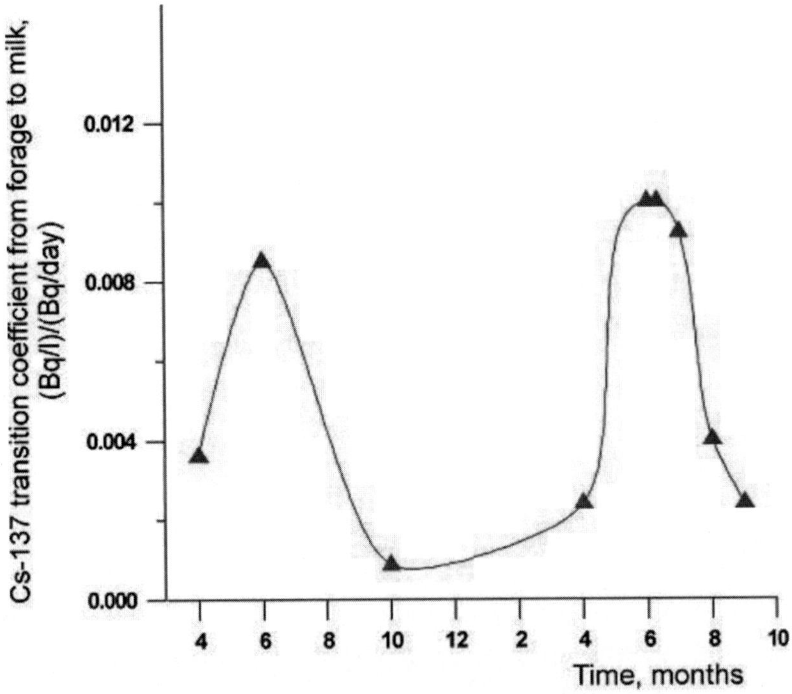

Figure 2. Seasonal dynamics of ^{137}Cs transfer coefficients from daily ration to milk

High variability of the part of ^{137}Cs transiting from the forage to milk, from 0.16 to 3.5% was detected. For the first time, this fact was determined in the studies under real production conditions, whereas under experimental conditions this value, on average, equaled 1% [25]. Apparently, such differences were reasoned by different digestibility of radionuclides from forage of different composition.

MODELING OF RADIONUCLIDE BEHAVIOR IN AGRICULTURAL AND NATURAL ECOSYSTEMS

After the Chernobyl accident special attention was devoted to mathematical modeling of radionuclide behavior in agricultural and natural ecosystems. First of all, this is stipulated by the necessity to solve applied tasks related to prediction of radionuclide concentration in natural objects and foods forming the ration of the population. Radioecological information obtained in the post-accident period allowed for modernizing approaches to development of migration models and broaden their application sphere. Generalization and analysis of the experimental data in the framework of the models allowed for quantitative estimation of radionuclide migration processes with respect to fallout properties and characteristics of ecosystems.

The features of emergency emission formation brought up the problem of prediction behavior of radionuclides with respect to fallout characteristics. Experimental studies have determined differences in the rates of sorption and fixation of radionuclides in soils in the zone with predominance of the fuel component and the zone of aerial fallouts. A model of radionuclide migration was designed describing sorption and fixation of radionuclides in soils and including concentration of various forms of radionuclides as the key compartments, which characterize their biological mobility in the soil. This model adapted for the "nearest" zone conditions (the zone, where fuel particles dominate in the fallouts) was used for analyzing the factors defining dynamics of the biological availability of the radionuclide in the soil – plant system [15, 26]. The model describes and qualitatively estimates the effect of leaching fuel particles the mobility of radionuclides in soils. In the initial period after the accident, the main factor defining biological availability of ^{137}Cs is the character of fallouts and properties of soils – after relative stabilization of the radiation situation.

Based on the results obtained for different types of meadows, a model of ^{137}Cs vertical migration in soils with respect to the forms of this radionuclide in the soil was designed [20]. Parameters of the model were estimated for a wide selection of meadows and soils typical of the territories, exposed to contamination. This model helped in calculating the ecological half-life periods of ^{137}Cs concentration decrease in the root area of the soil.

Further modifications of the radionuclide migration models in the soil – plants system were performed in two directions:

- modeling of the migration features with respect to specific characteristics of ecosystems (for example, for meadow ecosystems on peaty soils);
- description of the mechanisms defining behavior of radionuclides in the soil and controlling nonlinear processes proceeding.

Realization of the first direction has designed a model of ^{137}Cs behavior in the peaty soil with respect to the season dynamics of plant biomass and increasing upper layer of peat-bogs due to dying-off and decomposition of plant residues [27]. In the framework of the second direction, the models based on identification of the mechanisms controlling physicochemical and biological processes, which define accumulation of radionuclides in the plants, were designed [28]. As these models were realized, the change of radionuclides behavior under the effect of agromeliorants used as protectors against accumulation of a radionuclide in the agricultural production was described. The model allowed for estimation of changes in the biological mobility of ^{137}Cs in the soil – plants system after application of potassium and dolomite powder.

To provide for decision information-analytical support on the agricultural production management on the territories contaminated by the accident, a complex of mathematical models of ^{137}Cs migration in agricultural ecosystems was developed [29]. The models describe dynamics of this radionuclide transfer under conditions of protective and rehabilitation measures applied to ploughed fields, hayfields and pastures, and in the natural meadow systems. The concentration of ^{137}Cs in the main types of agricultural products is predicted and the efficiency of countermeasures in the accident zone is estimated.

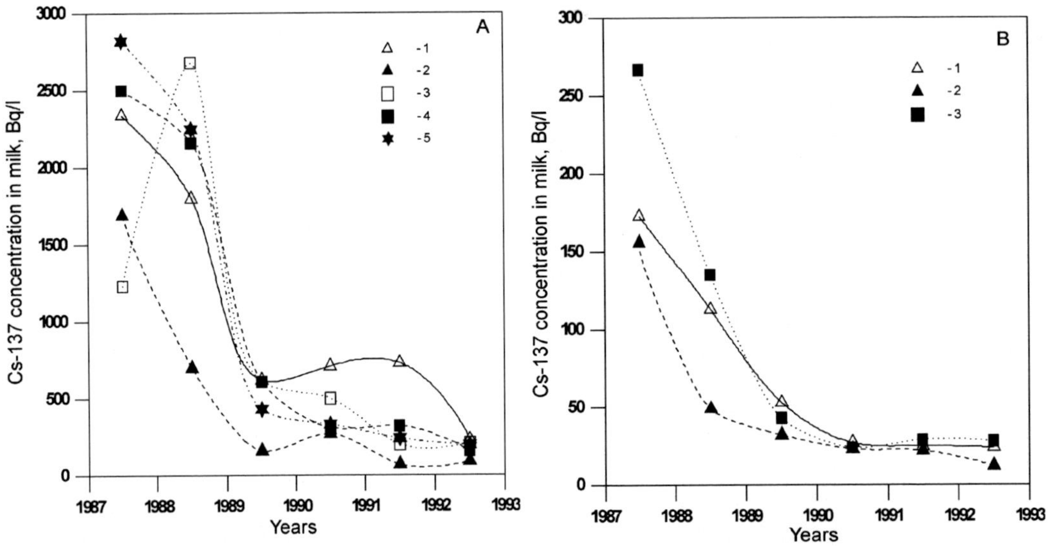

Figure 4. Dynamics of ^{137}Cs concentration in milk in the regions of Russia, exposed to radioactive contamination by the Chernobyl accident: A: Bryansk Oblast (regions: 1 – Krasnogorsk; 2 – Klimovsk; 3 – Klintsovsk; 4 – Novozybkov; 5 – Gordeev); B: Kaluga Oblast (regions: 1 – Ul'yanovsk; 2 – Zhizdrinsk; 2 – Khvastovich)

A great scope of experimental data obtained in the period of the Chernobyl accident allowed for extending the ideas about the main mechanisms, which control ^{137}Cs behavior in the forest components, and modifying the migration models for this radionuclide in forest ecosystems. The models designed for prediction ^{137}Cs accumulation in the main types of forest products (mushrooms, berries, forest animals and wood) are parameterized for forest ecosystems of various types [30]. A complex of these models was applied to justification of the optimal system of countermeasures in contaminated forests of Bryansk Oblast [31].

DYNAMICS OF AGRICULTURAL PRODUCTS CONTAMINATION ON THE TERRITORY OF THE RUSSIAN FEDERATION

The observed concentration of ^{137}Cs in the agricultural products was caused by its sorption by the solid soil phase, implementation of protective measures and radioactive decay [32]. The rate of ^{137}Cs concentration decrease in the agricultural products was significantly different for different areas (Figure 3).

This is because in the initial two years after the accident the main conditions for decreasing contamination of the agricultural products were focused on the areas with the highest contamination levels (Novozybkov, Krasnogorsk and Gordeev regions of Bryansk Oblast) [32 – 36]. In these areas, owing to large-scale soil fertility measures (radical and surface improvement of hayfields and pastures) and introduction of organizational measures a sharp (3 times, approximately) reduction of ^{137}Cs concentration in milk was observed. In contaminated regions of Kaluga Oblast, where protective measures were highly limited, ^{137}Cs concentration in milk was reduced much slower.

The highest reduction rate of ^{137}Cs concentration in milk and other agricultural products was observed in the first year after the accident. The average ^{137}Cs concentrations in the period of 1987 – 1989 were reduced 4 times in Krasnogorsk, Gordeev and Klintsov regions of Bryansk Oblast and more than 7 times in Novozybkov region of Bryansk Oblast [34]. In the further period, reduction of ^{137}Cs concentration in milk was slower and equaled, on average, 2 to 3 times in the period of 1989 – 1992 (Table 7).

Table 7. Environmental half-life periods of Cs-137 concentration reduction in milk, years

Regions	Average period, 1987 – 1994 (T_e^t)	First period, 1987 – 1989 (T_e^1)	Second period, 1990 – 1995 (T_e^2)
Bryansk Oblast			
Gordeev	1.7	1.2	2.6
Klimovsk	1.8	-	4.7
Klintsov	2.3	1.3	2.6
Krasnogorsk	1.8	1.0	3.5
Novozybkov	1.6	0.8	4.2
Kaluga Oblast			
Khvastovich	2.3	1.5	6.3
Ul'yanovsk	4.8	1.8	11.7
Zhizdrinsk	2.9	1.1	5.7

Analogous regularities are also typical of ^{137}Cs concentration reduction in the crop growing products. In the majority of cases, T_e values calculated for the main types of the crop growing products are higher than for milk, equaling for the regions of Bryansk Oblast: T^1_e - from 0.6 to 1.7 years, T^2_e - from 1.7 to 14 years [36].

For the regions of Kaluga Oblast, T^1_e varies from 1.0 to 2.9 years and T^2_e from 3 to 14 years or more. Thus, T_e^t for all types of crop growing products in the regions of Bryansk Oblast are much lower (1.0 – 2.8 years) than in Kaluga Oblast regions (2.0 – 6.9 years), which correlates with higher scope of countermeasures implemented in Bryansk Oblast and earlier initiation of their implementation after the accident.

Since 1995, the rate of ^{137}Cs concentration decrease in the agricultural products became much lower, and dynamics of this decrease is significantly different for the regions with different scopes of implementation of the protective measures in 1990 – 1992. Since 1995 till 1999, in the regions with the most intensive introduction of counter measures a tendency to increase contamination of the products was observed. In the areas with limited use of protective measures the increase of the contamination level of products due to reduced application of countermeasures was compensated by natural reduction of ^{137}Cs availability in the soil, and concentration of this radionuclide in the products was at the same level.

Organization of the Agricultural Production and Implementation of Protective Measures on Contaminated Territories

Rehabilitation of the territories contaminated by the Chernobyl accident is characterized by some specific features [37]. In various periods after the accident, it was necessary to design a system of protective measures, which would allow for maximal reduction of the radioactive contamination consequences in the agricultural production. Primarily, the stage-by-stage implementation of the protective measures was initiated after the Chernobyl accident. In the first period, general approaches to implementations of countermeasures in the zones with different contamination levels were designed. At the second stage, protective measures were developed with respect to the soil-climatic features of contaminated territories. In the remote period after the accident, rehabilitation strategies taking into account the specificity of agricultural enterprises were suggested. Systems of protective measures separate for each farm or settlement e.g. targeted protection was designed [38, 39].

The new directions of the studies in the field of scientific justification of the protective measure system are development of a strategy for estimating their effectiveness and composing the list of criteria for selection of the most optimal actions [40]. Usually, multiplicity of radionuclide concentration decrease in the products is the main criterion for the estimation of countermeasure effectiveness. However, this parameter is not sufficient. First of all, this relates to the estimation of the influence of protective measures in the agriculture on reduction of the collective dose of radiation among the population, because the application of intense techniques increases productivity of cultures and, consequently, total yield of radionuclides with the products [41]. Hence, total yield of ^{137}Cs depends on the soil conditions, scales of protective measures, and the features of agricultural production management. For the purpose of estimating the effectiveness of protective measures, a set of criteria including radiological, dose and economic ones has been suggested [42].

APPLICATION OF PROTECTIVE MEASURES IN THE FIRST YEAR AFTER THE ACCIDENT

The first period after the accident was determined as the period of iodine danger due to the presence of short-lived iodine radionuclides in the fallouts, mostly ^{131}I. The most acute conditions were observed in several initial weeks. The fallout happened, when the cattle had been moved to the pasture, and radioiodine was immediately included into trophic chains. In this period, milk was the critical food product. Prohibitive and restrictive measures were recommended: prohibition and restriction of milk consumption, prohibition for owning private dairy stock in the areas with the highest contamination levels [43]. The main protective measure in the agriculture was elimination of contaminated grass stand from the cattle forage, i.e. cattle stabling. Moreover, special processing of agricultural products was used (first of all, milk processing to condensed or powder milk), where ^{131}I concentration exceeded temporarily permissible concentrations (TPC = 3700 Bq/l). The most important was adoption of the life time monitoring method of the farm livestock for radionuclide concentration that stopped unreasonable slaughtering on the contaminated territories.

The second period after the accident is related to aerial contamination of agricultural lands. It lasted during the first vegetative period after radioactive fallout. At this stage, the main way of radionuclide delivery to plant growing and fodder cropping products was direct contamination of above-ground mass of crops. In this period, the main problems were induced by agricultural product contamination with ^{134}Cs and ^{137}Cs.

The main measures in the agriculture were either prohibitive or restrictive:

- prohibition of slaughtering in the areas with ^{137}Cs contamination level exceeding 555 kBq/m^2 or recommendation for 1.5 month cattle feeding by "pure" forage before slaughtering;
- changing types of processing of agricultural crops – reduction of the number of operations producing dust;
- limited use of contaminated dung;
- storage of haylage and silage instead of hay;
- mandatory dosimetric control;
- restricted milk consumption, produced by the private sector, mandatory processing of milk;
- in some regions, restrictions on owning dairy cattle and other livestock in the private sector were imposed.

Private livestock was evacuated from settlements with ^{137}Cs contamination above 555 Bq/m^2 (15 Ci/km^2) and population support by uncontaminated food was organized, which was produced in the national sector or imported from uncontaminated regions. These measures included about 200 settlements. Already in 1986, 8,813 pieces of cattle stock and over 15,000 pieces of small cattle and pigs were evacuated from these regions. Further on, in 1987 – 1988, 6,881 pieces of cattle stock and over 10,000 pieces of other livestock were withdrawn from these settlements, additionally.

In the period of field works and harvesting the main measures were aimed at prevention of secondary contaminations of seeds: reduction of cultivation operations, maximal use of chemical weeding and motorized methods of harvesting, etc. For the first time, a system of soil processing for winter crops was designed and adopted. It included ploughing with soil overturning by 4 – 5 cm deeper than usual, placement of lime to acid soil and increased portions of phosphorus-potassium fertilizers [44 – 46].

In this period, the methods of industrial processing of contaminated agricultural materials were designed. They allowed for reducing concentration of radionuclides in the processed products to temporarily permissible concentrations (TPC). In particular, milk processing to butter and beef to sausages was recommended. Alcohol production from contaminated grain was tested.

LONG-TERM PROTECTIVE MEASURES

The third period in development of the radiological situation in the agriculture started with the second vegetation period after radioactive fallouts (since 1987). The main way of radionuclides delivery to the agricultural migration chains was their root consumption by plants from the soil. In this period, the important factor for justifying measures is accounting for the influence of soil-climatic and geochemical features of contaminated territories, which cause difference in migration of radionuclides. This period may last dozens and hundreds of years.

Protective Measures in the Plant Growing

Plant growing is one of the main branches in agriculture of contaminated regions of Russia. After the accident the contamination zone included territories with different soil-climatic conditions and different techniques of cropping that brought up a task of optimizing the use of protective measures with respect to zonal features of agriculture management. The studies performed allowed for selecting three groups of methods, which may be applied to plant growing: land treatment, agrochemical and organizational [45, 46].

Organizational measures include selection of crops, characterized by low accumulation of radionuclides, or industrial crops, not used for food production. Accumulation of ^{137}Cs by more than 100 cultivar species of main crops [47] was studied. For the use on contaminated territories, zoned species and cultivar crops were recommended, which, on average, allowed for 3-fold reduction of ^{137}Cs accumulation in the products.

A set of agricultural methods, which may be used on croplands, includes standard ploughing 18 – 25 cm deep; ploughing with soil overturning by 4 – 5 cm deeper compared with usual depth; deep ploughing (by 50 – 70 cm) with soil overturning. Due to redistribution of radionuclides in the top soil, in the first year of application after fallout happened standard ploughing by 18 – 25 cm allowed for 1.3 – 3.0-fold decrease of radionuclide transition to plants [44, 48]. Special ploughing technique, which included increasing ploughing depth with the soil overturning by 4 – 5 cm and provided for reduction of radionuclide accumulation in plants by 5 – 7 times. Further on, usual ploughing was used that provided for burial of the contaminated layer. Such method appeared quite effective for croplands and reseeding of meadow pastures, repeated every 4 – 5 years.

Agrochemical measures include traditional methods (the use of fertilizers and agromelioration agents), however, for contaminated areas the most optimal doses and combinations of their application should be justified. The main agrochemical method for limitation of ^{137}Cs delivery from the soil to plants is application of potash fertilizers that reduces radionuclide delivery to plants 2 – 20 times [48 – 51]. Introduction of increased phosphorus fertilizer doses reduced ^{137}Cs delivery to plants on soddy-podzolic sandy-loam soils and sands by 2 – 2.5 times, and for leached humus – by 1.5 – 3 times [49, 52]. The effectiveness of application of nitrogan fertilizers depends on their forms. Introduction of fertilizers formed as ammonium salts increases ^{137}Cs accumulation in plants.

Table 8. The effectiveness of agrotechnical and agrochemical protective methods for reducing accumulation of radionuclides by plant products

Procedure	Changeable parameters	Effectiveness reduction factor of radionuclide accumulation in plants, times
Ploughing	Redistribution of radionuclide in the top soil	1.5 – 2.5
Ploughing with soil overturning	Mechanical moving of contaminated layer to subjacent soil horizons	up to 5 – 10
Lime application	Soil acidity change, saturation of soil absorbing layer with Ca	1.5 – 2.0
Introduction of increased amounts of phosphorus-potassium fertilizers	Soil acidity change, increase of positive ion K concentration, change of saturation with bases	1.5 – 2.0
Introduction of organic fertilizers	The change of carbon exchange scope and concentration	1.5 – 2.5
Application of clay minerals	The increase of sorption ability of soils and competitive influence of positive ions K and Ca emitted to the soil solution	On light soils – 1.5-3.0-fold decrease of ^{137}Cs accumulation in plants; on other soils no effect is observed
Complex application of melioration agents	The change of soil acidity and carbon content, increasing saturation with bases, increasing concentration of positive ions K and Ca	Up to 5.0 times

Based on the studies performed, the application of mineral fertilizers in the ratio N:P:K = 1:1.5:2 was justified [44, 46]. It is recommended to introduce nitric fertilizers in the form of nitric salts in doses optimal for current soil and predominantly before seeding.

In the most cases, introduction of organic fertilizers reduces ^{137}Cs delivery to plants by 1.5 – 3 times, whence the highest effect is observed for soils of light granulometric composition [49, 53]. For reduction of ^{90}Sr transfer, the efficiency of application of organic fertilizers depends on the crop species. Introduction of dung for potatoes, lettice and corn is the effective method and provides for a decrease of radionuclide accumulation in plants by 4.6 – 5.6 times.

Lime application is a widely used method for increasing productivity of acidic soils [46, 48 – 50]. In case of radioactive contamination, lime doses must be 1.5 – 2 times increased compared with the dose for simple soil acidity neutralization. 1.5 – 4.0-fold lime application reduces ^{90}Sr and ^{137}Cs transfer to agricultural products. Lime ($CaCO_3$), dolomite powder, calcareous tuff, lake lime, bog lime and peat-tuffs are used.

In the initial years after the accident, the question about application of clay minerals, representing aluminosilicates of alkaline and alkaline-earth metals, having sorption ability and being able to absorb ions, including radioisotopes, in both exchangeable and fixed form, was the subject of considerable discussion. Tests with zeolite indicated a dependence of the effectiveness of this method on the soil and crop fertility [50, 54]. The use of zeolite on low-fertility soils causes a decrease of ^{137}Cs transfer to plants, on average, by 1.5 times. The highest effectiveness is reached on 2 – 3 years after the introduction. The application of bentonite and zeolite reduced ^{90}Sr transition in plants by 1.9 – 2.4 times. However, in some cases, introduction of clay materials caused no effect, for example, if they were applied to high fertility soils.

Table 9. Effectiveness of protective measures for meadows

Countermeasure	Reduction factor of 137Cs content decrease in the grass stand, times
Removal of the upper contaminated layer of the soil	5 - 15
Ploughing: standard	1.8 - 3.2
with soil overturning	2.0 - 6.0
deep	8 - 16
Disking and rotary tillage	1.2 - 1.8
Radical improvement	2.7 - 6.2
Surface improvement	1.6 - 2.9
Drying	2.8
Drying and surface improvement	2.5 - 5.5
Drying and radical improvement	3 - 10
Introduction of clay minerals to the soil surface in the first period of the accident	1.5 - 2.0
Application of nontraditional melioration agents (zeolite, palygorskite, vermiculite, etc.)	1 - 2.5

Protective Measures in Forage Production

Under conditions of radioactive contamination of the territory, the arrangement of the forage supply is the most important component of cattle breeding. There are two groups of agrotechnical methods, traditionally used for forage grasslands: surface and radical improvement of hayfields and pastures. Implementation of traditional measures for the purpose of increasing productivity of grass stands is also effective from the viewpoint of radionuclide accumulation reduction in the grass stand [55 – 57].

Maximal effectiveness of melioration agents was observed for low-fertility acidic soils of light mechanical composition. For the meadows subject to permanent or short-term overwetting (floodplain, lowland and boggy meadows), the application of agrotechnical methods reduced radionuclide transition to plants more significantly than in dry valleys. From the point of view of decreasing ^{137}Cs transfer to the grass stand, the radical improvement of meadows, which includes ploughing, lime application and N:P:K = 1:1.5:2 introduction, is more effective than the surface improvement.

Protective Measures in Cattle Breeding

The system of countermeasures in cattle breeding includes four groups of methods: organizational, restrictive, veterinarian and zootechnical [46, 58 – 60].

Restrictive measures: prohibition for holding dairy cattle; prohibition for using unimproved pastures or hayfields for the dairy cattle; prohibition for pasture maintenance of the dairy cattle. In the first period after the accident, these measures adopted indicated the highest effectiveness. In the initial years after the accident, over 15,000 pieces of cattle stock and over 14,000 pigs were confiscated from the population of Bryansk Oblast.

Organizational measures: forage grassland inventory by contamination density; prohibition of forage grassland use, if standard quality products may not be produced on them; the change in structure of cultivated areas and pastoral cropping.

Veterinarian countermeasures include application of two main groups of substances providing for production of specified "pure" products of cattle breeding (milk, beef). The first group includes specific cesium-binding compounds: ferrocyn, ferrocyn-containing boli and cattle salt licks. The second group contains natural adsorbents: zeolites, vermiculite, various clays, rotten-stones and gaizes. In addition to their adsorptive properties, these adsorbents represent additional sources of macro- and trace element for animals.

Zootechnical measures: prudent management of meadows and pastures (arrangement of cow feeding according to "green conveyor" type; substitution of cow pasture management by stable-extensive management; observance rate of stocking on the pasture, ha/piece; forage selection for rations; prior to slaughter stock fattening by "pure" forage". The arrangement of cow feeding according to the "green conveyor" type using extra nutrition from sown grasses or complete substitution of natural pastures by artificial forage grasslands reduces ^{137}Cs transition to milk 2 times.

The substitution of cow pasture management by stable-extensive management reduces ^{137}Cs transfes to milk and beef. This system is based on the use of long-term cultivated pastures with a significant supply (up to 50%) of green forage due to seeding and then mowed for feeding cattle from feed troughs.

The observance of rate of cattle stocking and controlled grazing (not shorter than 10 cm) reduces ^{137}Cs transfes to milk. The enclosure pasturing with daily zoning per animal is recommended.

The cattle feeding during stable management by mixed or ensilage-concentrate forage consisting of fodders, produced on meliorated fodder lands, provides for low ^{137}Cs delivery to the organism. It is desirable to increase the part of cellular tissue in the cattle ration in order to reduce radiocesium consumption in the digestive system. The cattle feeding by rough forage as hay, silage and haylage allows reduction of ^{137}Cs concentration in milk.

Table 10. Reduction of ^{137}Cs concentration in cattle breeding products as a result of countermeasure application

Type of counter measures	Kind of animals	Type of product	**Reduction factor, times**
Restrictive	Cattle stock	Milk	8.3-8.5 8.4
Organizational	Cattle stock	Milk	4.0-4.1 4.1
	Cattle stock	Beef	3.3-3.5 3.4
Veterinarian			
Application of Cs-binders	Cattle stock	Milk	1.5-21.8 6.9
	Cattle stock	Beef	1.9-2.3 2.1
Application of adsorbents	Cattle stock	Milk	1.2-2.0 1.9
Zootechnical			
Prior to slaughter feeding by "pure" forage	Cattle stock	Beef	2.0-15.2 7.7
	Horses	Beef	1.9-9.5 5.1
	Lambs	Beef	2.8-76.4 27.2
Prudent management of hayfields and pastures	Cattle stock	Milk	1.3-10.4 4.5
Forage component selection for the ration	Cattle stock	Milk	1.7-2.5 2.1
	Cattle stock	Beef	32.6-41.8 37.2

As the cattle are fed by "pure" forage, ^{137}Cs is quite rapidly removed from the organism. The first stage corresponds to the period of contaminated feed mass removal from the gastrointestinal tract (T_1); the second stage coincides with the period of mobile (intercellular) ^{137}Cs removal from the organism (T_2); the third stage – to the period of low mobile (intracellular) ^{137}Cs removal from the organism (T_3). For the cattle stock, $T_1 = 6 – 7$ days, $T_2 = 8 – 9$ days, and T_3 is defined by pollution intensity of muscles. For horses and lambs, T_1 and T_2 equal 8 – 9, 7 – 8 and 12 – 13, 5 – 6 days, respectively. In young animals the rate of ^{137}Cs removal from the muscles is higher and decreases with the age. For saving consumption of "pure" forage, they should be used only for feeding milking cows or beef cattle in the prior to slaughter period. The nuclide concentration in the forage is the higher, the longer time remains before the cattle slaughter.

It was confirmed by practice that three feeding stages are required: initial, intermediate, and final. As the cattle is fed by "pure" forage at the final stage, the life-time monitoring of animals is mandatory in order to avoid their slaughter.

ARRANGEMENT OF AGRICULTURAL PRODUCTION IN THE ZONES WITH DIFFERENT CONTAMINATION LEVELS

The features of emergency emission led to formation of the territories with irregular type of contamination that brought up the task of organizing the zonal system of the agricultural production management with respect to radiation contamination level [43, 45, 46]. At the second year after the accident, realization of the zonal principle for the agricultural production indicated the presence of 4 contamination density zones: 37 – 185, 185 – 555, 555 – 1480, and above 1480 kBq/m^2. For each separated zone, a complex of protective measures was suggested and realized.

In the first zone (with ^{137}Cs contamination density within 37 – 185 kBq/m^2) crops are cultivated by techniques, accepted for current soil-climatic zone. Lime application to acidic soils is performed with respect to hydrolytic acidity. Fertilizers are introduced in doses providing for stable harvests. At hayfields and pastures, the radical improvement was implemented in the only case, if they are located on peat-bogs, because in this case exceeding of ^{137}Cs TPC in milk may be observed for the contamination density about 37 kBq/m^2.

In the second zone (with ^{137}Cs contamination density within 185 – 555 kBq/m^2) plants on the croplands are grown without any restrictions. Mineral fertilizers are introduced in doses providing for planned harvests. If there are risks of obtaining products with exceeded TPC, higher doses of phosphate-potassium fertilizers are used. The measures for pastures and hayfields depend on the type of soils. The surface improvement with introduction of phosphorus and potassium fertilizers and undersowing perennial grasses is recommended for floodplane meadows and dry valleys with loamy and clayey soils. All natural pastures and hayfields on peat-bogs and mechanically light soils are subject to radical improvement with introduction of 1.5-fold increased doses of phosphate-potassium fertilizers.

In the third zone (with ^{137}Cs contamination density within 555 – 1480 kBq/m^2) the agriculture management suggested full-scale countermeasures providing for obtaining products satisfying radiological standards. For farming, introduction of lime and increased doses of phosphorus-potassium fertilizers was recommended. Lime materials are introduced on the basis of 1.5 doses for hydrolytic acidity. On hayfields and pastures the root improvement is performed with annual introduction of increased (1.5 – 2.0-fold) doses of phosphorus-potassium fertilizers. Organics of animal origin are applied in this zone without any limits.

Agricultural lands of the fourth zone (with ^{137}Cs contamination density over 1480 kBq/m^2) were eliminated from the agricultural use. Agricultural production on the territories contaminated by ^{137}Cs in the dose above 2960 kBq/m^2 was terminated. Based on the data of the primary examination, performed in the period since autumn 1986 till spring 1987, it was decided to eliminate 17,300 ha from economic use in Krasnogorsk, Gordeev and Novozybkov regions of Bryansk Oblast. Further investigations indicated that the agricultural lands with ^{137}Cs contamination density above 1480 kBq/m^2equaled 6,300 ha, which are still the part of the restriction zone.

THE SCOPE OF PROTECTIVE MEASURES IN THE AGRICULTURE AFTER THE ACCIDENT

Since 1986 till 1988 the measures in the agricultural complex were undertaken and continuously increasing, and in the period of 1988 – 1992 they were implemented in optimal scopes. This allowed for a significant reduction of the production volume with contamination level above TPC: from 86% in 1986 to 1.7% in 1994 for milk, from 15.2% to 0.06% for beef, and from 78% to less than 0.01% for grain, respectively.

The main scope of protective measures was implemented in the southwestern region of Bryansk Oblast. However, in other contaminated areas they were also significant.

On contaminated territories, the main protective measure was introduction of increased doses of potassium fertilizers. Since 1993, the volumes of applied chemical agents and agromelioration measures were being reduced. The introduction of mineral fertilizers per 1 ha of ploughed field and for separate crops was decreased by 3 – 11 times (Table 12).

Table 11. The volumes of agromelioration agents applied in the Oblasts of the Russian Federation contaminated by the Chernobyl accident, thousand ha

Measures	1986 - 1990	1991 - 1995	1996 - 2000	2001 - 2004
Bryansk Oblast				
Lime application	335.8	145.0	24.5	34.8
Phosphorite application	183.8	130.6	23.0	14.1
Potassium application	1359.6	630.2	164.1	90.8
Kaluga Oblast				
Lime application	706.9	408.2	77.2	7.31
Phosphorite application	286.3	121.5	4.0	0.7
Potassium application	101.1	1.0	12.0	8.6
Orel Oblast				
Lime application	355	144.0	2.2	1.81
Phosphorite application	143	53.9	4.56	1.20
Potassium application	224	102.3	14.0	1.58
Tula Oblast				
Lime application	421.1	248.7	20.6	17.0
Phosphorite application	227.9	81.5	0	–
Potassium application	159.8	96.1	22.4	2.5

Deterioration of economic conditions of the farms does not allow for implementation of the required protective measures, which increases volumes of contaminated products. For example, in 1995 – 1996 in Novozybkov region the deficiency of potassium fertilizers for cereals and potatoes caused more than 2-fold increase of ^{137}Cs concentration in the products, and in 2003 – 2004 up to an order of magnitude compared with the period, when they were used in optimal volumes (1991 – 1992).

Table 12. Volumes of introduced mineral fertilizers to ploughed lands of contaminated areas in Bryansk Oblast, kg/ha

Regions	1981-1985	1986-1990	1991-1995	1996-2000	2001-2004
Gordeev	-	207	99	57	43
Zlynkovsk	-	298	176	45	31
Klimovsk	189	247	143	31	22
Klintsov	184	273	129	55	50
Krasnogorsk	180	200	85	43	43
Novozybkov	261	352	178	65	37
Starodub	187	223	131	57	59
On average:					
by contaminated regions	200	257	134	50	41
by Oblast	164	209	117	38	33

On the territory of contaminated Oblasts, the main problem was represented by production of forage and animal products [61, 62]. For example, in Bryansk Oblast on the farms of the southwestern zone in hays of about 20% fodder lands ^{137}Cs concentration exceeds the standards varying from 650 to 66,000 Bq/kg. Cs-137 concentration in the pasture grass and rough forage stipulates specific activity of produced milk within the range of 80 – 5,000 Bq/l and for beef – from 800 to 14,000 Bq/kg.

In the field of forage production and meadow management, one of the main methods providing for a significant reduction of forage contamination was the radical improvement of low-productivity hayfield-pasture lands (root mat, backset, grass seeding, lime application and fertilizer introduction). Radiological value of the root melioration of contaminated hayfield-pasture lands is stipulated by their significance in production of critical dose-forming products (milk, first of all).

During initial 5 – 8 years after the accident, the radical improvement was widely used for all regions exposed to contamination (Table 13).

Table 13. The scope of radical improvement of meadows and pastures in Oblasts of the Russian Federation, contaminated by the Chernobyl accident, thousand ha

Oblast	1986-1990	1991-1995	1996-2000	2001-2004
Bryansk	97.6	30.8	12.8	20.79
Kaluga	11.6	8.5	2.0	-
Orel	-	12.2	-	-
Tula	-	4.7	-	-

In the recent years (2001 – 2004), application of the radical improvement as a countermeasure for reducing contamination of the grass stand on pastures and hayfields was applied only to seven southwestern regions of Bryansk Oblast.

For the cattle breeding, the method of prior to slaughter feeding of animals by "pure" forage (3 – 4 weeks before slaughter all animals were fed by forage with low ^{137}Cs concentration) has proved its high effectiveness. The scale of application of these measures, which were corrected on the basis of life-time determination of ^{137}Cs concentration in muscles of animals, reached 5,000 to 20,000 heads of cattle stock.

Since 1993, in contaminated regions of Bryansk Oblast, both the collective sector and private farms initiated the large-scale introduction of ferrocyn-containing compounds (Table 14).

Table 14. The volumes of ferrocene-containing compounds in contaminated regions of Bryansk Oblast in 1993 – 2004

Compound	1993-1995	1995-2000	2001-2004
Bifege, tons	24.1	77.7	23.65
Ferrocyn, tons	13.7	2.8	5.6
Boli, thous. pcs.	13.1	19.8	-
HJ-90, tons	0	4.0	-

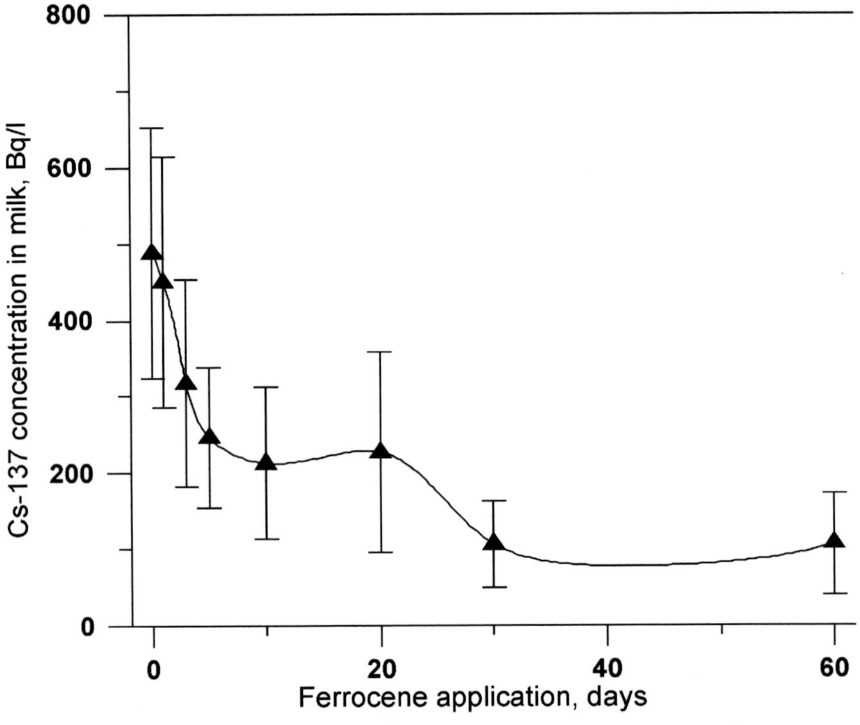

Figure 5. Dynamics of ^{137}Cs concentration decrease in cow milk with ferrocyn application

There are four types of compounds used: ferrocyn powder, bifege, lick boli salt containing ferrocyn; ferrocyn-containing boli. Ferrocyn application is one of the most effective methods and leads to 4 – 6-fold decrease of ^{137}Cs concentration in milk (Figure 5).

The application of ferrocene-containing compounds, which reached from 500,000 to 1 million heads of cattle stock per year, allowed for preserving minimal volumes of cattle breeding products with ^{137}Cs concentration above TPC in the period of low-intensive agrotechnical and agrochemical measures. Annually, application of such compounds reduces contamination of more than 30,000 tons of milk and 5,000 tons of beef of the slaughter weight.

For processing branches of the agricultural sector, contaminated milk processing to butter and cheeses is the most popular way. In the case of iodine danger, milk (concentrated or powdered) was stored until decay of short-living radionuclides. Obtaining of radionuclide-free products from plant raw material is guaranteed by their processing to oil, alcohol, starch and sugar. Special technology of potato processing which reduced ^{137}Cs concentration 6 – 7 times was adopted. Taking into account high production volumes in the contamination zone, this technique allowed for a significant decrease of collective doses of radiation, perceived by the population.

The Results of Adapting the System of Protective Measures in Agriculture

Reduction of Agricultural Product Volume with ^{137}Cs Exceeding Norms

Radiological examination of agricultural lands and production, performed in the first period after the accident, indicated increasing standards for radionuclide concentration in products in four Oblasts only, which are Bryansk, Kaluga, Orel and Tula. In five most contaminated regions of Bryansk Oblast (Gordeev, Novozybkov, Krasnogorsk, Klintsov and Klimovsk), above 80% of produced cereals, milk and forage were under TPC levels in relation to radionuclide contents, assumed for the primary period after the accident (TCP-86). In Kaluga Oblast (Zhizdrinsk, Khvastovich and Ul'yanovsk regions) 70% of cereals displayed exceeding of the standards. Orel and Tula Oblasts showed somewhat smaller amounts of products exceeding the radiation standards (up to 40% in Bolkhov region of Orel Oblast and 15% in Plavsk region of Tula Oblast, respectively).

Contamination levels for lands in Kaluga, Orel and Tula Oblasts was found much lower than in Bryansk Oblast. Moreover, in Tula and Orel Oblasts, on heavy mechanical compositions, the possibility of obtaining products with the contamination level above TPC was much lower than for light sandy and sandy-loam soils typical of Bryansk Oblast. In this relation, as well as due to protective measures performed, exceeding of the norms in Tula Oblast was observed in 1987 exclusively (0.7% by cereals), whereas in Orel Oblast due to the measures undertaken all products fitted the standards. Further on, targeted application of protective measures in these Oblasts, applied in scientifically justified volumes, allowed for avoiding production with exceeded radiation standards. In Kaluga Oblast, exceeding of radiation standards in the agricultural products (cereals and potatoes) was observed in 1988. At the same time, in forage control levels are exceeded still, while by a small value (below 10%).

The highest levels of agricultural products contamination and, consequently, volumes of products with exceeded standards were observed in Bryansk Oblast (Figure 6).

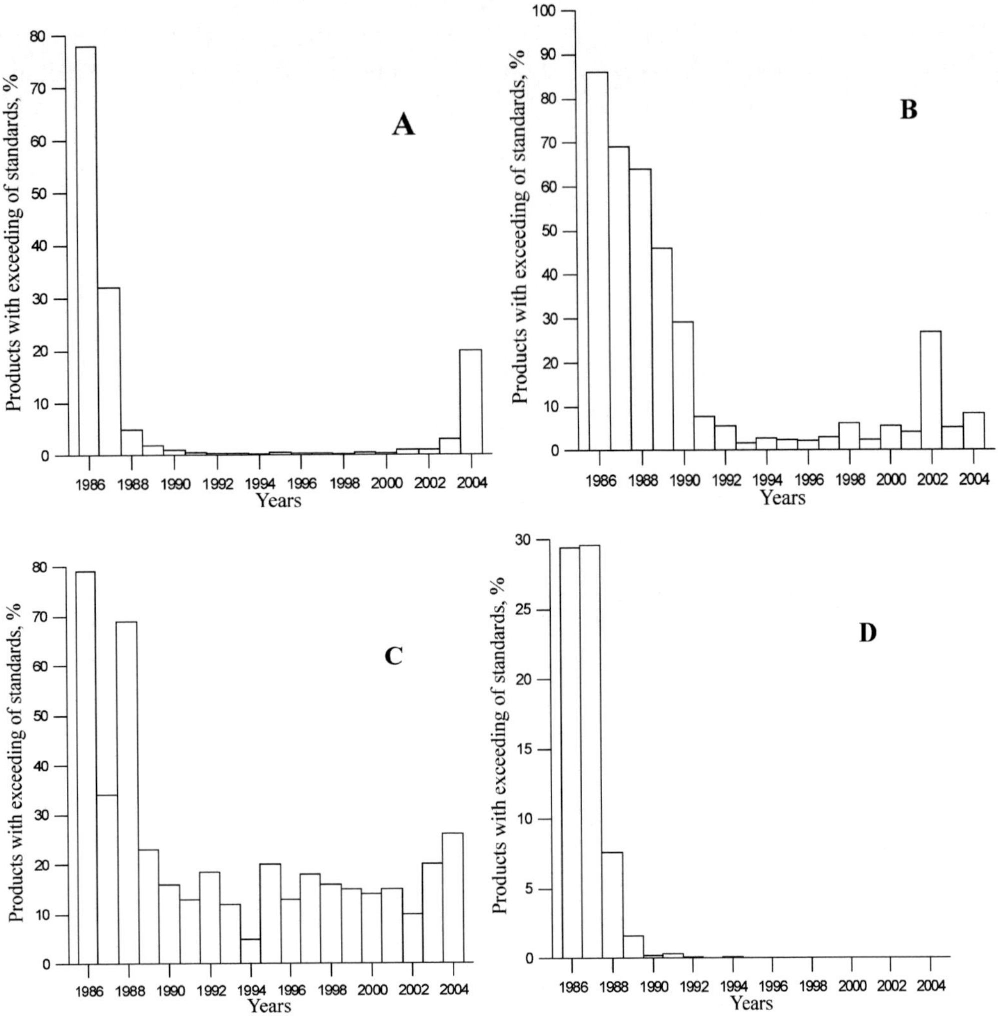

Figure 6. Reduction of production volumes with exceeding radiation standards in contaminated regions of Bryansk Oblast in 1986 – 2004: A – cereals; B – milk; C – hay; D – beef

Owing to adoption of special measures to the cattle breeding (animal feeding by "pure" forage and arrangement of stable–range management of animals) contamination levels of the cattle breeding products was always decreasing, and since 1991, the part of products exceeding the radiation standards is below 10% of total production volume in contaminated regions. The highest effect was reached for providing production of cereals and potatoes pursuant to radiation standards. For example, in 1991 cereals production with ^{137}Cs concentration below 370 Bq/kg (Figure 6) was decreased to extremely minimal volume (below 0.1%). However, at transition to SanPiN 2.3.2-1078-01, the part of cereals inadequate to new stricter standards (160 Bq/kg) was sharply increased to 20% in 2004. Moreover, it should be noted that recently, the volume of applied mineral fertilizers for cereals was abruptly decreased, as well as the scope of protective measures, which, in turn, led to an increase of ^{137}Cs accumulation in grain.

Production of milk according to current radiation standards is still the most complicated question. By 1995, owing to adoption of the protective measure complex and intensive application of ferrocyn-containing compounds production of milk inconsistent with the current standards was reduced to the minimum and preserved at this level by now.

Contribution of Protective Measures into Reduction of Agricultural Product Contamination

A decrease of agricultural products contamination was defined by three groups of factors:

- natural biogeochemical processes determining reduction of biological availability in the soil–plants system;
- protective measures;
- radioactive decay.

The contribution of each of these factors into reduction of agricultural product contamination in Bryansk and Kaluga Oblasts was estimated (Table 15).

Table 15. Contributions of factors determining reduction of ^{137}Cs concentration in the agricultural products

Factors	Regions with intensively applied countermeasures (Bryansk Oblast)		Regions with partly applied countermeasures (Kaluga Oblast)	
	milk, beef	potatoes, cereals	milk, beef	potatoes, cereals
Natural bio-geochemical processes	**0.33**	**0.36**	**0.60**	**0.73**
Countermeasures	**0.61**	**0.57**	**0.28**	**0.12**
Radioactive decay	0.06	0.07	0.12	0.15

Contamination of agricultural products in Bryansk Oblast was reduced quicker than in Kaluga Oblast due to more intensive application of the protective measures. In the period of intensive application, the contribution of protective measures into reduction of the product contamination (0.57 – 0.61) considerably exceeded the contribution of natural processes in contaminated regions of Bryansk Oblast. In Kaluga Oblast, in the areas with limited application of protective measures, the main factor providing for reduction of radionuclide transition to the agricultural products was a complex of natural biogeochemical processes, which defined strength of ^{137}Cs fixation in soils and its biological mobility.

CONCLUSION

The analysis of 20-year experience in liquidation of the Chernobyl accident consequences in the branch of agricultural production reliably indicates that realization of scientific programs of rehabilitation measures in the agriculture has radically improved radiological and social-psychological situation in the country and considerably recovered the potential of the agricultural sector in the region impacted by the accident.

However, despite a significant improvement of the radiation conditions and successful application of rehabilitation measures, the question of full radiation safety of the population living on the territories, contaminated due to the accident, is not completely resolved yet. Contamination levels of agricultural lands, high rates of radionuclide transition from low-fertility soils to cultivated crops and the abrupt reduction of the scope of protective measures undertaken provide for a considerable contribution of internal irradiation into the total dose perceived, from 46 to 90%. The main dose-forming agricultural product is milk, the contribution of which to the internal radiation dose reaches 75%, as a maximum.

In Russia, there are 6 critical areas among contaminated regions of Bryansk Oblast. At present, they include 3,929 settlements, among which 181 settlements (with 33,500 citizens) possess the average annual radiation dose above 1 mSv. Citizens of these settlements own about 4,000 dairy cattle pieces and use about 12,000 ha of haylands and pastures. They produce agricultural products with the highest ^{137}Cs concentration. For milk, the standards adopted in SanPiN 2.3.2.1078-01 are exceeded 20 times, on average, (30 – 60 times in some periods), and 3 – 4 times for beef.

The feature of the modern time after the accident is transition to norms typical of pre-accident period or uncontaminated territories, i.e. stiffening the norms on MPC for radionuclides in the products. Analysis of the radiation control data indicates impossibility of obtaining forage and cattle breeding products conforming to the standards without application of special protective measures in 23 collective farms and 123 settlements. For 11 farms, the excess of SanPiN 2.3.2.1078-01 norms will lasts for long time, and a half of milk produced will not conform to the radiation standards. Forecasting of the situation indicates that milk conforming to the norms in these settlements may be produced not earlier than in 2025 – 2030. If protective measures are not applied, reduction of the annual radiation doses for population to 1 mSv will take a long time – by 2062 for “critical” settlements.

A separate problem in the complex of liquidation of the Chernobyl accident consequences in Russia is return of lands, temporarily eliminated from the land-use management. According to Rosgidromet forecasting, due to radioactive decay by 2139 maximal ^{137}Cs concentration will fall below 1480 kBq/m^2 for all contaminated territories, i.e. the restriction zone will be eliminated. However, a significant part of these territories may be restored for the economic use much earlier owing to changing radiation situation and application of specially designed protective measure systems. In the remote period after the accident, one of the main tasks is rehabilitation of lands temporarily eliminated from the land-use management.

In the remote post-accident period, to provide for the radiation safety of the population from contaminated territories, the following tasks must be solved in the branch of agriculture management:

- continuous monitoring of the radiation conditions;
- modernization of the radiation monitoring and control systems;
- formation of radiological passports for every collective farm and settlement to provide correct forecasting of changes in the radiation conditions and radiation doses

perceived by the population and, consequently, development of targeted rehabilitation strategies;

- continuing implementation of protective measures, development and adoption of new techniques and methods providing for reduction of radionuclide concentration in agricultural products;
- justification, development and adoption of targeted strategies for rehabilitation of cultivated lands;
- justification of the rehabilitation measure system aimed at returning lands, temporarily eliminated from the land management, to use.

REFERENCES

[1] Israel Yu.A., Kvasnikova E.V., Nazarov I.M., and Fridman Sh.D., "Global and regional radioactive contamination by cesium-137 in the European part of the former USSR", *Meteorologia i Gidrologia*, 1994, No. 5, pp. 5 – 9. (Rus)

[2] Kozubov G.M. and Taskaev A.I., *Radiobiological and Radioecological Investigations of Woody Plants*, St. Petersburg, Nauka, 1994, 256 p. (Rus)

[3] Alexakhin R.M., Buldakov L.A., Gubanov V.A. *et al.*, *Large Radiation Accidents: Consequences and Protective Measures*, Moscow, IzdAT, 2001, 752 p. (Rus)

[4] Geras'kin S.A., Dikarev V.G., Zyablitskaya E.Ya. *et al.*, *J. Environ. Radioact.*, 2003, vol. 66, pp. 155 - 169.

[5] Shavchenko V.A., Kal'chenko V.A., Abramov V.I. *et al.*, *Radiats. Biol. Radioekol.*, 1999, vol. 39(1), pp. 162 – 176. (Rus)

[6] Zyablitskaya E.Ya., Spirin E.V., Sanzharova N.I., and Alexakhin R.M., "Genetic and biological effects of chronic irradiation of fall rye crops by radioactive fallouts from the Chernobyl accident", *Radiobiologia*, 1990, vol. 30(3), pp. 291 – 295. (Rus)

[7] Suvorova L.I., Spirin D.A., Martyushov V.Z. *et al.*, In: *Radiation Aspects of the Chernobyl Accident*, vol. 2, St. Petersburg, Gidrometeoizdat, 1993, pp. 321 – 325. (Rus)

[8] Smirnov E.G. and Suvorova L.I., In Book: *The Effect of Radioactive Contamination on Surface Ecosystems in the Chernobyl Accident Zone. Proc. Komi Scientific Center UrO RAS No. 145*, vol. 2, Syktyvkar, 1996, pp. 27 – 37. (Rus)

[9] Shershunova V.I. and Zainullin V.G., *Radiats. Biol. Radioekol.*, 1995, vol. 35(5), pp. 690 – 695. (Rus)

[10] Alexakhin R.M., Sarapul'tsev I.A., Spirin E.V., and Udalov D.B., *Doklady RAN*, 1992, vol. 323(3), pp. 576 – 579. (Rus)

[11] Shevchenko A.S., Vakulenko A.D. and Isamov N.N., *Doklady VASKhNIL*, 1990, No. 11, pp. 55 – 58. (Rus)

[12] Budarkov V.A., Zenkin A.S., Arkhipov N.P. *et al.*, *Radiobiol.*, 1992, vol. 32(3), pp. 451 – 458. (Rus)

[13] Korneyev N.A., Povalyaev A.P., Alexakhin R.M. *et al.*, "The sphere of agricultural production – radiological consequences of the accident at Chernobyl nuclear power

plant and basic protective measures", *Atomnaya Energia*, 1988, vol. 65(2), pp. 129 – 134. (Rus)

[14] Sanzharova N.I., Fesenko S.V., Nedbaevskaya N.A., and Alexakhin R.M., "Aerial contamination of agrarian ecosystems after the Chernobyl accident", *Radiats. Biol. Radioecol.*, 2001, vol. 41(4), pp. 408 – 414. (Rus)

[15] Sanzharova N.I., Fesenko S.V., Alexakhin R.M., Anisimov V.S, Kuznetsov V.K., and Chernyaeva L.G., "Changes in the forms of ^{137}Cs and its availability for plants as dependent on properties of fallout after the Chernobyl nuclear power plant accident", *The Science of Total Environment*, 1994, vol. 154, pp. 9 – 22.

[16] Fesenko S.V., Sanzharova N.I., Alexakhin R.M., and Spiridonov S.I., "The change of biological availability of ^{137}Cs after the Chernobyl accident", *Pochvovedenie*, 1995, No. 5, pp. 508 – 513. (Rus)

[17] Arkhipov A.N., "The behavior of ^{90}Sr and ^{137}Cs in agrarian ecosystems of the Chernobyl accident restriction zone", *Candidate Dissertation Thesis on Biology*, Obninsk, 1995, 26 p. (Rus)

[18] Sanzharova N.I., Fesenko S.V., Lisyansky K.B., Kuznetsov V.K., Abramova T.N., and Kotik V.A., "The forms of occurrence in soils and dynamics of accumulation of ^{137}Cs radionuclide in cultivated crops after the Chernobyl accident", *Pochvovedenie*, 1997, No. 2, pp. 159 – 164. (Rus)

[19] Sanzharova N.I., Kotik V.A., Arkhipov A.N., Sokolik G.A., Ivanov Yu.A., Fesenko S.V., and Levchuk S.E., "Quantitative parameters of the vertical migration of radionuclides in the soils of various types of meadows", *Radiats. Biol. Radioekol.*, 1996, vol. 36(4), pp. 488 – 497. (Rus)

[20] Fesenko S.V., Spiridonov S.I., Sanzharova N.I., and Alexakhin R.M., "Estimation of half-reduction periods for ^{137}Cs concentration in the rooting zone of meadow ecosystems", *Radiats. Biol. Radioekol.*, 1997, vol. 37(2), pp. 267 – 280. (Rus)

[21] Sanzharova N.I., Fesenko S.V., Alexakhin R.M., Abramova T.N., and Kuznetsov V.K., "The assessment of factors defining dynamics of agricultural product contamination by ^{137}Cs after the Chernobyl accident", *Radiats. Biol. Radiekol.*, 1995, vol. 35(3), pp. 307 – 315. (Rus)

[22] Nedbaevskaya N.A., Blinova L.D., Kyishev I.I., Sanzharova N.I., and Alexakhin R.M., "Dynamics of radionuclide concentration in fallouts, pasture vegetation and milk in Leningrad Oblast after the Chernobyl accident", *Atomnaya Energia*, 1991, vol. 70(1), pp. 29 – 31. (Rus)

[23] Isamov N.N., Sirotkin A.N., Fesenko S.V., Sokolova E.A., and Sidorova E.V., "Migration regularities of industrial contaminants in trophic chains of lactating cows", *Ekologia*, 1998, No. 6, pp. 441 – 446. (Rus)

[24] Sirotkin A.N., Sanzharova N.I., Isamov N.N., Loy V.I., and Chetokin A.M., "Migration of ^{137}Cs in the food chain of lactating cows after the Chernobyl accident", In Coll.: *Liquidation of consequences of Kaluga Oblast contamination with Radionuclides, Caused by the Chernobyl Accident*, Kaluga – Obninsk, 1992, pp. 170 – 176. (Rus)

[25] Alexakhin R.M., Sanzharova N.I., Fesenko S.V., Kurganov A.A., and Mosharov V.N., "Basic results of the works on liquidation of the Chernobyl accident consequences in the field of agricultural production", In: *Chernobyl: 15 Years After*, Ed. N.V. Gerasimova, Moscow, 2001, pp. 105 – 141. (Rus)

[26] Fesenko S.V., Spiridonov S.I., Sanzharova N.I., and Alexakhin R.M., "Dynamics of ^{137}Cs bioavailability in a soil-plant system in areas of the Chernobyl Nuclear Power Plant accident zone with a different physicochemical composition of radioactive fallout", *J. Environmental Radioactivity*, 1997, vol. 34(3), pp. 287 - 313.

[27] Fesenko S.V., Spiridonov S.I., Sanzharova N.I., Anisimov V.S., and Alexakhin R.M., "Modeling of ^{137}Cs migration in the soil–plants system in peaty soils contaminated as a result of the Chernobyl accident", *Ekologia*, 2002, No. 3, pp. 185 – 192. (Rus)

[28] Spiridonov S.I., Fesenko S.V., and Sanzharova N.I., "Modeling of ^{137}Cs migration in the soil–plants system after melioration agent application", *Radiats. Biol. Radioekol.*, 2001, vol. 41(3), pp. 334 – 337. (Rus)

[29] Shubina O.A. and Fesenko S.V., "Modeling of ^{137}Cs migration in agrarian ecosystems in case of applied protective and rehabilitation measures", *Radiats. Biol. Radioekol.*, 2004, vol. 44(5), pp. 591 – 602. (Rus)

[30] Spiridonov S.I., Fesenko S.V., Avila R., and Gontarenko I.A., "Modeling of ^{137}Cs behavior in forest ecosystems and forecasting of its accumulation in the forest products", *Radiats. Biol. Radioekol.*, 2001, vol. 41(2), pp. 217 – 225. (Rus)

[31] Spiridonov S.I., Fesenko S.V., and Gontarenko I.A., "Development of protective measure strategy aimed at liquidation of radioactive contamination consequences in forest ecosystems", *Radiats. Biol. Radioekol.*, 2004, vol. 44(1), pp. 104 – 112. (Rus)

[32] Fesenko S.V., Alexakhin R.M., Sanzharova N.I., and Spiridonov S.I. "Analysis of the factors determining reduction of ^{137}Cs availability for inclusion into agricultural food chains", *Doklady AN*, 1995, vol. 343(5), pp. 715 – 718. (Rus)

[33] Fesenko S.V., Alexakhin R.M., Spiridonov S.I., and Sanzharova N.I. "Dynamics of ^{137}Cs concentration in agricultural products in areas of Russia contaminated as a result of the accident at the Chernobyl nuclear power plant", *Radiation Protection Dosimetry*, 1995, vol. 60(2), pp. 155 – 166.

[34] Fesenko S.V., Alexakhin R.M., Sanzharova N.I., and Spiridonov S.I., "Regularities in ^{137}Cs concentration change in the cattle breeding products on the territory of the Russian Federation contaminated by the Chernobyl accident", *Radiats. Biol. Radioekol.*, 1995, vol. 35(3), pp. 316 – 327. (Rus)

[35] Alexakhin R.M., Fesenko S.V., Sanzharova N.I., Spiridonov S.I., Vorob'ev G.T., and Yakovleva N.A., "On reduction of ^{137}Cs concentration in plant growing products contaminated by the Chernobyl accident", *Doklady RASKhN*, 1995, No. 3, pp. 20 – 21. (Rus)

[36] Fesenko S.V., Colgan P.A., Sanzharova N.I., Lissianski K.B., Vazquez C. and Guardans R., "The dynamics of the transfer of caesium-137 to animal fodders in areas of Russia affected by the Chernobyl accident and resulting doses from the consumption of milk and milk products", *Radiation Protection and Dosimetry*, 1997, vol. 69(4), pp. 289 – 299.

[37] Alexakhin R.M. "Protective measures for the agricultural production at a radiation accident", *Atomnaya Energia*, 1992, vol. 72(2), p. 208. (Rus)

[38] Panov A.V., Fesenko S.V., and Alexakhin R.M. "The efficiency of measures aimed at reduction of radiation doses perceived by the population of the countryside settlements in a remote period after the Chernobyl accident", *Radiats. Biol. Radioekol.*, 2001, vol. 41(6), pp. 682 – 694. (Rus)

[39] Panov A.V., Fesenko S.V., and Alexakhin R.M. "The use of computerized systems at adoption of protective measures to agriculture after the Chernobyl accident", *Radiats. Biol. Radioekol.*, 2005, vol. 45(1), pp. 19 – 29. (Rus)

[40] Fesenko S.V., Panov A.V., and Alexakhin R.M., "Methodical approach to justification of protective measures for countryside settlements in the remote period after the Chernobyl accident", *Radiats. Biol. Radioekol.*, 2001, vol. 41(4), pp. 415 – 426. (Rus)

[41] Alexakhin R.A., Firsakova S.K., Rauret G., Arkhipov N.P., Vandecasteele C., Ivanov Yu.A., Fesenko S.V., and Sanzharova N.I., "Fluxes of radionuclides in the agricultural environment - main results and still unsolved problems. Radiological consequences of the Chernobyl accident", *Proceedings of the First International Conference*, Minsk, Belarus, 18-22 March 1996, p. 39 – 48.

[42] Panov A.V., Pakhomov A.Yu., Fesenko S.V., and Alexakhin R.M., "The measures for decreasing concentration of radionuclides in the products of collective farms after the Chernobyl accident", *Vestnik RASKhN*, 2005, No. 1, pp. 55 – 58. (Rus)

[43] *Agricultural Radioecology*, Ed. N.A. Korneyev and R.M. Alexakhin, Moscow, 1992, 400 p. (Rus)

[44] Ratnikov A.N., Alexakhin R.M., Zhigareva T.L., Sanzharova N.I., and Popova G.I., "Efficiency of a complex of agromelioration measures for reduction of ^{137}Cs accumulation in the plant growing products in the Chernobyl accident zone (at the territory of Russia)", *Agrokhimia*, 1992, No. 9, pp. 112 – 116. (Rus)

[45] *A Manual for Agriculture Management under Conditions of Radioactive Contamination of a Part of RSFSR, Ukrainian SSR and Belorussian SSR for the Period of 1988 – 1990*, Moscow, 1988. (Rus)

[46] *Recommendations for Agriculture Management under Conditions of Radioactive Contamination of the Territory as a Result of the Chernobyl Accident for the Period of 1991 – 1995*, Moscow, 1991. (Rus)

[47] Kuznetsov V.K., Sanzharova N.I., Kalashnikov K.G., and Alexakhin R.M., "The accumulation of ^{137}Cs in the plant growing products with respect to species and varietal features of cultivated crops", *Sel'skokhozyaistvennaya Biologia*, 2000, No. 1, pp. 64 – 70. (Rus)

[48] Zhigareva T.L., Ratnikov A.N., Popova G.I., Sanzharova N.I., Petrov K.V., and Belous N.M., "The effectiveness of mineral fertilizers on radiation-contaminated territories", *Khimia v Sel'skom Khozyaistve*, 1996, No. 1, pp. 35 – 38. (Rus)

[49] Belous N.M., Moiseenko F.V., and Sanzharova N.I., "The effectiveness of agrochemical measures aimed at reduction of ^{137}Cs accumulation in agricultural harvests in the area of Bryansk polesye", In Coll: *Proc. Scientific Session of Rosselkhozakademia (27 – 29 June, 2002). Problems of Industrial Impact on the Agricultural Sector and Rehabilitation of Contaminated Territories,* Moscow, 2003, pp. 165 – 183. (Rus)

[50] Sanzharova N.I., Kuznetsov V.K., Aksenova S.P., and Kotik Zh.A., "The accumulation of ^{137}Cs in cultivated crops on sands and sandy-loam soils of Belarus Polesye, caused by the effect of various melioration agents", *Sel'skokhozyaistvennaya Biologia*, 1996, No. 5, pp. 55 – 60. (Rus)

[51] Sanzharova N.I., Belova N.V., Yurikov P.I., Vorob'eva L.A., Moiseenko F.V., and Spiridonov S.I., "Transition of ^{137}Cs to plants from sod-podzol soil with respect to potassium dose and its mobility", *Agrokhimia*, 2004, No. 7, pp. 58 – 66. (Rus)

[52] Kuznetsov V.K., Sanzharova N.I., Anisimov V.S., and Abramova T.N., "The effect of phosphate fertilizers on ^{137}Cs accumulation by agricultural products", *Agrokhimia*, 2001, No. 9, pp. 47 – 53. (Rus)

[53] Belova N.V., Draganskaya N.V., and Sanzharova N.I., "The effect of organic fertilizers on biological mobility of cesium-137 in the soil", *Plodorodie*, 2004, vol. 20(5), pp. 35 – 36. (Rus)

[54] Kuznetsov V.K., Sanzharova N.I., Aksenova S.P., and Kotik Zh.A., "The reduction of ^{137}Cs accumulation in cultivated crops affected by melioration agents", *Agrokhimia*, 1995, No. 4, pp. 74 – 79. (Rus)

[55] Sanzharova N.I., Fesenko S.V., Kotik V.A., and Spiridonov S.I., "Behavior of radionuclides in meadows and efficiency of countermeasures", *Radiation Protection Dosimetry*, 1996, vol. 64(1/2), pp. 43 – 48.

[56] Prister B.S., Belli M., Sanzharova N.I., Fesenko S.V., Bunzl K., Petriaev E., Sokolik G. A. *et al.*, "Behavior of radionuclides in meadows including countermeasure application. The radiological consequences of the Chernobyl accident", *Proceedings of the First International Conference*, Minsk, Belarus, 18-22 March 1996, pp. 59 – 68.

[57] *Behavior of Radionuclides in Natural and Semi-natural Environments, Final Report*, Ed. M. Belli and F. Tikhomirov, Part Y11.5. EUR 16531 EN, 1996, p. 147

[58] Korneyev N.A. and Sirotkin A.N., *The Fundamentals of Radioecology of the Farm Animals*, Moscow, Energoatomizdat, 1987, 208 p. (Rus)

[59] Isamov-Jr. N.N., Sanzharova N.I., and Kuznetsov V.K., "Protective technological methods in fodder production and cattle breeding in case of radioactive contamination of agricultural lands", *Dostizheniya Nauki i Tekhniki v APK*, 2004, No. 7, pp. 30 – 32. (Rus)

[60] Fesenko S.V., Sanzharova N.I., Pasternak A.D., Goryainov V.A., Vorob'ev G.T., Novikov A.A. *et al.*, *A Manual for Cattle Breeding Management for the Remote Period after the Chernobyl Accident*, VNIISKhRAE Library, 2001. (Rus)

[61] Fesenko S.V., Pakhomov A.Yu., Pasternak A.D. *et al.*, "Regularities of cesium-137 concentration change in milk in the remote period of the Chernobyl accident", *Radiats. Biol. Radioekol.*, 2004, vol. 44(3), pp. 336 – 345. (Rus)

[62] Isamov N.N., Rudakov A.P., Isamov-Jr. N.N., Pasternak A.D., Goryainov V.A. *et al.*, "About radionuclide contamination of forage, raw materials and products of cattle breeding", *Sel'skokhozyaistvennaya Biologia*, 2004, No. 6, pp. 29 – 32. (Rus)

In: 20 Years after the Chernobyl Accident
Editors: E. Burlakova and V. Naidich, pp. 231-246
ISBN 1-60021-249-2

Chapter 13

RADIATION DAMAGE OF PLANTS IN THE CHERNOBYL NUCLEAR ACCIDENT IMPACT ZONE

D. M. Grodzinsky and I. N. Gudkov
Institute of Cell Biology and Genetic Engineering, NAS of Ukraine
National Agricultural University of Ukraine, Kiev

ABSTRACT

Somatic and genetic radiobiological effects, observed for cultivated and wild plant species in the 30-kilometer exclusion zone of Chernobyl Nuclear Power Plant and other radionuclide-polluted territories. Special attention is paid to remote reactions of plants: weakened immunity, occurrence of mutations, changes in phytocoenosis. The question about possible adaptation of plants to increased radiation conditions is discussed.

Keywords: Chernobyl nuclear accident, plants, radiobiological effects

Even many radiobiologists have a strict prejudice that plants are supposedly extremely resistant to radiation, and on radiation-polluted territories this component of biocoenosis is of importance only as the primary stage of radionuclide transfer via trophic chains to the man. So far as concerns possible radiation damage of plants, this probability is frequently neglected or ignored. On the one hand, as some damage of a plant as the organism is assumed, it is suggested that this will cause no harm to the environment, as it takes place in the case of damaging of an animal or a man. On the other hand, somehow it is considered to be that plants may easily adapt to the increased levels of external and internal radiation.

Actually, the overwhelming majority of plant species possess higher resistance to radiation compared with mammals. For example, for mammals semilethal doses of acute infrequently ionizing radiation equal 2 – 8 Gy, whereas for many higher plant families they are 20 – 30 Gy. There are also higher plant families, for example, pinaceous, liliaceous, leguminous etc., for some representatives of which semilethal doses equal 2 – 8 Gy. Their radiation damaging may cause significant changes in phytocoenoses and the entire biocoenosis. Destruction (death) of about 600 ha of the notorious "Red Forest" at the Chernobyl Nuclear Power Plant in autumn 1986 – spring 1987 is the clear demonstration of this fact.

However, death is the utmost, the most severe damage of the living organism, observed at the Chernobyl nuclear accident in plants only under conditions of extremely high dose rates in the first weeks and months after its initiation on comparatively narrow territories neighboring Chernobyl Nuclear Power Plant. Other forms of damage have covered much broader regions and are also observed in remote periods. The most clear radiobiological effects related to growth suppression, morphological changes, increasing amount of cells possessing chromosomal aberrations, etc. occurred in the initial years after the accident. Even five, ten and up to the present days various responses of plants to increased radiation suppression are observed in the accident effect zone, not yet limited by the 30-km zone around Chernobyl NPP. Not always easily explainable phenomena are observed in this case:

- Despite the fact that the level of occurrence of various radiobiological effects usually exceeds the spontaneous level, it does not correlate with the γ-radiation dose rate.
- A deviation from linear dependence between the level of damages and the radiation dose is frequently observed, especially in more remote periods after the accident;
- Mismatch of manifestation of the registered radiobiological effects, both somatic and genetic, and radiation doses – they correspond to the levels, at least, by an order of magnitude higher than those calculated by usual methods or estimated by dosimeters – dose accumulators.

Undoubtedly, it is generally senseless to relate the absorbed dose 5 – 10 years after radioactive emission, as well as at present, to the background radiation or estimate it for the whole plant. At present, the dose absorbed by all organisms, including plants, is by 70 – 95% due to the internal irradiation of incorporated radionuclides. Similar to the majority of common elements, their basic feature is the ability for irregular distribution in organism's tissues and accumulation in some places, forming extremely high local doses of radiation. Usually, they are actively metabolizing and permanently regenerating tissues, in which important physiological and biochemical processes proceed, accompanied by the cell fusion. Basing on the analysis of radioautographs of variously originated radioactive compound distribution in the plant, we have outlined for many times that discrepancy between the suggested absorbed dose and radiobiological effect is induced by the concentration of radionuclides in critical organs of plants – in meristems, entire growth zones, forming generative organs, where the highest activity of cells requiring many different substances takes place [1, 2].

It was proved by A.N. Mikheev [3] in his special experiments. He grew corn and greens as the aqueous culture on radioactive strontium and cesium salt solutions and radiation-polluted Chernobyl soils and detected strontium predominant accumulation in the root extension zone and cesium in meristem – in its 3-mm tip. In the latter case, ^{137}Cs concentration in meristem exceeded its concentration in the rest root by more than an order of magnitude. On this basis, the author logically concludes that radioactive cesium induces the dose, absorbed by meristem, at least an order of magnitude higher than by other tissues.

At present and in future years, of special meaning is the study of the effect of dose levels formed in the accident effect zone on remote radiobiological effects, firstly, such as the states of immunity, adaptation abilities, genetic consequences, and changes in populations. Under the conditions of the majority of radionuclide-polluted area, this concerns radiobiology, which deals with the low dose effects.

GROWTH AND DEVELOPMENT

In 1986, in autumn, many countries in the Central Europe got significantly higher than usual harvests of agricultural crops. For example, in some northern regions of Ukraine cereals harvests by 10 – 30% exceeded the average by previous 20 years. Hungarian scientist A. Sabo [4] calculated total doses of external radiation, accumulated by plants during the vegetation period in 1986, and demonstrated that for many species they approached the stimulating level. This is the reason for acceleration of plant growth and their biomass accumulation in that dramatic year.

In 1986, during the vegetation period many scientists determined extremely high stimulation of growth of one-year timber species shoots [5] both in the 30-kilometer exclusion zone around Chernobyl NPP and outside it.

Figure 1. Pine tree species transplanted in 1988 at the site of the former "Red Forest" (July 2004; at the left Chernobyl NPP contours are seen; author made photo)

The increase of germinating ability and power of seeds, force and rate of plant growth, intensification of some physiological and biological processes, the cell fusion in meristems, and, finally, productivity increase were observed by many authors who studied plant growing processes at test sites in the accident effect area in later time.

There is no doubt that these phenomena, which may be qualified as manifestation of the radiation stimulation or radiation hormesis, must not be taken for as positive events, because they represent one of the effects of ionizing radiation – the specific reaction, the consequences of systematic manifestations of which in phytocoenoses may be hardly foreseen.

Frequently on polluted territories, especially in the closest 10-kilometer zone of Chernobyl NPP suppression of growth of separate organs and the whole plants was observed for many species. This was accompanied by intensity decrease of photosynthesis, transpiration, synthesis of some metabolites, many proteins, in particular, and other processes. Deceleration of plant development was also observed: delay of occurrence of some phases of ontogenesis and increased duration of the vegetation period.

Suppression of growth processes and development of plants is still observed for coniferous species growing at the place of buried "Red Forest", where soil pollution level reaches dozens of thousands kBq per square meter by both ^{137}Cs and ^{90}Sr (Figure 1).

MORPHOLOGICAL CHANGES

At the background of growth and development process deceleration, multiple morphological (teratogenic) changes in plants are clearly observed. Multiple works are devoted to variations in sizes of separate organs (increase or decrease), their shape (wrinkledness, curling, curvatures, asymmetry, jointing and division, thickening and thinning, etc.), color, disturbances of branching and leaf arrangement (phyllotaxis), premature fall of leaves (needles), flowers, blossom clusters, occurrence of tumors, and many other processes. Various morphological changes in coniferous species are most widely described in the cited works (Figure 2). Some of them may be observed still (Figure 3).

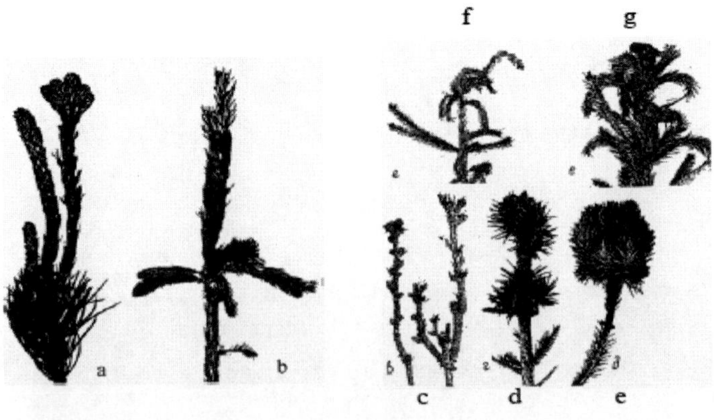

Figure 2. Anomalous vegetative shoots on pine trees (a, b) and spruce (c – g), formed in the effect zone of the Chernobyl nuclear accident in 1987 – 1988 [6, 7]

Figure 3. Morphological changes, currently observed in pine trees: needle loss, apical domination removal, dichotomy (July 2004; author made photo)

Among infrequent morphological changes in plants in the natural state, tumor formation should be mentioned. The cases of tumor-like growth induced by some types of viruses, bacteria, fungi, nematodes, insects, mites, and ionizing radiation are known. Nevertheless, one must admit that they are rather scanty, and origination of these neoplasms in plants is poorly studied, including the case of induction by radiation.

The most detailed studies of the occurrence of tumors on radiation-polluted territories were implemented by M.A. Dmitrieva *et al.* [8] among populations of natural species of plants. Occurrence of tumor-like neoplasms were infrequently observed for wall hawkweed (*Hieracium murorum L.*) and narrow-leaved hawkweed (*H. umbellatum L.*), raspberry (*Rubus idacus L.*), and blackberry (*R. caesius L.*). Hence, on the fifth year after the accident, when γ-radiation background decrease by three times, down to 2.3 mR/h, about 80% of species in populations of corn sowthistle (*Ciresium arvence L. Scop.*) obtained tumor-like outgrowths on the main and side vegetative shoots.

For plants affected by chronic radiation, these authors have also detected more intensive formation of cecidia on English oak (*Quercus robur L.*) leaves. Cecidia were formed on a significant part of the leaf, up to five per each lamina. The authors suggest that the abundance of cecidium formation testifies about plant immunity suppression, because to their opinion cecidia are formed due to typical damages of plants by viruses, bacteria and other organisms rather than by direct impact of ionizing radiation.

Chromosome Aberrations

The frequency of chromosome aberrations in meristem cells is one of the indices of radiation effect, most frequently used on radiation-polluted territories. Hence, cytogenetic analysis is usually applied to 2 – 4-day old plantlets, grown up in laboratories from seeds, gathered from plants developed under radiation pollution conditions. In this case, the test long time ago accepted by radiobiologists appeared very suitable and informative.

E.Ya. Zyablinskaya *et al.* [9] have studied dose dependencies of the yield of cells with chromosome aberrations in fall rye plantlets, grown up from seeds, gathered in 1986 at five sites in the 30-kilometer zone possessing different levels of total radionuclide pollution density in the upper 0 – 20-centimeter layer of the horizon. The authors carefully calculated the absorbed dose of external β– and γ–radiation (the external radiation is the factor which defined the radiation damage at that most acute period) that allowed for composing quite strict quantitative dose dependence of cytogenetic damages in the root meristem.

Highlighting the possibility of a significant error in estimation of absorbed doses, nevertheless, the authors pay special attention to nonlinearity of the dependence obtained and the presence of two dose ranges, within which the quantity of cells with aberrations does not increase. They explain this by the features of cell response to a weak external impact, during formation of which the model of gene expression is changed and inducible repair systems are activated.

To some extent, this is confirmed by observation of a noticeable increase of germinating ability of seeds and acceleration of plantlet root growth at the background of 2 – 3-fold increase of aberrant cell quantity and even some decrease of mitotic activity in meristems at 1.3 – 7.7 Gy doses. These processes were suppressed at maximal dose of 12 Gy only.

Even several years after the accident, despite clear tendency to decrease with time, significantly increased quantity of cytogenetic damages depending on the level of radiation pollution of soils was observed. Hence, the decrease rate of their level was significantly lower than that of radiation pollution of the locality, and analysis of genetic variability in three consecutive generations of fall rye and wheat demonstrated increased yield of damages in the second and the third generations.

Basing on these studies, the authors conclude that chronic low-dose radiation in regions, radionuclide-polluted by radiation catastrophes, may induce hereditary destabilization genetic structures, which, in particular, is manifested by increasing yield of cytogenetic damages and karyotypic variability of irradiated organism descendants. In its turn, the latter is the reflection of cytogenetic adaptation processes in populations of plants and animals [10].

Growing large horse beans (*Vicia faba L.*) of Belorussian and Russian black cultivars, which possess extremely high radiosensitivity compared with other kinds of herbage plants (LD_{50} = 3 – 4 Gy), at the sites in the 10-km zone southward from Chernobyl NPP, where the surface pollution of soils equaled $2 \cdot 10^6$ Bq/m^2 by ^{90}Sr and $4 \cdot 10^6$ Bq/m^2 by ^{137}Cs, and the absorbed dose during three months of plant growing reached 0.15 – 0.4 Gy, the authors registered 2 – 3-fold increase of the quantity of cells with chromosome aberrations in meristems of outgrown seeds, formed in them. As the pollution level and corresponding dose increased to 0.5 – 2.5 Gy, the latter parameter showed 7-fold increase [11].

For the purpose of studying cytogenetic effect of chronic radiation on narrow-leaved hawk's-beard (*Crepis tectorum L.*) population in the 30-km zone, long-term investigations were carried out by V.A. Shevchenko and L.I. Grinikh [12]. It has been shown that in the first two years after the accident chronic radiation of 5 – 10 mR/h dose rate by γ-radiation induces up to 10 – 15% increase of chromosome aberration quantity in the first mitosis metaphases of root meristem in plantlets, outgrown from seeds gathered in the 30-km zone, compared with 0.3 – 0.5% for the control at 0.02 – 0.03 mR/h dose. In this case, cells possessing several aberrations were frequently observed. It has been found that the frequency of cells with aberrations in nine populations under study varies in much narrower ranges compared with the dose rate.

On the second year after the accident, beside the plantlets possessing aberrations in some cells, there were ones having all aberrant cells or a definite part of them possessing the same aberration e.g. these plantlets attained the changed kariotype of cells. The latter testifies about active microevolution processes proceeding in chronically irradiated populations of plants.

On the third and the fourth years the authors gathered seeds totally and separately from each plant. The latter operation allowed for implementing cytogenetic analysis of each plant progenitors. In these experiments a correlation between the frequency of cells with chromosome aberrations and the number of plantlets with nonidentical structural changes in kariotype, among which pericentric inversions and reciprocal translocations dominated, were observed at all radiation rates. In these tests 9-chromosome kariotypes of several heterozygous plants were also observed, which were liberated from homozygous plantlets with changed kariotype [13]. These data much more clearly indicate active mutation processes proceeding in plants growing on radiation-polluted territories.

MUTATIONS

The occurrence of mutations in plant populations on radiation-polluted territories is of special interest, because formation of remote consequences of chronic irradiation may be realized via genetic damages. During the period after the accident many investigators studying various agricultural and wild species of plants have observed the increase of the quantity of cells with chromosome aberrations and the frequency of chlorophyll [6] and waxy-mutations [2], and morphological mutations. In 1987 – 1989, V.V. Morgun and V.F. Logvinenko [14] indicated 2 – 23-fold increase of visual mutation frequency for various sorts of fall wheat grown in the 30-km zone compared with the level of spontaneous variability before the accident.

Since 1986, V.A. Shevchenko *et al.* [15] have studied mutation process proceeding in populations of some herbaceous, frutescent and ligneous species using indices of quantitative estimation of cells with chromosome aberrations in meristems and frequency of embryonic lethal mutations. The results obtained unambiguously indicate the genetic load, induced by radionuclide pollution of the pollution of the territory, being increased during the initial 2 – 3 years. In subsequent generations the frequency of mutations decreased, but the rate of genetic effect manifestation in descendants was and is still lower than the rate of dose decrease, which fell by three orders of magnitude during ten years.

Figure 4. Various ear shapes for Mironovskaya 808 wheat descendants, outgrown from grains gathered in autumn 1986

Since 1986, the authors of this review in cooperation with K.D. Kolomiets and L.A. Burdenyuk [16] have been studying kinetics of mutational variability of four fall wheat genotypes in consecutive 15 generation, impacted by radiation in the 10-km restriction zone (by August 1986 the pollution level at test sites equaled by ^{137}Cs $n{\cdot}10^6$ Bq/kg for soils and (1

– 2)·10^5 Bq/kg for plants (in this case, ^{137}Cs gave 10% or lesser of total radioactivity). In the first vegetation the quantity of plants with morphological anomalies equaled 60 – 80%, whereas in the second vegetation in the same zone at natural seeding it reached 60%. Families with abrupt deviations of features from the initial forms were clearly observed. The main among these deviations was ear sterility. Plants of the third vegetation proceeded in the 30-km zone also display predominance of partial or full sterility, grain alteration, shortened ear, etc. The plants grown on pure soils since that generation demonstrated 2 – 3-fold lower sterility, but morphological anomalies were high-frequent still. Changes in dimensions and shapes of plants and separate organs, the number of organs, color, bushiness, leaf wax bloom occurrence, etc. were observed. Determination of a tendency to preserve mutagenesis level, induced by the presence of two generations under chronic radiation, in the following time both in the 30-km zone and outside it e.g. on relatively pure selective field, was of special importance. Its intensity altered with time according to wave dependence, but in the following period no significant deviation of this process was observed. By 1999, 1,342 mutant lines of wheat were detected, characterized by different resistance to splitting intensity in subsequent generations (Figure 4).

Thus, the mutagenic effect on plants from radionuclide-polluted medium, as well as consequences of acute ionizing radiation impact during many years may be realized in many generations. Very often increasing quantity of chromosome aberrations, chlorophyllous and other mutations is explained by radiation-induced genome instability, which not only mutagenic process intensification, but also disturbance of typical gene depression norm is taken for.

IMMUNITY

The experience of operations on radionuclide-polluted territories testify about decreasing phytoimmune potential of plants under these conditions and, simultaneously, some disease-producing factors become more aggressive and the probability of occurrence of new phytopathogenes and races with enhanced virulence possessing high migration rate by the population increases. In particular, this is confirmed by studies of the population structure culm smut producing factor *Puccinia graminis*, carried out by A.P. Dmitriev *et al.* [17] on 11 species of wild and agricultural plants. The development stage of this disease in the 30-kilometer exclusion zone around Chernobyl NPP (the 30-km zone) is extremely high. The authors isolated 22 monoisolates of this factor, all of which were found virulent. Besides, only one race was widespread, whereas the others were infrequent, among which one race was found extremely virulent.

For the stationary conditions in the 10-km restricted zone, these authors have also indicated fall wheat vulnerability by mildew is 2 – 2.5 times higher than for control plants. Similarly, vulnerability by brown rust increased with radionuclide pollution of soils.

A decrease of plant resistance to phytopathogenic fungi is preserved for plantlets outgrown from seeds, sampled from polluted soils. For example, fall wheat resistance to mildew was twice lower than for plants grown on unpolluted soil. Increased sensitivity to brown rust is also preserved.

Wheat and rue grains produced under radionuclide pollution conditions possessed 33 – 55% lower activity of proteinase inhibitors. These inhibitors form stable complexes with proteolytic enzymes of phytopathogenic fungi with the result that activity of the latter is suppressed.

ADAPTATION

Some authors present data on increasing vital capacity of plants, grown under conditions of the radiation pollution, and their higher resistance to additional irradiation impact in laboratory. This is explained by different reasons: induction of repair systems, increasing concentration of natural substances possessing radioprotector properties, radiation stimulation, and selection of more resistant specimens. Sometimes, this phenomenon is called adaptation to ionizing radiation.

It should be mentioned that these works are rather scanty, but they always attract much attention and are actively discussed, because they touch upon the urgent problem, many points of view on which are often speculative.

V.N. Pozolotina [18] has studied carefully the works on the possibility of adaptation mechanism formation in plants impacted by increased radiation doses. Basing on a complex of cytological and morphological criteria in the experiments on provocative additional gamma-radiation of dandelion (*Taraxacum officinale Wigg.*) seeds, gathered at different sites of the accident effect zone with the background gamma-radiation changing within the range of 0.2 – 19 mR/h, she has indicated their increased radiation sensitivity. Comparing these results with the works indicating the increase of radiation resistance, she concludes that such opposite effects demonstrate high genetic and morphological instability of plants, induced by long-term impact of low-dose radiation. Finally, the author concludes that it is beyond the reason to speak about formation of genetically fixed enhanced radioresistance of the living organisms subject to higher radiation impact.

Only in the single case V.V. Shevchenko and L.I. Grinikh [12, 13] have observed a 1.5-fold increase in resistance to radiation of hawk's-beard (*Crepis tectorum L.*) seeds, gathered on the fourth year in the Chernobyl accident effect zone and, therefore, subject to definite doses of β– and γ–radiation. For all other populations of plants, studied in different years in the 30-km zone of the Chernobyl accident, radiation-polluted regions of the Bryansk Oblast and in the Kyshtym accident zone, no changes in resistance to radiation were observed.

O.N. Popova *et al.* [19, 20] described elevation of resistance to radiation for cenopopulations of two wild plants from the Chernobyl accident effect zone – ribwort (*Plantago lanceolata L.*) and morning violet (*Viola matutina Klok.*) – due to their self-autopurification from unfavorable genetic damages at the generation change. For many years they were observing seemingly paradoxical phenomenon: on the one hand, absolutely logical decrease of the seed productivity of plants with increasing radiation dose (elevation of the soil pollution level), and on the other hand, the decrease of sterile pollen-grain yield. Basing on careful, nearly single-piece analysis of phytocoenoses, the authors conclude that relatively high radiation doses induce the change of genetic status of a definite cenopopulation due to elimination of genetic damages during seed progeny formation.

Similarly, S.A. Dmitrieva *et al.* [21] consider that natural populations become radiation adapted via their division into radioresistant and radiosensitive fractions. At additional irradiation, the second fraction may be eliminated, but may also be preserved and represent the source of elevated frequency of mutations.

Thus, increased resistance to radiation for plants is not the result of their adaptation to elevated radiation doses, but is the selection consequence. Moreover, it is beyond the reason to speak about possible formation of specific radioadaptation mechanisms in plants or any living organisms during their ontogenesis.

Though sometimes the ability of radiated or still being radiated plants to accumulate some metabolites, flavonoids, in particular, including anthocyanin, carotinoids, fats, amino acids, especially containing SH-groups, and other radioprotectors is taken for the radioadaptation. However, this ability is usual nonspecific in relation to ionizing radiation action and represents a general response to damaging factor. Moreover, the unambiguous relation between endogenic concentration of these substances in plants and their radioresistance was not observed.

THE INFLUENCE ON PHYTOCOENOSES

Finally, the question of preservation of natural autochthonous vegetation on radiation-polluted territories is extremely urgent. Obviously, firstly, this requires preservation of the reproductive ability damage below a definite critical level for the species subject to radiation impact. Since the Chernobyl accident, the species composition of phytocoenoses and agrocoenoses in the 30-km exclusion zone have changed drastically. However, in general, these changes are stipulated by changed type of management, manifested in abrupt limitation of agricultural operations down to their complete conservation and manifold reduction of the anthropogenic influence. At such background of intensive change of vegetation even on highly radiation-polluted territories particular influence of the radiation factor on seral processes can be hardly determined.

The study of the seed quality of plants, several generations of which grew on polluted territories, generally testifies about stability of the production process. O.N. Popova *et al.* [19, 20] present data on the seed progeny of more than 20 species of plants from several cenopopulations subject to the radiation impact of dose rates between 0.1 and 800 mR/h (background γ-radiation in 1986). It has been found that for the majority of species both weight and germinating ability of 1,000 tested seeds equaled to the control. However, for some species, ribwort, in particular, the seed productivity was found noticeably lower.

Tracking the state of orchard grass (*Dactilis glomerata L.*) growing in the 30-km exclusion zone during five years after the accident, V.I. Shershunova and V.G. Zainullin [22] have found that at the sites with high radiation pollution of soils this plant falls out the phytocoenosis. For analogous conditions, P.G. Sidorenko *et al.* [23] have detected up to 30% increase of the quantity of nonviable pollen for white clover (*Trifolium repens L.*), rose bay (*Chamaenerium angustifolium L.*) and white campion (*Melandrium album L.*). These data testify about that seral processes may also be stipulated by weakening of species positions after the radiation impact.

Basing on long-term observations carried out not only in the Chernobyl accident impact zone, but also in the East-Ural radioactive trail, Semipalatinsk Nuclear Test Site, V.A. Shevchenko *et al.* [15] consider genetic effects the most important factors for changes of coenoses, affected by ionizing irradiation of radioactive precipitates. It is their opinion that these genetic effects are the main reason for ecological changes in the compositions of phyto- and zoocoenoses. They also demonstrate changes in biocoenoses at different dose rates of chronic radiation.

As the dose rate increases, the following effects gradually increase in coenoses: biochemical changes, chromosome aberrations visible mutations, changes in genetic structure of the population, elimination of radiosensitive species, coenosis depletion and, finally, coenosis degradation, which is the strongest among these effects. The authors outline the range of dose rates, at which genetic effects become obvious – 10^{-6}–10^{-4} Gy/day. For example, for the dose rate equal $2 \cdot 10^{-5}$ Gy/day genetic effects are clearly observed in coniferous species. At 0.1 Gy/day or higher all genetic effects are observed in the majority of flora and fauna species. Elimination of radiosensitive species of plants and animals happens at dose rate of 10^{-2} Gy/day or higher. Dose dependencies taking into account genetic effects are convex or linear shaped and possessed no threshold. This one more time confirms the threshold-free concept and testifies about high probability of coenosis vulnerability for ionizing radiation.

The advantages of such approach for estimation of changes in coenoses are obvious, because registration of genetic changes in chronically irradiated coenoses allows for operating with tests much more sensitive than somatic effect criteria. In accordance with this approach, radiobiological effects at the phytocoenological level are initiated at the dose rate, which damages the most radiosensitive species, pine tree, for example.

One of the works [24] presents results of long-term radiation-genetic monitoring of pine trees (*Pinus sylvestris L.*). In the initial mitoses of root meristems in plantlets, outgrown from seeds gathered in 1986 and 1987 at radiation-polluted sites (giving from 0.1 to 20 Gy doses absorbed by plants), the number of cells with chromosome aberrations was 1.5 – 7-fold higher than in control. In 1993, at the sites with high (0.1 – 1 Gy) and low (1 – 5 Gy) damage of trees the value of cytogenetic effect approached the control level, but at the sites with moderate damage level (6 – 10 Gy) it remained at the level observed in 1986 – 1987. In 1997 – 1998, the quantity of cells with aberrations decreased. The mutation frequency of 20 enzymatic loci in all radiated populations was 4 – 17-fold higher than in control. By and large the use of genetic criteria and radiosensitive vegetative object, which is pine tree, in the experiment allowed for estimation of the radiation dose effects in the accident zone by 1 – 2 orders of magnitude lower than those, estimated with the help of traditional radiobiological criteria.

Using one of the most radiosensitive annual grasses – horse beans (*Vicia faba L.*), we succeeded in recording an increase in the quantity of cells with chromosome aberrations in primary root meristems of germinating seeds, gathered in the Chernobyl accident effect zone at the dose rate in the range of 10^{-3} – 10^{-2} Gy/day [25]. It is noteworthy that at these doses parental plants demonstrate radiation stimulation, and in plantlet meristems the increase of the quantity of mitotic cells with the quantity of cells with aberrations was observed, which also indicated the radiation stimulation.

COMBINED ACTION OF RADIATION AND OTHER FACTORS

Under long-term irradiation of biocoenoses of the highest urgency becomes the problem of combined action of various types of radiation and non-radiation factors. It is not inconceivable that this is the main reason for anomalously strong action of low doses. Unfortunately, this sort of data for plants, animals and man is very short both for the Chernobyl accident zone and general.

With regard for the sad experience in extinguishing the burning reactor and consecutive pollution of large squares with leads, V.A. Kryukov *et al.* [26] have demonstrated that addition of 64 mg or more of lead nitrate per 1 kg of soil, polluted with Chernobyl's radioactive cesium, at lead nitrate natural content of 20 mg/kg induced progressive increase of the number of embryonic lethal mutations in mouse-ear cress (*Arabidopsis thaliana (L.) Heynh.*). The authors pay not much attention to synergism of radiation and lead action; however, analysis of the broad primary material presented in the work indicates the action of these agents exceeding the additive level for some combinations. Moreover, the work describes a very interesting effect: separate action of radiation and lead does not practically induce chlorophyllous mutations of plants, whereas their combined action e.g. lead injection to radiation-polluted soil produces such mutations.

Zinc and cobalt salt injection into sod podzol in the 10-km zone of the accident in amount of kilograms per ha in terms of metal weakened the radiation impact on plants, the intensity of some morphological disturbances in horse beans (*Vicia faba L.*) and the number of cells with chromosome aberrations in the their plantlet root meristems decreased [11]. This means that in conditions of Polesye soils, depleted with all trace elements and obtained the highest radiation pollution level after the accident, antagonistic interactions between ionizing radiation and low concentrations of heavy metal salts may take place. In turn, this produces a radioprotective effect.

REPAIR

To-date, the ability of higher plants to repair is studied for acute radiation impact only. In case of chronic radiation, when DNA damages are not abundant, the state of repair systems may differ from their natural state. The authors in cooperation with Bubryak [27] studies repair processes in male gametophyte cells of birch pollen grains. For quantitative estimation of the repair processes common method of unscheduled DNA synthesis complementary DNA with labeled thymidine inclusion was used. Pollen grains were gathered at the sites possessing different radiation pollution. Total dose of pollen grains radiation was calculated by their radioactivity that reached 14.8 kBq/kg. Total doses obtained in the experiment equaled: 0.005 – 0.01, 0.04 – 0.08 and 0.5 – 1 Gy. Unscheduled synthesis was studied after acute radiation impacting the pollen grain by γ-radiation from ^{137}Cs in the dose range between 1 and 1,000 Gy, where potential ability of the repair systems may be determined. The data obtained allow for a conclusion that, generally, excisional DNA repair takes place, but it is low effective.

Hence, our investigations carried out in cooperation with V.D. Naumenko [28] indicated intensification of unscheduled DNA synthesis in lupine plantlets, outgrown from seeds, which were formed in the 30-km zone at the background approaching 40 mR/h. After γ-radiation in

10 Gy dose and processing with hydroxy-urea in concentration inhibiting replicative DNA synthesis, in meristem cells of such plantlets two-fold increase of ^{3}H-thymidine was observed.

In 1986 – 1987, for the populations of several wild species of plants, including relatively radiosensitive ones from the pea family, namely, tufted vetch (*Vicia cracca L.*), sampled from the sites around Chernobyl NPP with different radionuclide composition and pollution level, A.B. Syomov *et al.* [29] determined increasing quantity of single-stranded breaks in DNA of plants sampled from the sites with predominant quantities of γ– (^{134,137}Cs, ^{144}Ce) and α– (^{239,240}Pu) irradiating radionuclides. At the same time, in populations from the sites mostly polluted by β–irradiating radionuclides (^{90}Sr) the quantity of DNA breaks equaled the spontaneous level, which was related to their repair at this infrequently ionizing radiation type.

Chromatin and DNA breaks and probability of their repair were persuasively shown by V.N. Reshetnikov *et al.* [30] for 1–4-day plantlets and tylosis tissues of seed embryos from triticale and spring barley, produced in the 30-km restriction zone at the sites with 555, 1,480 and 2,220 kBq/m^2 radiation pollution level (the authors give no indication, but, apparently, by ^{137}Cs).

Regenerative Reduction

Reduction processes are provided by repair, repopulation of cells in meristems, lateral bud regeneration, and secondary meristem formation. After acute irradiation of perennial woody plants in the first months after the accident, the radiation damage was developing during several years. After 1990, repair processes began proceeding. They are very slow, but, nevertheless, obvious [7], though they are significantly masked by natural forest recovery processes.

Conclusion

It is obvious that up-to-date no endangered vegetation species or significant changes in phytocoenoses were determined in the Chernobyl accident zone even on the territories with high radiation pollution level, apparently, except for the former “Red Forest” area and several sites of much smaller square. At the same time, obvious and indisputable facts of manifestation of definite radiobiological effects are observed, which testify about radiation damage of some species. Among these effects genetic ones are most urgent due to their origination from genome stability break. They become the reason for mutations, stratification of populations, elimination of some species and, apparently, other effects, which can hardly be foreseen now. This shows no contradiction, because cumulative radiobiological phenomena last for a long time and, apparently, remote consequences of the radiation impact have been not completely realized yet. Moreover, adverse radiation action is opposed by the system of repair processes appealed to provide for stability of biocoenoses.

REFERENCES

[1] Grodzynsky D.M., Bulakh A.A., and Gudkov I.N., *The Chernobyl Catastrophe*, Kiev, Naukova dumka, 1995, pp. 293 – 311. (Rus)

[2] Grodzynsky D.M. and Gudkov I.N., *Chernobyl. The Restriction Zone*, Kiev, Naukova dumka, 2001, pp. 325 – 377. (Ukr)

[3] Mikheev A.N., *Radiats. Biol. Radioekol.*, 1999, vol. 39(6), pp. 663 – 666. (Rus)

[4] Szabo A.S., J. *Radioanal. Nucl. Chem.: Lett.*, 1987, vol. 119(6), pp. 503 – 511.

[5] Kal'chenko V.A., Fedotov I.S., Igonina E.B. *et al.*, *Radiats. Biol. Radioekol.*, 2000, vol. 40(5), pp. 607 – 614. (Rus)

[6] Grodzynsky D.M., Kolomiets K.D., Kutlakhmedov Yu.A. *et al.*, *Anthropogenic Radionuclide Anomaly and Plant*, Kiev, Lyibid', 1991, 160 p. (Rus)

[7] Kozubov G.M. and Taskaev A.I., *Radiobiological Investigations of Coniferous in the Chernobyl Accident Zone*, Moscow, IPTs "Design. Information. Mapping", 2002, 272 p.(Rus)

[8] Dmitrieva S.A and Parfenov V.I., *Radioctive Pollution of Vegetation in Belarus*, Minsk, Navuka i tekhnika, 1995, pp. 275 – 337. (Rus)

[9] Zyablitskaya E.Ya., Geras'kin S.A., Udalova A.A., and Spirin E.V., *Radiats. Biol. Radioekol.*, 1996, vol. 36(4), pp. 498 – 505. (Rus)

[10] Geras'kin S.A., Dikarev V.G., Udalova A.A. *et al.*, *Radiats. Biol. Radioekol.*, 1998, vol. 38(3), pp. 367 – 374. (Rus)

[11] Gudkov I.N., Kitsko V.E., Grisyuk S.N *et al.*, *Radiats. Biol. Radioekol.*, 1999, vol. 39(2-3), pp. 349 – 353. (Rus)

[12] Shevchenko V.V. and Grinikh L.I., *Radiobiologia*, 1990, vol. 30(6), pp. 728 – 734. (Rus)

[13] Shevchenko V.V. and Grinikh L.I., *Radiats. Biol. Radioekol.*, 1995, vol. 35(5). (Rus)

[14] Morgun V.V. and Logvinenko V.F., *Mutational Selection of Wheat*, Kiev, Naukova dumka, 1995, 624 p. (Rus)

[15] Shevchenko V.A., Abramov V.I., Kal'chenko V.A. *et al.*, *Radiats. Biol. Radioekol.*, 1996, vol. 36(4), pp. 531 – 545. (Rus)

[16] Grodzinsky D.M., Kolomiets K.D., and Burdenyuk L.A., *Bull. Ecological State of the Restriction Zone*, 2001, No. 16, pp. 50 – 54. (Ukr)

[17] Dmitriev A.P., Kryizhanovskaya M.R., and Gushcha N.I., *Nauch. Vestnik NAU*, 2001, No. 45, pp. 69 – 76. (Ukr)

[18] Pozolotina V.N., *Remote Consequences of Radiation Impact for Plants*, Ekaterinburg, Akademkniga, 2003, 186 p. (Rus)

[19] Popova O.N., Frolova N.P., and Taskaev A.I., *Radioecological Monitoring of Natural Environmental Systems*, Syktykvkar, Komi NTs RAN, 1993, pp. 64 – 79. (Rus)

[20] Popova O.N., Frolova N.P., and Taskaev A.I., *Radiats. Biol. Radioekol.*, 1994, vol. 34(6), pp. 872 – 876. (Rus)

[21] Dmitrieva S.A., Prfenov V.I., and Davidchik T.O., *Intern. Conf. "The Questions of Radiation Genetics at the Edge of the Century", Thes. Rep.*, Moscow, Ros. Univ. Druzhby Narodov, 2000, pp. 264 – 265. (Rus)

[22] Shershunova V.I. and Zainullin V.G., *Radiats. Biol. Radioekol.*, 1995, vol. 35(5), pp. 690 – 695. (Rus)

[23] Sidorenko P.G., Kordyum E.L., Pryadko E.I. *et al.*, *Radiobiological Congress, Thes. Rep.*, Part 3, Pushchino, 1993, pp. 919 – 920. (Rus)

[24] Kal'chenko B.A., Shevchenko V.A., Rubanovich A.V. *et al.*, *Radiats. Biol. Radioekol.*, 1995, vol. 35(5), pp. 702 – 707. (Rus)

[25] Gudkov I.N., Gural'chuk Zh.Z., and Ivanova E.A., *Nauchn. Vestnik NAU*, 2001, No. 45, pp. 62 – 69. (Ukr)

[26] Kryukov V.i., Shishkin V.A., and Sokolenko S.F., *Radiats. Biol. Radioekol.*, 1996, vol. 36(2), pp. 209 – 218. (Rus)

[27] Bubryak I.I. and Grodzinsky D.M., *Radiobiologia*, 1989, vol. 29(5), pp. 589 – 594. (Rus)

[28] Naumenko V.D. and Grodzinsky D.M., *Intern. Conf. "The Questions of Radiation Genetics at the Edge of the Century", Thes. Rep.*, Moscow, Ros. Univ. Druzhby Narodov, 2000, pp. 167 – 168. (Rus)

[29] Syomov A.B., Pitsina S.N., and Sergeeva S.A., *Sci. Total Environ.*, 1992, vol. 112(1), pp. 1 – 8.

[30] Reshetnikov V.N., Lapteva O.K., Sosnovskaya E.F., and Roshchenko M.V., *Radiats. Biol. Radioekol.*, 1996, vol. 36(4), pp. 567 – 572. (Rus)

In: 20 Years after the Chernobyl Accident
Editors: E. Burlakova and V. Naidich, pp. 247-264
ISBN 1-60021-249-2

Chapter 14

CYTOGENETIC INDICES OF ENGLISH OAK (*QUERCUS ROBUR L.*) SEMINAL PROGENY SUBJECT TO RADIOACTIVE RADIATION IN THE CHERNOBYL NUCLEAR DISASTER AND GROWING ON TERRITORIES WITH DIFFERENT LEVELS OF ANTHROPOGENIC CONTAMINATION

V. G. Artyukhov and V. N. Kalaev*
Voronezh State University, Voronezh

ABSTRACT

Cytogenetic characteristics (mitotic activity, cell mitosis stage propagating time, level and spectrum of mitosis disturbances, and nucleolar characteristics) of English oak (*Quercus robur L.*) seminal progeny subject to radioactive radiation in the Chernobyl nuclear disaster and growing on territories with different levels of anthropogenic contamination have been studied. Despite the normalizing radioactive background in the investigated territories, the changes in cytogenetic characteristics of English oak seminal progeny compared with the control are observed. The results obtained are accounted for by genome radiation-induced instability and wave kinetics of the mutation process. The questions of relevancy and expediency of the use of English oak seminal progeny subject to radioactive impact in forest recreation are discussed.

Keywords: English oak (Quercus robur L.), territory contamination, mitotic activity, remote effects, mitosis pathology, Chernobyl disaster, nucleolar characteristics

---

* Addressee: 1 Universitetskaya Ploshchad', 394006 Voronezh, Russia; Genetics, Selection and Evolution Theory Cathedral, Biological and Soil Department Voronezh State University; Tel.: (7-4732) 20-88-76; E-mail: Dr_Huixs@mail.ru

The Chernobyl nuclear disaster caused radioactive contamination of extremely wide forest squares. This has set up the problem of scientific justification of the forest management on radiation-polluted lands [1]. Despite a significant decrease of the radiation background on polluted territories due to partial decomposition of radioactive species (especially short-living ones) and their penetration deep into the soil, the following question remains unsolved: what radiation-induced cytogenetic effects and of what origin will be observed in the seminal progeny of trees growing on these territories. Solution of this question is of greatest importance, because it will allow for science-based recommendations for forest recreation using seminal progeny, gathered from previously polluted territories and adequately assessment of cytogenetic monitoring data of the areas polluted with radiation during the Chernobyl nuclear accident using seminal progeny of trees as the test-object. In relation to the above-mentioned, an attempt was made to determine cytogenetic indices (mitotic activity, cell mitosis stage propagating time, level and spectrum of mitosis disturbances, frequency of occurrence of cells with 2 or more plasmosomes in the nucleus) for English oak seminal progeny subject to radioactive impact in the Chernobyl nuclear accident in 1986, in remote periods (10 – 12 years after the radiation impact) on the territories with different levels of anthropogenic pollution (namely, significant car emissions and without significant anthropogenic loads).

English oak is one of the main and the most valuable forest-forming species in Russia. In the literature analyzed, we have found no works studying remote cytogenetic consequences of radioactive pollution effect in the seminal progeny of English oak in the regions, where the background gamma radiation have already been normalized. Previously, the seminal progeny of English oak growing in Bransky area of broad-leaved forests subject to radioactive pollution, caused by the Chernobyl nuclear accident, where the background gamma radiation at the moment of investigation varied within the range of 0.4 – 0.5 mR/h, displayed a significant alteration of mitotic activity (however, the authors of the cited works had not mentioned time of material fixation that may have an effect on reliability of the results obtained, because this parameter is changing during the day [2, 3]), an increase of the cell mitotic anaphase proceeding time and the number of chromosomal disturbances, among which bridges were the most frequent [4].

MATERIALS AND METHODS

For the test territories possessing different levels of anthropogenic pollution, we have selected two separate sites. The test site with a significant level of anthropogenic pollution is located at the 9th km of Zadonsky highway (between Moscow and Voronezh). According to the data obtained by Nuclear Physics Department of Voronezh State University, in 1986, radiation increase (to ~200 μR/h) was observed at this site. The main pollutants of the environment in this area were car emissions. Materials for cytogenetic monitoring were gathered in autumn 1994 from 8 test sites and in 1996 and 1998 from one test site. At each test site 25 acorns from phenotypic normal trees, having no visible damages by depredators and fungal diseases were gathered. The background gamma radiation at the test sites during the acorn gathering varied from 12 to 18 μR/h. It seems to us that cytogenetic effects observed on the current sites should be considered as the result of combined impact car emissions and remote consequences of radioactive radiation.

Cytogenetic effects of radioactive impact on the sites with insignificant anthropogenic pollution affecting the seminal progeny of English oak were measured in the area of Staryi Kurlak village, located in the Anninsky District of Voronezh Region. In 1986, the soils of this zone were polluted with cesium-137 up to 3 Ci/km^2 level [5].

Staryi Kurlak village is located 100 kilometers southwest from Voronezh City, on the Oka-Don Plain, nearby the Bityug River, which is the left confluent to the Don River. This area is characterized by the moderate continental climate with the average temperatures equal −9.5°C in January and +20.7°C in July. The average annual rain capacity is 550 – 560 mm (with the maximum in the spring-summer season). The vegetation period (with the average temperature above 5°C) lasts 190 days. The soils are represented by typical black earth soil. The current wood belongs to the Khrenovsky forestland.

The seeds were gathered from three test sites: Kashara (1996, 1998; 1997 – the lean year); Kushelevo, the site of natural hydrological memorial of the Voronezh Region – Chernoe Lake [6] (1996, 1997, 1998); Yasnaya Polyana (1998). These test sites are located 5 – 7 km north from Staryi Kurlak village. The distance between the test sites was 1.5 – 2 km. The background gamma radiation during gathering have not exceeded 14 – 17 μR/h. According to the data of Nuclear Physics Department of Voronezh State University, the site pollution with cesium-137 equaled:

Site	Year	Level, Ci/km^2
Kashara	1995	0.74
	1998	0.15
Kushelevo	1995	0.15
	1998	0.07
Yasnaya Polyana	1998	0.23

Possible decrease of soil pollution level with cesium may be caused by its migration down to deeper occurrences and simultaneous partial decomposition. At present, the sites under consideration are not subject to any anthropogenic impact. Therefore, cytogenetic effects observed in the seminal progeny of English oak may be considered as the remote consequences of the radioactive pollution. For testing the plants, 25 acorns from phenotypic normal trees having no visible damages by depredators and fungal diseases were gathered at each site. The seeds were taken from 4 phenotypic normal trees from each test site. The trees were approximately aged as 100 years. Besides oak, the test sites possessed common pine (*Pinus silvestris*). All test sites had abundant acorn yield.

The control was performed on seeds gathered in Usmansky forest (near "Venevitinovo" biological station of Voronezh State University). This site is located about 30 km westward from Voronezh City, and its soil and climatic conditions are the same as for the test territories. Pollution of the control site with chemical substances and radionuclides equaled MPC or lower [7]. The background gamma radiation varied within the range of 10 – 14 μR/h, and β-radiant flux density was below 1.2 β-particle/min·cm^2, which falls within the permitted range. Seminal material gathered at the control and the test sites was identical. (Due to bad harvest of acorns in 1997 and 1998 the control values to be compared with the tests data of cytogenetic indices for English oak seminal progeny were taken for 1995 and 1996 season yields, and statistically reliable differences in these indices were not observed.) The investigations in these years were stipulated by optimal conditions for this oak species in the mentioned area, because in this period the yield of acorns was quite abundant. Years 1998 and 1999 had bad harvest or the seminal material was seriously damaged by fungal diseases and insect pests. It is the authors' opinion that acorn gathering at the ecologically safe site in the years favorable for the current species and allows for judging about cytogenetic indices of the oak in norm. The comparison with cytogenetic of acorn plantlets from polluted sites in the years favorable for fruiting will allow for decreasing or neglecting the climate component in possible changes of cytogenetic indices and the genotype (resistance to diseases).

The investigation material was English oak acorn plantlet roots. Acorns were germinated in green sand at room temperature. As the roots became 2 – 3 cm long, they were fixed in a mixture of 96% ethyl alcohol and glacial acetic acid (3:1). The plantlet roots were fixed during 22 h (in winter), when mitotic activity and mitosis pathology peaks were observed in control samples [3, 8]. In 1994, acorn plantlet roots, gathered on the site neighboring Zadonsky highway, were placed into acetoalcohol at 4 p.m. For the comparison, the control group was fixed at the same time. The material was stored at +4°C in a refrigerator. Microslides were prepared according to the previously described technique [9].

From 7 to 10 microslides were prepared from oak acorn plantlet roots gathered on each test and control site. The slides were studied on a light microscope Laboval-4 (Carl Zeiss, Jena) at 40×1.5×10 and 100×1.5×10 magnifications. For every slide total number of tracked cells (700 or more), the number of fusing cells existing in one mitosis phase or another, the number of pathological mitoses, and the number of meta-, ana- and telophases with persistant nucleoli were considered. Basing on the data obtained, the mitotic index (the part of fusing cells, %), the part of pathological mitoses among the total number of fusing cells and meta-, ana- and telophase (the mitosis studies, where disturbances might be detected by the method used), the part of metha-, ana- and telophases with persistant nucleoli in the total amount of fusing cells at the mentioned stages, and cell distribution by mitosis stages (anaphase and telophase cells were counted together, because in case of dyeing with acetohematoxylin these stages may hardly be distinguished) were determined. Pathological mitoses were classified according to Alov [10].

The occurrence of one type of disturbances or another was expressed in percents of their total number. The part of cells with different amounts of nucleoli in the nucleus (%) was considered.

Statistic treatment of the data was performed on IBM PC/AT by *Stadia* statistical software. The data aggregate procedure and their processing are discussed in the work by Kulaichev [11]. Cytogenetic characteristics of English oak were compared by the following criteria: the frequency of persistant nucleolus and pathological mitosis occurrence by nonparametric X-criterion of Van-der-Varden ranks, because the distribution of these indices does not obey normal law, mitotic index and the number of cells with n nucleoli in the nucleus by Student's parametric t-criterion (the distribution of these indices obeys the normal law), and variance by Fischer's F-criterion. The comparison of the fractions of cells possessing different types of pathological mitoses using their angular transformation and introducing the Yates correction was carried out according to Lakin [12]. The correlation dependence was determined with the help of parametric correlation index (r) and Spirman's rank correlation index (r_S).

RESULTS AND DISCUSSION

Cytogenetic Parameters of English Oak Seminal Progeny Growing on the Territories with Significant Level of Anthropogenic Pollution

The investigation of plantlet roots of acorns, gathered in 1994, showed the increase of mitotic index in cells of apical meristem compared with the control (Figure 1). It is also found that in the test oak plantlets the part of root meristem cells in the mitosis prophase is increased. For example, for the plantlets of acorns gathered on 8 test sites of the studied territory a part of cells showing the mitosis prophase in the total amount of fusing cells varied within the range of 77.7 – 90.1% (Table 1) , whereas for control plantlets this value equaled 68%. Apparently, the increased mitotic index for test plantlets is explained by increasing fraction of cells in the mitotic prophase.

Table 1. The number of prophase cells (in percents to total number of fusing cells) in the apical meristem of English oak plantlets sampled in 1994 at the test sites near the 9th km of Zadonsky highway

Site number	Percentage of prophase cells
1	80.3
2	78.4
3	79.0
4	82.2
5	75.7
6	88.4
7	90.1
8	77.7

For 1996 and 1998, the data on cytogenetic characteristics of English oak seminal progeny from the Zadonsky highway zone are shown in Table 2.

In 1996, test plantlets of English oak showed no increase of the mitotic index compared with the control ((11.2 ± 1.2)% for the test and (13.5 ± 0.6)% for the control), whereas in 1998 a sharp increase to (21.1 ± 2.4)% was observed (reliability of the difference with the control is $p < 0.01\%$). However, no increase of the cell fraction at the prophase stage was observed. Vice versa, it decreased to (64.1 ± 4.1)% (with reliability of the difference from the control equal $p < 0.05$). In turn, the decrease of prophase cell fraction reflects upon fractions of metaphase and anaphase-telophase cells, which were increasing, but showed no reliable differences with the control. Therefore, this phenomenon may be considered as the stimulation of mitotic activity rather than a delay in passing the mitosis stages by cells. According to Fischer's F-criterion, despite similar values of mitotic indices in the test implemented in 1996 and in the control, the values of this index variance were reliably different from the control ($p < 0.05$, 11.9 for the test and 5.7 for the control). In 1998, the variance was reliably higher in the test ($p < 0.01$). High variance in the test selection of oak, obtained in 1996 and 1998 by the *mitotic index* parameter, allows for speaking about genetic inhomogeneity of the seminal progeny of the trees growing on the studied territory by sensitivity to anthropogenic pollution and the presence of the forms sensitive and resistant to the effect of pollutants in the selection under study. Hence, the part of these forms is constant in the mentioned selection, because no reliable differences in variances by years were observed. Apparently, the selection of pollution-resistant maternal trees producing great amounts of stable forms will allow creation of protective forestation, highly resistant to the pollutant impacts. For oak progeny in 1994, the frequency of pathological mitoses did not exceed control values (Figure 2), which is probably associated with a small number of cells existing in metaphase, anaphase or telophase, when mitosis pathologies may be detected. Hence, according to Alov [10], the increase of the prophase fusing cell fraction may be considered as the mitosis pathology associated with chromosomal aberrations and the absence of a possibility for highly damaged cells to transit to the next mitosis stage. As informed by us before [9], the oak plantlets studied in 1996 displayed increasing amount of mitosis disturbances compared with the control of 1996 at the estimation by Wilcoxon's *U*-criterion ($p < 0.05$); at the estimation by Van-der-Varden's *X*-criterion no reliable differences were displayed. Therefore, the question about reliability of the differences remained unsolved. The comparison of test values with generalized control data showed no reliable differences. The absence of differences in indices of pathological mitoses in tests and control is confirmed by close values of pathological mitosis numbers, calculated without consideration of prophase cells. In 1998, the difference between tests and the control are reliable similar to comparison of pathological mitoses both with ($p < 0.01$) and without accounting for ($p < 0.05$) prophases. In all years of the studies, no micronuclei were observed in the meristem of oak plantlet roots. The increase of mitotic index and the number of mitosis disturbances testify about root meristem adaptation to pollution. The increase of the number of fusing cells compensates their death due to fusion disturbance. The comparison of data for 1996 and 1998 shows the reliable ($p < 0.05$) increase of mitosis anomalies.

Table 2. Cytogenetic characteristics of English oak seminal progeny, grown in 1996 and 1998 near the 9th km of Zadonsky highway

Cytogenetic indices, %	Year of investigation		Control
	1996	1998	1995 + 1996
Mitotic index	11.8 ± 1.2 [2]	21.1 ± 2.4 **	13.5 ± 0.6
Mitotic index with no respect to prophases	3.9 ± 0.5 [1]	21.1 ± 2.4 **	3.6 ± 0.2
Mitosis pathologies	1.8 ± 0.3 [1]	7.6 ± 1.4 **	1.4 ± 0.3
Mitosis pathologies with no respect to prophases	6.9 ± 2.0 [1]	10.2 ± 1.6 *	3.1 ± 0.3
Occurrence frequency of cells with two nucleoli in the nucleus	2.5 ± 0.7	10.2 ± 1.6 *	3.1 ± 0.3
Occurrence frequency of persistant nucleolus in metaphase	14.4 ± 3.7	11.8 ± 4.2	13.6 ± 2.8
Prophase cell pool	68.2 ± 3.4	64.1 ± 4.1 *	72.3 ± 2.1
Metaphase cell pool	16.6 ± 3.0	18.1 ± 2.9	13.9 ± 1.1
Anaphase-telophase cell pool	13.2 ± 2.4	27.8 ± 1.0	13.7 ± 1.4

Notes: * Reliable differences with the control ($p < 0.05$);
** reliable differences with the control ($p < 0.01$);
[1] – reliable differences for 1996 and 1998 ($p < 0.05$);
[2] – reliable differences for 1996 and 1998 ($p < 0.02$)

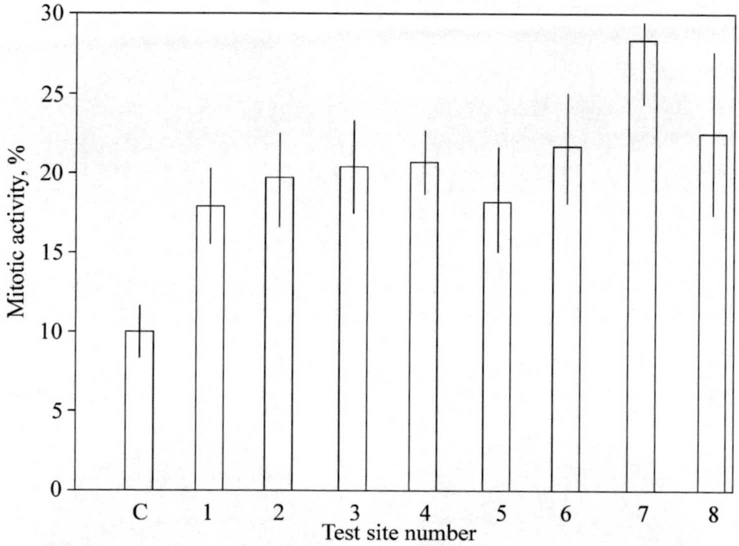

Figure 1. Mitotic activity of root meristem cells of English oak plantlets sampled in 1994 at the site near 9th km of Zadonsky highway; C is control

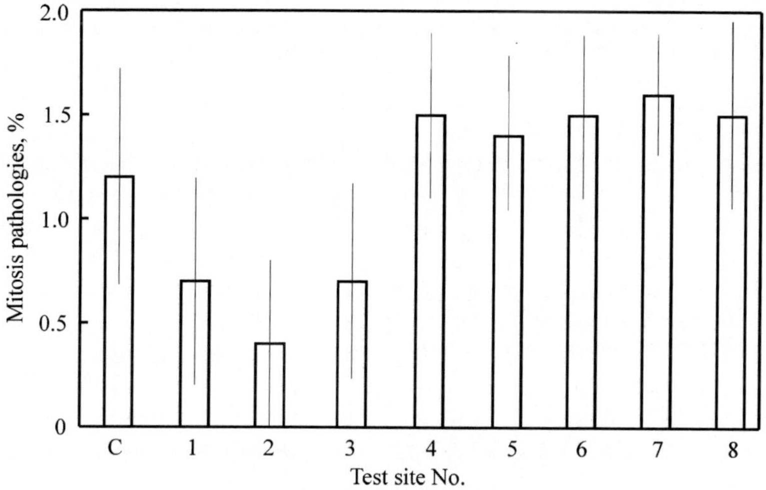

Figure 2. The occurrence frequency of mitosis pathologies in root meristem cells of English oak plantlets, sampled in 1994 at a site near the 9th km of Zadonsky highway; C is control

The increase of prophase cell fraction in oak plantlets in 1994 and the fraction of pathological mitoses in 1998 was, apparently, related to a synergic aftereffect of radiation and car emissions. Such effects were observed by many authors [13 – 16] for some species of herbs and pine trees growing on the territories, subject to radioactive pollution in the Chernobyl nuclear accident. The increase of cytogenetic anomalies in plants from the territories neighboring highways, at natural background radiation was indicated. The frequency of chromosomal aberrations in plants was the same as on the territories with high radiation pollution level.

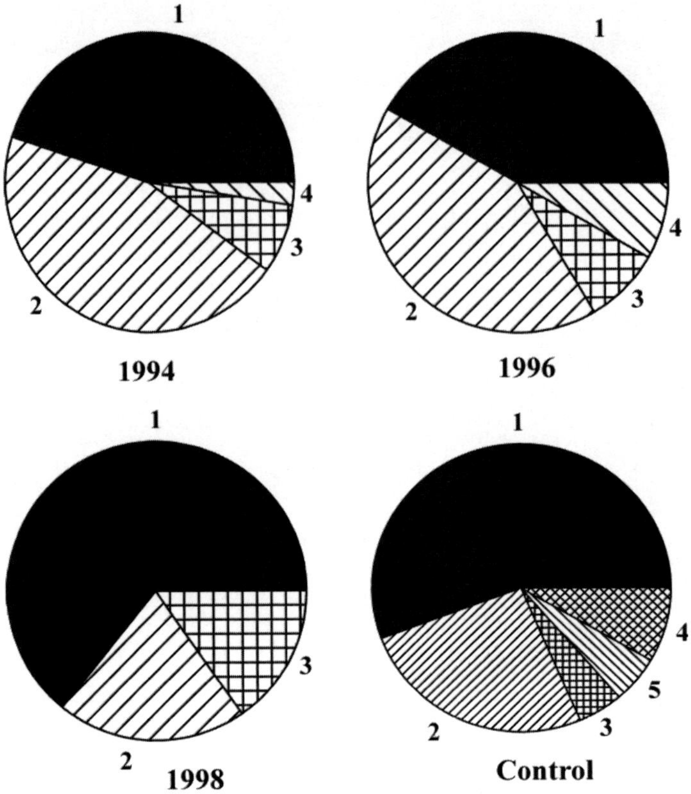

Figure 3. The spectrum of pathological mitoses in the root meristem of English oak plantlets, sampled in 1994, 1996 and 1998 at a site near the 9th km of Zadonsky highway: 1 – chromosome delay in metakinesis (1994 – 45.0%; 1996 – 41.6%; 1998 – 64.1%; control – 55.6%); 2 – bridges (1994 – 45.0%; 1996 – 41.6%; 1998 – 20.5%; control – 25.5 %); 3 – chromosome delay in anaphase (1994 – 7.5%; 1996 – 8.3%; 1998 – 15.4%; control – 4.9 %); 4 – "*infrequent*" pathology (1994 – 0%; 1996 – 0%; 1998 – 0%; control – 5.3%); 5 – agglutination (1994 – 2.5%; 1996 – 8.3%; 1998 – 0%; control – 8.4 %)

Normalization of the prophase cell fraction, mitotic index and mitosis pathologies in 1996 and the following change of cytogenetic characteristics of the oak seminal progeny in 1998, which led to mitosis disturbances increase and prophase cell decrease, may be the consequence of irregular pollution from motor transport in different years and decrease of post-radiation component in the initiation of changes of these indices.

The spectrum of pathological mitoses on the studied territory is shown in Figure 3. The tests showed bridges, chromosome delay in metakinesis and anaphase, agglutinations. Interestingly, the fraction of bridges is disturbed in the spectrum. Moreover, in 1994, the differences with the control indices are reliable ($p < 0.05$), whereas in 1996, despite a big fraction of bridges, they are uncertain ($p < 0.07$). In 1998, the fraction of bridges decreased to 20.5%, which is somewhat lower (but not reliably) than in the control. Hence, the fractions of bridges in 1994 and 1996 are quite different ($p < 0.05$). The increase of bridge fraction in the general spectrum of disturbances and small number or the absence of agglutinations indicate that in plantlets under study in the mentioned years "*break – rejoining*" processes and formation of dicentric chromosomes dominate compared with the norm, in which chromosome agglutination dominates [8]. In 1998, the decrease of bridge fraction was probably related to the fact that at fusing cell pool increase the "*break – rejoining*" processes are masked by increasing pool of agglutinations, caused by endogenic processes. Another possible explanation is the decrease of post-radiation component in formation of disturbances. As shown before [17, 18], the decrease in the spectrum of fragment disturbances and the bridge domination reflect the repair abilities of the objects after radiation impact. The greater the number of bridges is, the better the repair mechanisms work.

In 1994, the pool of cells with two nucleoli in test oak plantlets increased only at one test site (Figure 4); in 1996 and 1998, no difference in cell fractions with different numbers of nucleoli in the test and the control was observed. Maximal number of nucleoli in the nucleus observed in the experiment equaled 4 (the pool of cells with 3 – 4 nucleoli gave less than 0.01%). Similar values were observed in the control.

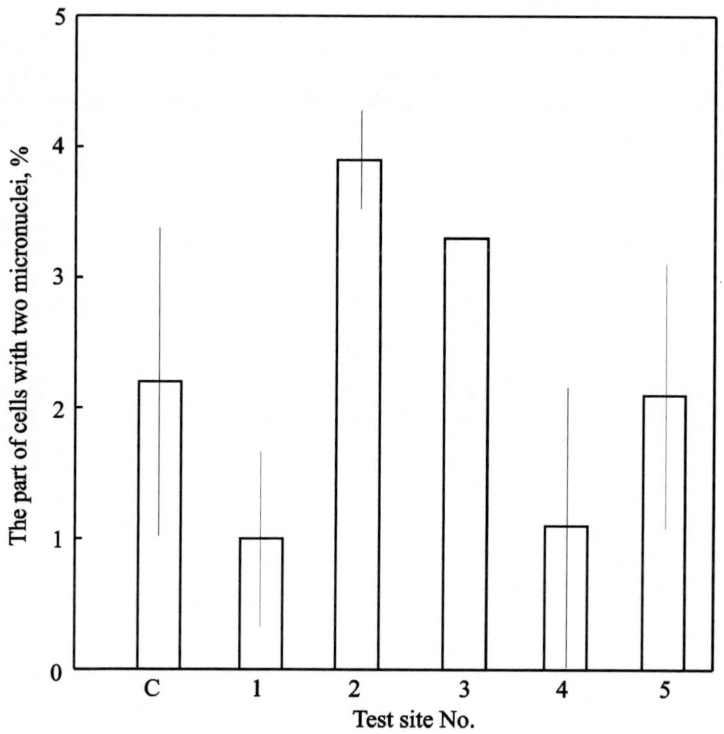

Figure 4. The occurrence frequency for cells having two nucleoli in the nuclei of English oak plantlet root cells, sampled in 1994 at a site near the 9th km of Zadonsky highway; C is the control

For the oak seminal progeny, at metaphase and anaphase-telophase stages a persistant nucleolus was observed. For the first time, this phenomenon was detected for the second-year plantlets growing at the site with increased background radiation, induced by consequences of the Chernobyl nuclear accident [19, 20]. Further on, persistant nucleoli in the cells of English oak plantlet roots occurred under non-stress conditions at the highest mitotic activity [3]. This phenomenon is stipulated by puffing of ribosomal genes, which provide for normal transcription in case of some unique gene functions suppressed by radiation [19].

In 1996 and 1998, no reliable differences in the occurrence of metaphase cells with persistant nucleoli in test and control pools were observed. In 1994, persistant nucleoli were not observed at all, which is most probably reasoned by the material fixation time. Though according to our previous investigations the change in the number of nucleoli is the most sensitive index of the stress impact, used in cytogenetic monitoring, the absence of differences in this index between test and control sites in 1994, 1996 and 1998 may be explained by low effectiveness of this index for the types of plants, which cells have small number of nucleoli in the nucleus (for example, English oak with 1 – 2 nucleoli in the nucleus). According to Arkhipchuk [21], the accurate estimation of the stress impact for such species requires determination of nucleolus sizes, but not their number. To our point of view, the change in cytogenetic indices of the oak (the mitotic index and the level of pathological mitoses), more resistant to the stress impact, may indirectly confirm this suggestion. It seems to us that the absence of differences in the frequency of persistant nucleolus occurrence at the mitosis metaphase indicates that the stress impact has exceeded a definite level, which initiated suppression of metabolic processes.

Cytogenetic Indices of English Oak Seminal Progeny Growing on the Territories with Insignificant Level of Anthropogenic Load

The results of investigations, carried out near Stary Kurlak village, Voronezh Oblast, radiation-polluted in 1986, show the change in cytogenetic indices of English oak seminal progeny growing on the studied sites in different years of investigation (Table 3). As indicated, the mitotic activity of the seminal progeny varied with years. Despite the decrease of mitotic activity compared for 1996 and 1998, which may be related to the decrease of Cs-137 concentration in the soil, we may not speak about the existence of dependence between the mentioned parameters, because in 1997, on the Kushelevo test site the mitotic index increased to (15.2 ± 1.4)% (though it is statistically uncertain), compared with the control value (13.5 ± 1.6)% and the value of 1996 (13.8 ± 0.7)%, and no correlation ability between the mitotic index at the test sites and Cs-137 concentration in the soil was observed. In 1998, at all test sites the mitotic index reliably decreased, compared with the control and previous years ($p < 0.05$). However, this process may not be considered as depression of mitotic activity, because the decrease of mitotic index fell within the range of this parameter variation in norm [3, 8]. Most likely, in this year a shift of daily peak of mitotic activity took place: possible changes of fusion pulses under the effect of factors possessing different etiology was discussed in a series of articles [22 – 24]. As mentioned above, analysis of the data obtained in different years indicated their changes, but comparison of mitotic index values obtained for different sites in 1996 and 1998 showed their good similarity. This is our opinion that this fact testifies about the absence of feedback in remote processes after the radiation impact on test sites and significant modulating influence of other factors (climatic, apparently).

Table 3. Cytogenetic indices of English oak seminal progeny growing in the area, radiation-polluted in 1986 in the Chernobyl nuclear accident

Cytogenetic indices, %	Test sites						Control 1995 + 1996
	Kushelevo			Kashara		Yasnaya Polyana	
	1996	1997	1998	1996	1998	1998	
Mitotic index	13.8 ± 0.7	15.2 ± 1.4	11.8 ± 0.5 [1]	15.0 ± 1.3	11.5 ± 0.4 [1]	11.6 ± 0.4 [1]	13.5 ± 0.6
Mitotic index with no respect to prophases	4.8 ± 0.2 [2]	6.9 ± 1.3 [1]	6.2 ± 0.3 [3]	5.4 ± 0.4 [2]	6.3 ± 0.4 [3]	6.1 ± 0.3 [3]	3.6 ± 0.2
Mitosis pathologies	1.8 ± 0.4	3.3 ± 0.8 [1]	2.6 ± 0.7 [1]	3.1 ± 0.6 [2]	3.0 ± 0.6 [2]	2.8 ± 0.4 [2]	1.4 ± 0.3
Mitosis pathologies with no respect to prophases	5.7 ± 1.2	8.6 ± 3.5	5.1 ±1.5	8.4 ± 1.6 [1]	5.3 ± 1.0	5.2 ± 0.8	5.9 ± 1.4
Occurrence frequency of cells with two nucleoli in the nucleus	1.6 ± 1.6 [1]	3.1 ± 0.9	6.5 ± 1.3 [1]	2.5 ± 0.5	4.3 ± 0.5 [1]	3.4 ± 0.5	3.1 ± 0.3
Prophase cell pool	62.5 ± 2.1 [1]	56.2 ± 5.6 [1]	47.2 ± 1.7 [3]	63.1 ± 3.1 [1]	45.8 ± 2.4 [3]	47.1 ± 2.5 [3]	72.3 ± 2.1
Metaphase cell pool	16.6 ± 0.9 [1]	19.6 ± 0.6 [1]	21.4 ± 1.3 [2]	15.8 ± 1.6	25.3 ± 1.5 [3]	23.4 ± 1.9 [3]	13.9 ± 1.1
Anaphase-telophase cell pool	18.0 ± 2.2 [1]	24.0 ± 5.9 [1]	30.5 ± 1.8 [3]	21.3 ± 2.3 [1]	30.7 ± 1.4 [3]	30.1 ± 1.6 [3]	13.7 ± 1.4

Notes: [1] – reliable differences with the control ($p < 0.05$);
[2] – reliable differences with the control ($p < 0.01$);
[3] – reliable differences with the control ($p < 0.001$)

Table 4. The "mitotic index" parameter variance for English oak seminal progeny growing at the test sites near Staryi Kurlak village

Test sites	Years of investigations		
	1996	1997	1998
Kushelevo	2.9	10.2	3.2
Kashara	13.8 *	–	1.7 *
Yasnaya Polyana	–	–	2.3 *
Control		5.7	

Note: * Reliable differences with the control ($p < 0.05$).

The variations of "*mitotic activity*" index variance are noteworthy (Table 4). In 1996, Kashara test site displayed variance increase, whereas in 1998, similar to Yasnaya Polyana, another test site, variance decreased below the control value. At Kushelevo site, no difference in this index with the control was observed. This site is the most pure by cesium content on the soil. Taking into account that at Kashara site the maximal Cs-137 concentration was detected in 1996, and in 1998 – at Yasnaya Polyana site, one may suggest the existence of relation between the radioactive pollution level of the site and genetic heterogeneity of the seminal progeny: the higher the pollution level at the site, the more significant variations of heterogeneity by *mitotic index* in different years. To put it differently, waves may occur in the population: either many quickly propagating (with high mitotic index) or slowly propagating (with low mitotic index) forms will occur or frequency of their occurrence will decrease to normal variability or even narrower range than in norm.

Data on the mitotic index calculated without consideration of prophase cells are also interesting. At all test sites this index exceeded the control level. This is stipulated by decreasing prophase cell pool and increasing metaphase and/or anaphase-telophase cell pools (Table 3). Since variability of the mitotic index calculated with neglected prophase pool falls outside its usual range, then basing on the investigations by Alov [10, 25], Kazantseva [26], and Kotlovaya [27] the increase of metaphase and anaphase-telophase cell pools may be considered a disturbance, related to cell cleavage spindle formation, cytotomy proceeding and start-up of genetic material checkpoint-repair system. Therefore, remote effects of radiation on the seminal progeny of English oak were manifested by changes in metabolic processes, which were expressed in disturbances of the mitotic apparatus formation.

Considering the changes in passing various mitosis stages by cells, the question about frequency of pathological mitoses in the root meristem of acorn plantlets with respect to the method of their calculation are noteworthy. The comparison of test and control values of pathological mitoses, calculated with consideration of prophases, indicated their reliable increase with time at all test sites, except for Kushelevo test site in 1996. At the first glance, this testifies about a significant increase of the control level of mitosis disturbances and allows for consideration of the current phenomenon as the remote radiation effect. However, the analysis of pathological mitoses, calculated with neglected prophase cell pools, in which no disturbances were observed, showed exceeding of the control level at one test site (Kashara) in 1996 only. The pathological mitosis values both with and without consideration of prophase cell pools by years indicated their increase in 1997 at Kushelevo test site

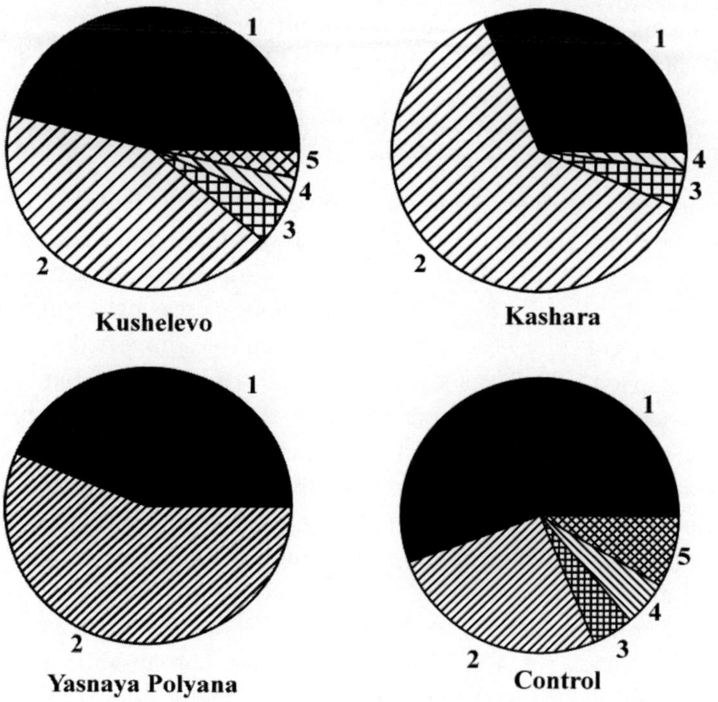

Figure 5. The spectrum of pathological mitoses in the root meristem of English oak plantlets, sampled in 1996 – 1998 at test sites near Staryi Kurlak village: 1 – chromosome delay in metakinasis (Kushelevo – 46.0%; Kashara – 31.3%; Yasnaya Polyana – 43.8%; control – 55.6%); 2 – bridges (Kushelevo – 42.9%; Kashara – 62.5%; Yasnaya Polyana – 56.2%; control – 25.2%); 3 – chromosome delay in anaphase (Kushelevo – 4.8%; Kashara – 4.2%; Yasnaya Polyana – 0%; control – 4.9%); 4 – "*infrequent*" pathologies (Kushelevo – 3.2%; Kashara – 2.1%; Yasnaya Polyana – 0%; control – 5.3%); 5 – agglutination (Kushelevo – 3.2%; Kashara – 0%; Yasnaya Polyana – 0%; control – 8.4%)

compared with 1998 values ($p < 0.05$) and decreasing disturbances at Kashara site in 1998 compared with 1996 ($p < 0.05$). Hence, comparison of the amount of pathological mitoses at different test sites in the same year did not show reliable differences and, therefore, it may be hardly considered that the change in percentage of pathological mitoses is stipulated by dynamics of Cs-137 concentration change in the soil. Basing on the above-said, one may agree with the opinion of Dmitrieva *et al.* [28] that remote radiation effect induces instability of cytogenetic indices of the seminal progeny of plants. They become, compared with non-irradiated selection, more sensitive to the effect of environmental factors. The wave kinetics observed 10 – 12 years after the Chernobyl nuclear disaster was stipulated by the presence and behavior of radiosensitive fraction in the seminal progeny. Under optimal or similar environmental conditions, the radiosensitive fraction is preserved and stipulates increased mutation frequency, determined in population gathers of seeds. As climatic conditions are unfavorable, it is eliminated, which is displayed in decreasing the frequency of induced cytogenetic disturbances down to control values [29]. The application of cytogenetic parameter of the *mitosis pathology* calculated with account for prophase cells and without it is of interest. It is our opinion that calculation of mitosis pathologies with respect to prophases is the rapid test method of integral estimation of the mitotic apparatus state, because it allows for detecting changes in its functioning (or the increase of mitosis pathologies and/or the change in cell mitosis duration, stipulated by disturbances in cleavage spindle formation or cytotomy). For more accurate determination of the factor, which change induced a disturbance of the mitotic apparatus, the calculation of mitosis pathologies neglecting prophase cell pool and cell distribution by mitosis stages should be implemented.

Figure 5 presents total spectrum of mitosis disturbances (data in each point are generalized for all years of investigation). As observed in the Figure, all test sites showed increasing bridge fractions and decreasing chromosomal delays in metakinesis compared with the control. Such spectrum of disturbances is detected in tissues with active systems repairing genetic damages in the cell [17, 18]. This testifies about increasing repair abilities of English oak plantlet root meristem cells, obtained from maternal trees impacted by radiation in remote periods. The comparison of mitosis disturbance spectrum for different sites indicated no significant differences. It is the authors' opinion that this gives one more proof to the supposition about the absence of direct relations between cytogenetic effects and soil pollution with radioactive cesium. Apparently, the increased repair abilities and changed cytogenetic parameters of the oak seminal progeny are reasoned by raise of genetic system of the organism to a new trigger level after the stress impact.

Experimental materials from the test sites showed significant variations in pools of cells having two nucleoli in the nucleus. The change of this index may indicate activation of latent nucleolar-organizing zones (in norm, the oak has two chromosome pairs with nucleolar organizers [30]) and metabolic activity increase in cells. Many authors [31 – 35] consider this phenomenon as the organism adaptation to unfavorable factors. Table 3 presents data on the occurrence frequency of cells having two nucleoli in the nucleus, obtained from the test sites. Acorn plantlets sampled from Kushelevo test site in 1996 showed reduced pool of cells having two nucleoli in the nucleus compared with the control ((1.6 ± 0.5)% in the test and (3.1 ± 0.3)% in the control ($p < 0.05$)), whereas in 1998 their fraction increased ((6.5 ± 1.3)% in the test ($p < 0.05$)). In the same year two plantlets from acorns sampled at this site had 3 cells with 5 nucleoli in the nucleus, which is untypical for English oak in norm. In 1998, samples from Kashara test site showed increasing pool of cells with two nucleoli ((4.5 ± 0.5)% in the test ($p < 0.05$)). Though in the control sampling the occurrence frequency of cells with two nucleoli showed no correlation with other cytogenetic characteristics, a negative relation of this index to mitotic index ($r = -0.325$; $p < 0.05$) and the prophase pool ($r = -0.369$; $p < 0.001$) occurred in the test. In the same test positive relation to "*pathological mitoses*" index calculated with respect to prophase cells ($r_S = 0.351$, $p < 0.01$) and without it ($r_S = 0.185$, $p < 0.05$) was obtained, simultaneously. According to Rostova [35], the relation between the indices becomes weaker under favorable conditions and, vice versa, becomes stronger under stress conditions. Basing on the above-mentioned correlation relations, one may suggest that English oak has two mechanisms of adaptation to the stress impact: the first mechanism is expressed in changing the prophase cell pool (it is related to positive correlation with pathological mitoses ($r_S = 0.231$, $p < 0.05$)); the second mechanism is related to activation of nucleolar-organizing zones (apparently, "*sleeping*" ones as for coniferous plants), because the number of cells with two nucleoli in the nucleus shows negative dependence on the prophase cell pool and positive dependence on "mitosis disturbance with no respect to "*mitosis disturbance with no respect to prophases*" index. Hence, one may suggest that these adaptation mechanisms are related to one another and compensate each other. Apparently, in remote periods after radiation impact on maternal trees of English oak, previously nucleolar-organizing zones become active in their seminal progeny.

Seminal progeny of the oak species under consideration had a persistant nucleolus in metaphase and anaphase-telophase cells. The occurrence frequency of this nucleolus and the number of plantlets analyzed, in which it was observed, are shown in Table 5. Clearly, in 1996 and 1997, the occurrence frequency of the persistant nucleolus at the mitosis metaphase stage was below the control level or it was not observed at all, whereas in 1998 no difference with the control was observed. This may be explained by ribosomal gene puffing decrease and considered as the remote radiation aftereffect.

Thus, basing on the investigations performed, it has been found that the seminal progeny of English oak maternal trees in remote periods on the sites subject to radiation impact and not subject to anthropogenic pollution showed changes in cytogenetic indices: mitotic activity (seminal progeny variability by this index is confirmed by high climate-sensitive oscillations) and cell mitosis stage duration, which is stipulated by disturbances in cytotomy processes and cleavage spindle formation. Despite normalization of the background radiation, the seminal progeny displayed increased repair activity of meristem cells. Apparently, "*sleeping*" nucleolar-organizing zones have been activated. In some years, the level of pathological mitoses was increased. The variation data correlate with the concept of the mutation wave kinetics in plants in remote periods after radiation impact. To our point of view, all these facts testify about undesirability of the application of English oak seminal progeny, sampled from the sites polluted with radiation, for forest recreation or it should be used very accurately (only after checking cytogenetic indices), because such seminal material may deteriorate the quality of forestation.

As the cell mitosis stage duration was different for all test sites, one may suggest it (this difference) as the most sensitive criterion of the stress impact for species possessing the greatest number of nucleolar-organizing chromosomes.

Table 5. Occurrence frequency of persistant nucleolus (PN) at metaphase stage of mitosis in English oak plantlet roots (in which it has been detected), sampled at a site near Staryi Kurlak village

Test site	Number of slides with PN/number of tracked slides			PN occurrence, %		
	Year of investigations					
	1996	1997	1998	1996	1997	1998
Kushelevo	0/7	2/7	15/15	0	6.2 ± 4.2*	10.0 ± 2.0
Kashara	2/8	–	11/11	3.0 ± 2.0 *	–	13.8 ± 3.1
Yasnaya Polyana	–	–	13/14	–	–	12.8 ± 2.0

Note: * Reliable differences with the control ($p < 0.05$).

Discussion

Basing on the studies, implemented at the sites with different anthropogenic pressing, one may consider that the seminal progeny of English oak trees growing at the sites polluted with radiation from the Chernobyl nuclear accident, even after normalization of the background radiation, displayed in stability of cytogenetic indices.

In the case of high pollution with car emissions, the seminal progeny of trees previously impacted by radiation show changes in cytogenetic characteristics and increasing genetic heterogeneity of the seminal progeny in accordance with the "*mitotic index*" parameter. In the case of pollution, the seminal progeny of oak may show mitotic index increasing both due to prophase cell growth and the rate increase of the cell mitotic cycle. The change of mitotic characteristics in apical root meristem may be considered as the additive response to the stress impact of pollutants, aimed at the compensation of cell death due to pathological mitoses and, therefore, at support of proliferative cell pool.

Since the seminal progeny of English oak had varying cytogenetic characteristics and heterogeneity of it increased by the "*mitotic index*" parameter simultaneously with the number of non-lethal mitosis disturbances and chromosome damages, we may not recommend acorn gathering for the purpose of forest recreation from the site near automobile roads, because this may give poor quality planting. The facts of cytogenetic parameter changing in English oak seminal progeny at a site polluted with automobile emissions testify about significant rearrangements in the genetic apparatus and occurrence of the mutant forms, which are able not only decrease quality of the planting, but also, probably, completely destroy the reproductive structure of artificially recreating planting.

The seminal progeny of English oak maternal trees in remote periods subject to radiation impact, growing on the sites not subject to anthropogenic pollution also showed changes in cytogenetic indices: mitotic activity and cell mitosis stage duration, which is stipulated by disturbances in cytotomy processes and cleavage spindle formation. Despite normalization of the background radiation, the seminal progeny showed increased repair activity in the meristem cells. Apparently, "*sleeping*" nucleolar-organizing zones were activated. In some years, the level of pathological mitoses increased. The changes correlate with the concept of mutation wave kinetics in plants in remote periods after radiation impact.

In the studied literature the reasons for radiation-induced instability of cytogenetic characteristics and the factors promoting its sharp increase or abrupt decrease in different years of tests are not completely disclosed yet. It is assumed that pursuant to short-term acute or long-term low ionizing radiation impact genome rearrangements leading to its radiation-induced instability appear in the plant cells [37]. The author of the cited work outlines two main mechanisms of its occurrence:

- Hereditary variability induced by intensification of mobile genetic component induction;
- Conformational rearrangements of genome changing activity of genes, which regulate cell differentiation processes.

At formation of remote radiation effects, plants may display different responses (the absence or the presence of genetic effect), realized in accordance with the external impact intensity and environmental conditions, both biotic and abiotic [37].

The current study is implemented in the framework of the Grant of the President of the Russian Federation supporting young Russian scientists and leading scientific schools of Russia "Cytogenetic responses of woody plants (on the example of deciduous species) to stress conditions, induced by anthropogenic pollution, and estimation of environmental genotoxicity for man by cytogenetic indices of woody plants" (grant MK–2587.2004.4).

REFERENCES

[1] Ipat'ev V.A., Tarasenko V.P., and Kholodilova L.V., *Oak – the Species of the Third Millennium*, in Collection of Proceedings of the Forest Institute, NAS of Belarus, Minsk, 1998, vol. 48, pp. 9 – 22. (Rus)

[2] Grif V.G. and Machs E.M., *Tsitologia*, 1994, vol. 36(11), pp. 1069 – 1084. (Rus)

[3] Butorina A.K. and Kalaev V.N., 'Diversity and adaptation in oak species', *Proc. Conf. IUFRO Working Party 2.08.05*, Oct. 12-17, 1997, College of Agricultural Sciences, The Pennsylvania State University, University Park, Pennsylvania, USA, Pennsylvania, 1998, pp. 46 – 48.

[4] Shlapakova S.A., 'Growth, reproductive ability and prospects of English oak reproduction in radiation-polluted forestry of Bryansky Region of broad-leaved forest zone', *Candidate Dissertation Thesis*, Bryansk, Bryansk State Engineering and Technological Academy, 1999, 20 p. (Rus)

[5] *Ecological Information Bulletin on the State, Problems and Forecasting of the Oblast Environmental Safety*, Voronezh, Izd. Voronezh State University, 1993, 129 p. (Rus)

[6] *Landscape Protection Areas and Special Nature Reserves of Vorogezh Oblast*, Voronezh, Tsentral'no-Chernozemnoe Knizhnoe Izd., 1983, 175 p. (Rus)

[7] Shchetinkina N.A., Pashkov A.N., Nemyikh V.N. *et al.*, *The State and Problms of the Usmansky Forest Ecological System*, Voronezh, 1992, Iss. 2, pp. 209 – 219. (Rus)

[8] Kalaev V.N., 'Cytogenetic environmental pollution monitoring applying plant test-objects', *Candidate Dissertation Thesis*, Voronezh, Voronezh State University, 2000, 25 p. (Rus)

[9] Butorina A.K., Kalaev V.N., Vostrikova T.V., and Myagkova O.E., *Tsitologia*, 2000, vol. 42(2), pp. 196 – 201. (Rus)

[10] Alov I.A., *Vestnik AMN SSSR*, 1965, No. 11, pp. 58 – 66. (Rus)

[11] Kulaichev A.P., 'Methods and data mining tools in Windows OS. Stadia 6.0', *Informatika i Komp'yutery*, 1996, 257 p. (Rus)

[12] Lakin G.F., *Biometry*, Moscow, Vysshaya Shkola, 1990, 352 p. (Rus)

[13] Dmitrieva S.A., *1st All-Union Congress on Radiobiology: Thes. Rep.*, vol. 3, Pushchino, 1989, pp. 589 – 590. (Rus)

[14] Sidorov V.P. and Tikhomirov F.A., *1st All-Union Congress on Radiobiology: Thes. Rep.*, vol. 2, Pushchino, 1989, pp. 530 – 531. (Rus)

[15] Butorina A.K., Kalaev V.N., Lunyak A.I. *et al.*, 'Burning questions in ecology and protection of natural environmental systems of southern and central regions of Russia', *Proc. Interrep. Sci.-Appl. Conference*, Krasnodar, 1996, pp. 166 – 167. (Rus)

[16] Butorina A.K. and Evstratov N., 'The first detected case of amitosis of pine', *Forest Genetics*, 1996, vol. 3(3), pp. 137 – 139.

[17] Akopyan E.M., *Genetika*, 1967, vol. 3(5), pp. 45 – 51. (Rus)

[18] Simakov E.A., *Radiobiologia*, 1983, vol. 23(5), pp. 703 – 706. (Rus)

[19] Butorina A.K. and Isakov Yu.N., *Doklady AN SSSR*, 1989, vol. 308(4), pp. 987 – 988. (Rus)

[20] Butorina A.K., Kosichenko N.E., Isakov Y.N. *et al.*, *Cytogenetic Studies of Forest Trees and Shrub Species*, Zagreb, 1997, pp. 211 – 226.

[21] Arkhipchuk V.V., 'The use of nucleolar characteristics in biotesting', *Tsitologia i Genetika*, 1995, vol. 29(3), pp. 6 – 9. (Rus)

[22] Generalova M.V., *Genetika*, 1969, vol. 5(6), pp. 48 – 51. (Rus)

[23] Machs E.M. and Grif V.G., 'Structure of the cellular cycle and mitosis rate in the plant meristem', *Tsitologia*, 1996, vol. 38(8), pp. 842 – 853. (Rus)

[24] Butorina A.K. and Kalaev V.N., *Ekologia*, 2000, No. 3, pp. 206 – 210. (Rus)

[25] Alov I.A., *Tsitologia*, 1976, vol. 3, pp. 40 – 66. (Rus)

[26] Kazantseva I.A., *Mitosis Pathology in Human Tumors*, Novosibirsk, Nauka, 1981, 114 p. (Rus)

[27] Koltovaya N.A., 'The checkpoint-control in *Saccharomyces cerevisiae* yeast', *Preprint of United Institute of Nuclear Investigations*, Dubna, 2001, 38 p. (Rus)

[28] Dmitrieva S.A., Parfenov V.I., and Davidchik T.O., *Tsitologia*, 1999, vol. 41(12), pp. 1062 – 1063. (Rus)

[29] Dmitrieva S.A. 'Karyology of flora in Belarus', *Doctor Dissertation Thesis*, Minsk, V.F. Kuprevich Institute of Experimental Botanics, NAS of Belarus, 2000, 42 p. (Rus)

[30] Butorina A.K., 'Cytogenetics of forest woody plants (in relation to the questions of their evolution and selection)', *Doctor Dissertation*, Voronezh, 1990, 368 p. (Rus)

[31] Dubrova A.N., *Zh. Obshch. Biologii*, 1989, vol. 50(2), pp. 213 – 217. (Rus)

[32] Shafikova L.M., 'Cytogenetic features of common pine (*Pinus silvestris*) under industrial pollution conditions', *Candidate Dissertation Thesis*, Krasnoyarsk, Botanic Garden – Institute of Scientific Centre of RAS, 1999, 20 p. (Rus)

[33] Muratova E.N. and Sedel'nikova T.S., 'The methods for estimating conditions and resistance of ecological systems in forests', *Thes. Rep. Intern. Conference*, 8 – 13 August, 1999, Krasnoyarsk, 1999, pp. 116 – 117. (Rus)

[34] Kalashnik N.A. and Khaidarova T.G., 'The methods for estimating conditions and resistance of ecological systems in forests', *Thes. Rep. Intern. Conference*, 8 – 13 August, 1999, Krasnoyarsk, 1999, pp. 79 – 80. (Rus)

[35] Kalaev V.N. and Karpova S.S., *Forest Genetics*, 2003, vol. 10(1), pp. 11 – 18.

[36] Rostova N.S., *Botan. Zh.*, 1999, vol. 84(11), pp. 50 – 66. (Rus)

[37] Pozolotina V.N., *Remote Radiation Consequences for Plants*, Ekaterinburg, Akademkniga, 2003, 243 p. (Rus)

In: 20 Years after the Chernobyl Accident
Editors: E. Burlakova and V. Naidich, pp. 265-285
ISBN 1-60021-249-2

Chapter 15

RADIONUCLIDES IN COMPONENTS OF AQUATIC ECOSYSTEMS OF THE CHERNOBYL ACCIDENT RESTRICTION ZONE

D. I. Gudkov[1*], M. I. Kuzmenko[1], S. I. Kireev[2], A. B. Nazarov[2], V. G. Klenus[1], A. E. Kaglyan[1], A. V. Kulachinsky[2] and L. N. Zub[3]

[1] Institute of Hydrobiology, National Academy of Sciences of Ukraine, Kiev
[2] State Scientific and Production Enterprise
"Chernobyl Radioecological Centre",
Ukraine Ministry for Emergency Situations, Chernobyl
[3] I.I. Shmalgauzen Institute of Zoology,
National Academy of Sciences of Ukraine, Kiev

ABSTRACT

The investigation results for dynamics of ^{90}Sr, ^{137}Cs, ^{238}Pu, $^{239+240}$Pu and ^{241}Am content and distribution in components of the aquatic ecosystem located in the Chernobyl accident restriction zone. The main data massif was obtained in the period of 1992 – 2004. Specific radionuclide activity data are present for bottom sediments, water, seston, macrozoobenthos (including bivalve mollusks), gastropods, higher aquatic plants, and fishes. The species specificity of radionuclide concentration by hydrobionts is studied, and the role of various groups of aquatic organisms in distribution of radioactive substances by the main components of the lake biocoenoses is estimated.

Keywords: ^{90}Sr, ^{137}Cs, ^{238}Pu, $^{239+240}$Pu, ^{241}Am, the Chernobyl accident restriction zone, freshwater ecosystems, bottom sediments, seston, macrozoobenthos, gastropods, higher aquatic plants, fishes

---

* Main addressee: 12 Geroev Stalingrada Avenue, 04210 Kiev, Ukraine; Institute of Hydrobiology, National Academy of Sciences of Ukraine, Department of Radioecology; Tel: (044) 418-9183; Fax: (044) 418-2232; E-mail: digudkov@svitonline.com

Radioactive substances emitted to the atmosphere during the Chernobyl accident have hit the areas of watersheds and precipitated directly on the surface of water objects, in which radioactive fission fragments and neutron activation products were partly dissolved, where chemically interacted with natural water and mineral organic components in it. In accordance with geochemical subordination in the sequence of elementary landscapes connected by migration flows of the substances, aquatic landscapes represent peculiar "receivers" of the majority of chemical compounds, including radionuclides, delivered by the overland discharge to the hydrological network and secluded reservoirs, located in internal-drainage lowlands. The latter reservoirs retain radioactive substances for long, where they distribute and migrate by the components of aquatic ecosystems.

Discussed in the present review are features of concentration and distribution dynamics for ^{90}Sr, ^{137}Cs, ^{238}Pu, $^{239+240}$Pu and ^{241}Am in aquatic ecosystems of the Chernobyl accident restriction zone possessing different hydrological conditions and levels of radionuclide contamination. The species specificity of radionuclide concentrating in hydrobionts and the values of various groups of aquatic organisms at radioactive substance distribution in hydrobiocoenosis are analyzed.

MATERIALS AND METHODS

The investigations were performed in the period of 1989 – 2004 in Azbuchin Lake, Yanovsky (Pripyatsky) crawl, Chernobyl NPP cooling pond, Pripyat' River left-bank flood plain lakes – Glubokoe and Dalekoe-1, and rivers Uzh (Cherevach village) and Pripyat' (Chernobyl town) (Figure 1). Water samples (20 liters each) were taken from the surface layers of water reservoirs. Seston (small plankton-forming organisms and inorganic and organic particles suspended in water) were filtered by a "blue belt" filter. Bottom sediments were sampled in 1995 – 1999, in winter, from ice with 20 – 70 m step using sampling unit PD-100. Sampling points were adjusted and fixed with the help of surveying tools.

Radionuclide concentration in tissues of biological objects was determined and analyzed for 28 species of higher aquatic plants, 6 species of mollusks and 18 species of fishes. Floristic and geobotanic descriptions, as well as sampling of macrophytes, mollusks and macrozoobentos for the purpose of estimating biomass and determining radionuclide content were performed using methods, described in refs. [1 – 5]. The results were measured in Bq/kg at natural humidity.

The radionuclide distribution by components of lake ecosystems was estimated by measuring specific activity of radionuclides in water, seston and hydrobionts, as well as using quantitative data on biomass obtained in summer periods (late July – August) of 1998 – 1999. Currently shown parameters of radioactive substance concentration in hydrobionts (concentration factor – *CF*) were calculated from specific radionuclide activity in tissues related to their average specific activity in habitat waters within one year for aquatic animal organisms and within the vegetation period for higher aquatic plants. The mistake of estimated radionuclide concentration in the ecosystem components fell within 15 – 20%.

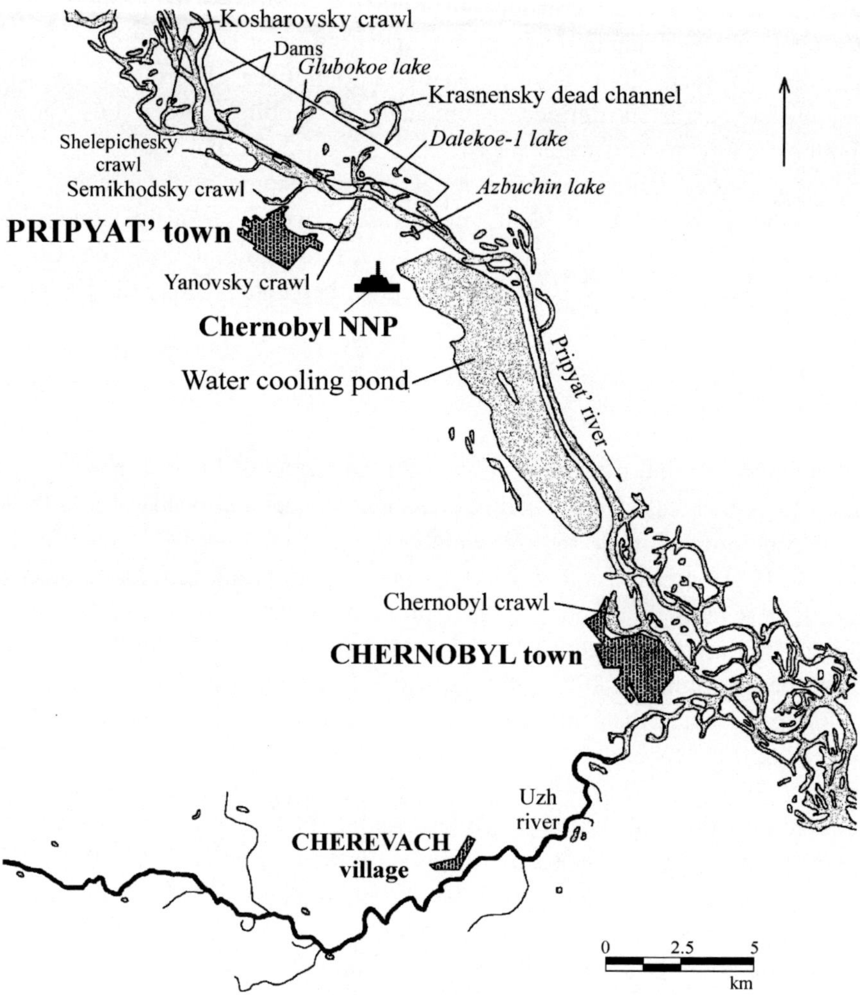

Figure 1. The map of main water bodies in the Chernobyl NPP restriction zone

RESULTS AND DISCUSSION

The modern level and composition of radionuclide contamination of aquatic ecosystems in the Chernobyl accident restriction zone is primarily provided by the intensity of radioactive substance delivery to a watershed area and hydrodynamic processes proceeding in water reservoirs. Hence, transformation of radionuclides in soils of the watershed areas and bottom sediments of water reservoirs, and their migration with water flows are of special importance.

The lowest specific radioactive activity is observed in river ecosystems, where bottom sediments were subject to natural decontamination (specifically, in high waters and spring floods) and, therefore, during years passed since the accident, these factors became insignificant as the secondary source of channel flow contamination. At present, most radionuclides are delivered to rivers from overflooding and inflow from higher contaminated water objects. At the same time, closed water reservoirs and, especially, lakes located in the nearest restriction zone, possess considerably higher radiation levels, provided by the water exchange limitation and comparatively high concentration of radionuclides, accumulated in bottom sediments. Hence, the concentration of radionuclides, determined for water of the majority of closed basins, is defined by intensity of mobile radionuclide form exchange between bottom sediments and water masses and, as well, external washout from the bailing area. Therefore, flooded landscapes of Pripyat' River become of the greatest significance. Impacted by the Chernobyl accident, these areas were seriously contaminated with radionuclides. Currently, they are the largest source of radionuclides, delivered with the overland flow to river systems of the restriction zone, then connected to the Dnepr basin.

Water

Maximal values of total β-activity of water in the mouth of Pripyat' river were registered during initial two weeks after the Chernobyl accident, reaching 10 kBq/l [6]. In this period, the main contribution was made by ^{131}I, which gave up to 70 – 90% of specific activity. By middle of May, 1986, after the end of aerosol emission, total radioactivity of water in Pripyat' river reduced to several thousand and then in June to several hundred Bq/l. In June, the contribution of ^{131}I into total radioactivity of water was below 30%. Since the end of October 1986 and till the beginning of 1987 total radioactivity of water in Pripyat' river infrequently exceeded 40 Bq/l. Hence, the main contribution into radionuclide contamination of aquatic objects was made by ^{90}Sr and ^{137}Cs radionuclides. Since 1988, the fraction of ^{90}Sr in total radioactivity of river water was significantly higher at the background of reduction of ^{137}Cs specific activity. At present, the ratio of these radionuclides in water bodies located in the Chernobyl NPP restriction zone is significantly different. For Uzh and Pripyat' rivers the average ^{90}Sr specific activity is 2.5 – 4-fold higher than that of ^{137}Cs, whereas in closed lakes this parameter may reach the level of 20-fold higher specific activity. Among water bodies located in the Chernobyl NPP restriction zone, the only exclusion is presented by the cooling pond of the Chernobyl NPP, in which ^{137}Cs concentration in water exceeds that of ^{90}Sr. As suggested, this is related to predominant delivery of high ^{137}Cs concentrations with contaminated reactor waters to the cooling pond during the accident period via discharge channel, but not atmospheric precipitations [7].

For all water objects, studied since the early 1990, located in the Chernobyl accident restriction zone, the general tendency in variations of radionuclide concentration in water of these objects is a decrease of ^{90}Sr and ^{137}Cs specific activity, which dynamics is firstly related to intensity of water exchange processes, with the only exclusion for closed water bodies of the left-bank flood plain of Pripyat' river, located at the territory of dammed area of Krasnenskaya flood plain, where recently, at the background of ^{137}Cs specific activity stabilization, a tendency to ^{90}Sr concentration increase was observed.

At present, the mean annual concentration in waters of test site water bodies is recorded in the wide ranges: from 0.11 Bq/l in Uzh river to 135 Bq/l or more in Glubokoe lake for ^{90}Sr and from 0.05 Bq/l in Pripyat' river to 8 Bq/l in Azbuchin lake for ^{137}Cs (Table 1). For the period of 1998 – 2003, the specific activity of transuranic elements, observed in water of some water objects is shown in Table 2.

Table 1. Dynamics of the mean annual concentration of radionuclides in water of some tested water bodies in the Chernobyl NPP restriction zone, Bq/l

Year	Cooling pond		Azbuchin lake		Glubokoe lake		Dalekoe-1 lake		Pripyat' river (Chernobyl town)		Uzh river (Cherevach vil.)	
	^{90}Sr	^{137}Cs	^{90}Sr	^{137}Cs	^{90}Sr	^{137}Cs	^{90}Sr	^{137}Cs	^{90}Sr	^{137}Cs	^{90}Sr	^{137}Cs
1992	4.8	5.2	–*	–	333	16	–	–	0.44	0.21	0.35	0.22
1993	4.4	6.3	–	–	281	18	–	–	0.85	0.21	0.81	0.20
1994	3.4	5.6	–	–	215	15	–	–	0.93	0.20	0.56	0.17
1995	3.2	4.8	229	22	159	11	–	–	0.33	0.11	0.31	0.13
1996	2.7	4.1	85	15	137	14	–	–	0.34	0.13	0.34	0.13
1997	2.2	2.8	85	13	100	13	45	4.5	0.25	0.16	0.26	0.08
1998	1.8	3.1	120	17	120	14	50	3.4	0.30	0.14	0.32	0.14
1999	1.9	3.1	190	23	120	14	45	2.8	0.50	0.15	0.25	0.10
2000	1.7	2.7	133	13	103	8	48	1.7	0.22	0.11	0.16	0.10
2001	1.5	2.1	110	10	79	7	35	2.6	0.23	0.12	0.18	0.09
2002	1.4	2.5	52	6	74	8	29	2.0	0.17	0.08	0.08	0.07
2003	1.7	2.4	49	8	102	7	40	2.3	0.15	0.05	0.11	0.06
2004	1.6	2.1	56	7	135	7	55	2.2	0.18	0.05	0.17	0.06

* – no measurements were performed

Table 2. The average specific activity of transuranic elements in water of some water bodies in the restriction zone in the period of 1998 – 2003, Bq/l

Water object	^{238}Pu	$^{239+240}Pu$	^{241}Am
Glubokoe lake	0.0050	0.0107	0.0061
Dalekoe-1 lake	0.0025	0.0068	0.0042
Chernobyl NPP cooling pond	0.00025	0.00071	0.00027
Pripyat' river (Chernobyl)	0.00017	0.00022	–*

* – no measurements were performed

Enhanced ^{90}Sr migration activity in the soils of contaminated areas comparing with that for ^{137}Cs provides for higher ^{90}Sr delivery to the surface discharge to river networks and subsequent withdrawal outside the restriction zone. Hence, firstly, ^{90}Sr discharge is defined by the annual discharge of river networks (Figure 2). In 2003, over 70% of ^{90}Sr discharged by Pripyat' river, was delivered to it by the surface runoff from the areas of the Chernobyl accident restriction zone. So far as concerns ^{137}Cs, in recent years, its specific activity in Pripyat' river upstream the restriction zone equaled that in the aligning of Chernobyl town. This testifies about that the most of ^{137}Cs is in low-soluble form. It is strongly fixed by soils that does not provide for its intensive discharge from drainage basin surfaces and occurrence in water flows. Totally, in the period of 1986 – 2004 Pripyat' river has discharged doses equal 158 TBq for ^{90}Sr and 126 TBq for ^{137}Cs.

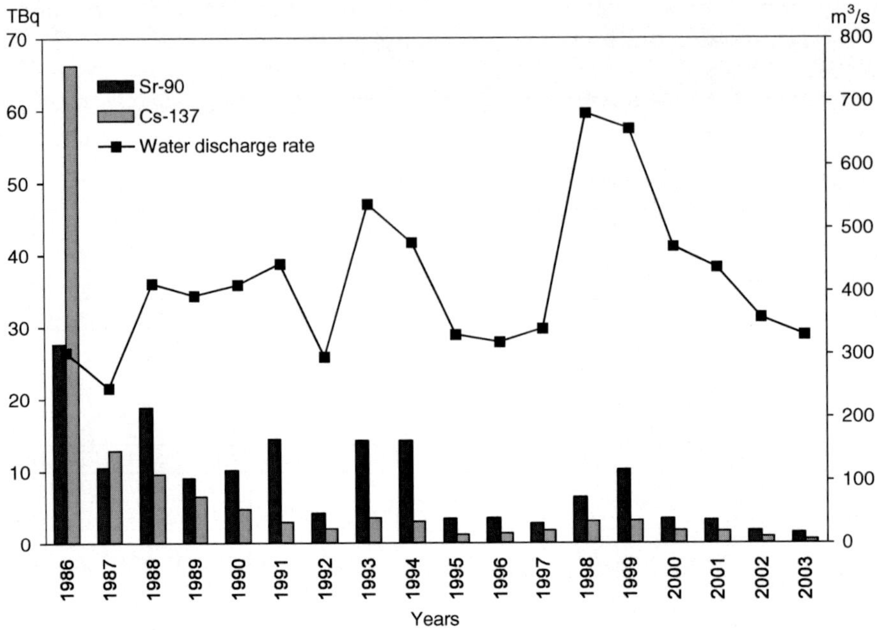

Figure 2. Radionuclides ^{90}Sr and ^{137}Cs withdrawal with Pripyat' river waters in through Chernobyl town section

Bottom Sediments

The main water objects of the restriction zone, in bottom sediments of which a noticeable quantity of radioactive substances is deposited, are closed or low flowage water bodies – lakes, cutoff meanders, crawls, abandoned irrigation channels, etc. Many crawls and cutoff meanders display so high contamination level of bottom sediments that to prevent radionuclide drainage during flood runoffs and spring tides were staked off from the truck of Pripyat' river by dead filling dams. Determinations of bottom sediment contamination level in dead water bodies, performed in the period of 1995 – 1999, allowed for specifying morphometry of the objects under study, their flat water square and water and water mass volume, as well as calculating radionuclide contamination density in bottom sediments (Table 3).

Among studied water objects, maximal contamination densities in bottom sediments were detected in Yanovsky (Pripyatsky) crawl, 2.5 km distant from the epicenter and on the formation way of the Northwest radioactive trail. In the current water body, localities with radionuclide concentration above 36 TBq/km^2 for ^{90}Sr and ^{137}Cs and above 1 TBq/km^2 for $^{238+239+240}Pu$ were detected. Within the zone of silt deposits, several areas (covering about 13% of the crawl area) are detected, where total density of radionuclide contamination equals 41 – 70 TBq/km^2, and total reserves of radionuclide activity are about 46%. The contamination densities for ^{90}Sr and ^{137}Cs exceeding 4 TBq/km^2 are usually confined to depths of 6 – 10 m, whereas concentrations above 37 TBq/km^2 – beneath 10 m. There is an area in the western region of the crawl, where anomalously high concentration of radionuclides was detected according to data, obtained in 1995: 307, 252, and 53 TBq/km^2 for ^{90}Sr, ^{137}Cs and $^{238+239+240}$Pu, respectively, which are approximately 20 times higher than the average for the whole crawl. Total reserve of radionuclides deposited in the bottom sediments of Yanovsky crawl equaled about 13.5 TBq for ^{90}Sr, 12.1 TBq for ^{137}Cs and 0.3 TBq for $^{238+239+240}$Pu, respectively.

During the accident, the cooling pond of Chernobyl NPP was also subject to intensive radionuclide contamination. Here, at depths beneath 11 m (up to 35% of the bottom square) reaches 100 cm at ^{137}Cs contamination density within the range of 19 – 133 TBq/km^2. At the depths of 3 – 11 m, the bottom is composed of primary soils, covered by silt 1 – 6 cm thick, where ^{137}Cs contamination density varies from 2 to 6 TBq/km^2 [8].

Table 3. Morphometric characteristics and concentration densities of radionuclides in bottom sediments of some water bodies in the restriction zone

Water body characteristics		Yanovsky crawl	Semikhodsky crawl	Shepelichesky crawl	Kosharovsky crawl	Azbuchin lake	Chernobyl crawl	Dalekoe-1 lake	Glubokoe lake
Distance from Chernobyl NPP*	azimuth, °	330	320	320	310	30	130	5	340
	km	2.5	4.5	7.0	9.5	2.0	13.0	4.5	6.5
Volume, million m^3		3.70	0.84	0.61	0.28	0.78	1.20	0.02	0.46
Water plane, km^2		0.84	0.41	0.23	0.16	0.27	0.61	0.01	0.17
Maximal depth, m		13.4	5.0	6.2	3.9	5.6	10.0	6.6	7.1
Average concentration densities of radionuclides, $\cdot 10^{10}$ Bq/km^2									
^{137}Cs		1,498.5	310.8	85.1	62.9	1,147.0	38.5	339.6	558.7
^{90}Sr		1,628.0	225.7	70.3	28.9	666.0	14.1	301.8	259.0
$^{238+239+240}$Pu		37.0	7.8	1.7	0.9	24.4	0.4	7.8	7.4
^{241}Am		31.1	7.0	1.4	7.4	21.8	0.4	7.7	6.3

* – to the point closest to Chernobyl NPP

According to assessments made in 2001 [7], bottom sediments of the cooling pond contains up to 160 TBq of ^{137}Cs, 24 TBq ^{90}Sr and 0.5 TBq $^{239+240}$Pu, respectively. About 30 – 50% of total radionuclides are accumulated in bottom sediments, precipitated at depths above 7 m.

At present, the radionuclide concentrations in bottom sediments of trunks in Uzh and Pripyat' rivers are just insignificantly higher than the pre-accident level. Much higher specific concentration of radionuclides is still observed in open crawls, cutoff meanders and other low flowage areas of rivers.

Higher Hydrophytes

Higher hydrophytes possess high production potential and an ability to actively absorb radioactive substances. Occupying littoral and partly sublittoral zones in the majority of water bodies and, hence, being one of predominant components of biocoenoses by biomass, hydrophyte coenoses may be of importance for self purification of aquatic ecosystems. The content of radionuclides in higher hydrophytes from water bodies of the Chernobyl NPP restriction zone is mostly determined by the radiation contamination level of water objects and neighboring lands, and the features of hydrochemical regime of water bodies affecting the forms of radionuclide presence in said water bodies and their biological accessibility.

Radionuclide ^{90}Sr and ^{137}Cs accumulation by macrophytes dwelling in the restriction zone possess signified species specificity. Species with relatively high ^{137}Cs concentration include aerial aquatic plants of Cyperaceae family, common reed (*Phragmites australis* (Cav.) Trin ex. Steud.), reed sweet-grass (*Glyceria maxima* (Hartm.) Holmb.), lesser reedmace (*Typha angustifolia* L.), and inherent aquatic plants – spiked milfoil (*Myriophyllum spicatum* L.) and water-soldier (*Stratiotes aloides* L.). The following species with floating leaves accumulate ^{137}Cs in low concentration: first of all, they are representatives of Nymphaeaceae family, which are white water lily (*Nymphaea candida* L.) and yellow water-lily (*Nuphar luteum* (L.) Smith), as well as frogbit (*Hydrocharis morsus-ranae* L.) [9]. Among species with comparatively high ^{90}Sr concentration, Potamogetonaceae family representatives are of special importance. It is obvious that this is related to Potamogetonaceae ability to intensively absorb calcium on their surface during photosynthesis processes, and this calcium is not washed off at standard sampling procedure [10]. Hence, calcium carbonate separated from may contain 7 – 20-fold high concentration of radioactive strontium than directly in tissues. The data on ^{137}Cs and ^{90}Sr concentration in macrophytes growing in water bodies of Chernobyl NPP restriction zone are shown in Table 4. The majority of higher aquatic plant species were characterized by increasing specific activities of ^{137}Cs and ^{90}Sr at the vegetation peak (the end of July – August) compared with spring and autumn periods.

In higher aquatic plants of lakes, located in the left flood plain of Pripyat' river, concentrations of $^{238+239+240}Pu$ and ^{241}Am were registered in the range of 0.4 – 65.6 (11.2) Bq/kg with *CF* within 23.6 – 4,175.2 and the range of 0.3 – 44.8 (11.0) Bq/kg with *CF* within 82.5 – 7,458.3, respectively. The highest *CF* was observed for lesser reedmace (*Typha angustifolia* L.). For this species, the values of this index were 5 – 7 times higher than average *CF* for other species of plants under study that allows for considering lesser reedmace as a specific accumulator of transuranic elements under conditions of water bodies, located in Chernobyl NNP restriction zone.

Within the period of investigation since 1989 till 2004, dynamics of the main radionuclide concentrations in tissues of the higher aquatic plants from river ecosystems of the restriction zone indicated a decrease of specific activity of ^{90}Sr and ^{137}Cs (Figure 3a, b). So far as concerns closed of low flowage water bodies, the most representative retrievals obtained in the period of 1993 – 2004 have shown that since the late 1990ies the higher aquatic plants related to different ecological groups are indicating a frank tendency to ^{90}Sr content increase in tissues (Figure 3c–f). Primarily, this tendency was observed for test materials, sampled in the period of 1993 – 1998 in Glubokoe lake [11], and then was confirmed by tests performed in 1998 – 2004 for Dalekoe-1 lake. So far as concerns ^{137}Cs, its specific activity in higher aquatic plants of the lakes under study either decreases or remains practically constant. Hence, in the middle 1990ies specific activity of ^{137}Cs in tissues of higher aquatic plants from Krasnenskaya flood plain was much higher than specific activity of ^{90}Sr in them, whereas in the late 1990ies these values became comparable, and at present specific activity of ^{90}Sr is much higher than that of ^{137}Cs, the concentration of which either decreasing or remaining practically constant.

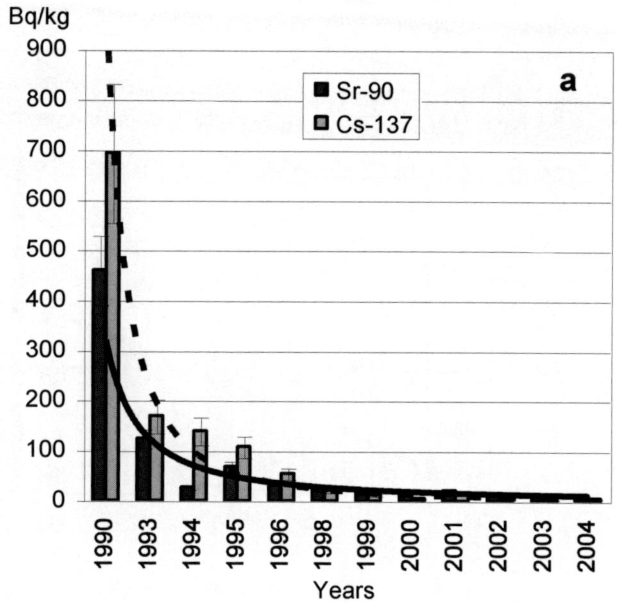
Bq/kg
900
800
700
600
500
400
300
200
100
0
a
Sr-90
Cs-137
1990
1993
1994
1995
1996
1998
1999
2000
2001
2002
2003
2004
Years

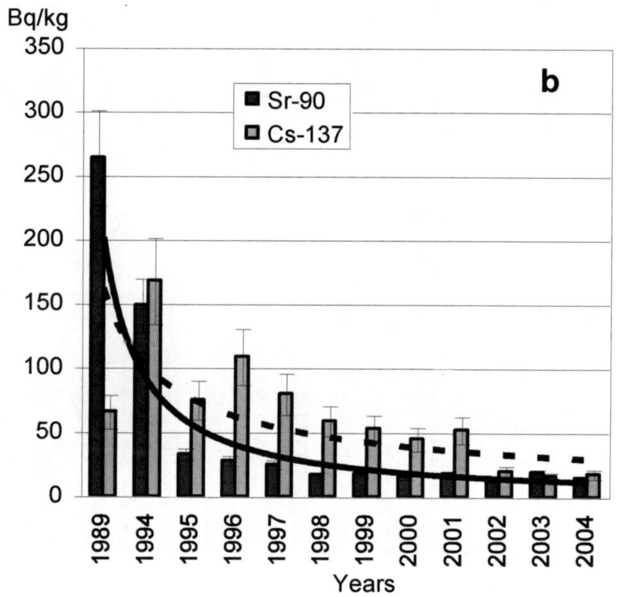
Bq/kg
350
300
250
200
150
100
50
0
b
Sr-90
Cs-137
1989
1994
1995
1996
1997
1998
1999
2000
2001
2002
2003
2004
Years

Figure 3(a, b)

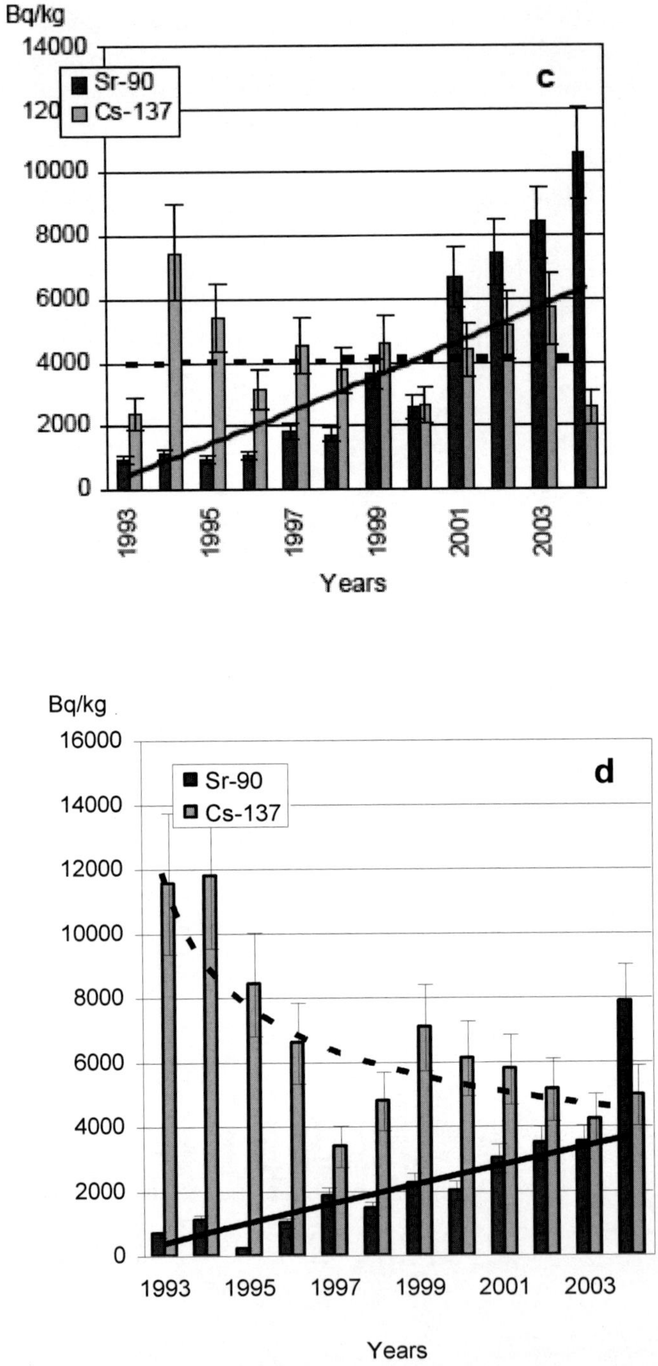
Bq/kg
14000
12000
10000
8000
6000
4000
2000
0
Sr-90
Cs-137
c
1993
1995
1997
1999
2001
2003
Years
Bq/kg
16000
14000
12000
10000
8000
6000
4000
2000
0
Sr-90
Cs-137
d
1993
1995
1997
1999
2001
2003
Years

Figure 3(c, d)

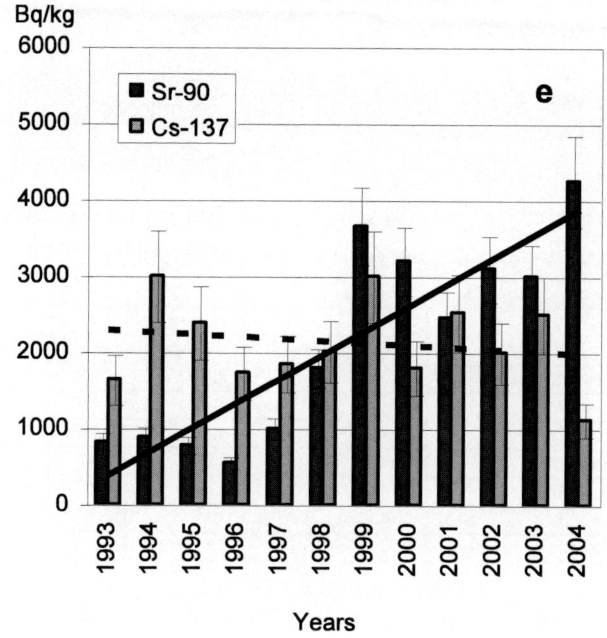

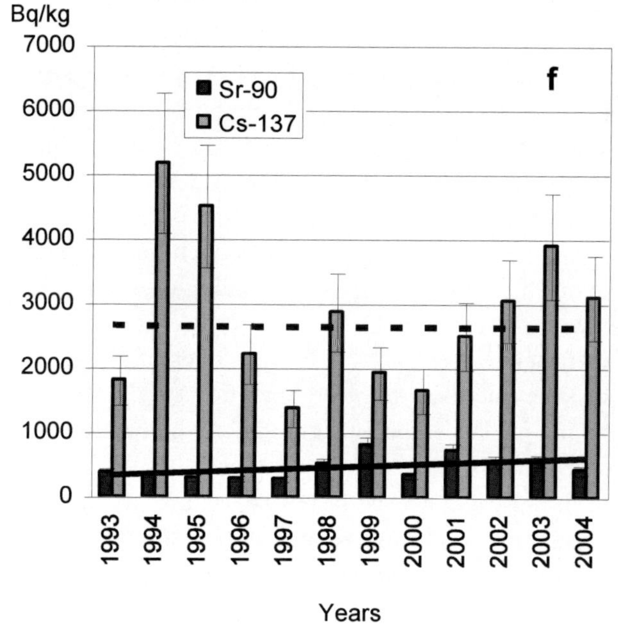

Figure 3. Dynamics of radionuclide content in higher aquatic plants of Chernobyl NPP restriction zone: a – lesser reedmace (Pripyat' river); b – fennel pondweed (Pripyat' river); c – water-soldier (Glubokoe lake); d – yellow water lily (Glubokoe lake); e – yellow water lily (Dalekoe-1 lake); f – common reed (Chernobyl NPP cooling pond). Continuous line denotes ^{90}Sr specific activity trend; dashed line denotes ^{137}Cs specific activity trend

In the cooling pond of Chernobyl NPP, due to more intensive water exchange and, at the same time, high contamination density of bottom sediments with ^{137}Cs, which defines the ratio between ^{90}Sr and ^{137}Cs contents in water, anomalous for water bodies of the restriction zone, dynamics of ^{90}Sr specific activity in tissues of macrophytes is not so clear, and ^{137}Cs specific activity varies within a broad range showing no clear linear dependence (Figure 3e).

Table 4. Specific activity of radionuclides in higher aquatic plants from water bodies in the restriction zone of Chernobyl NPP in the period of 1997 – 2001, Bq/kg

Water object	^{90}Sr			^{137}Cs		
	max	min	average	max	min	average
Glubokoe lake	14,060	67	2,210	36,500	1,220	8,730
Dalekoe-1 lake	5,100	200	1,810	19,470	1,170	5,170
Azbuchin lake	24,210	730	4,900	23,860	220	2,030
Cooling pond	1,600	36	218	3,130	310	1,380
Yanovsky crawl	2,200	110	779	1,702	38	814
Uzh river (Cherevach vil.)	234	3	10	185	4	42
Pripyat' river (Chernobyl)	357	5	15	164	5	30

It is suggested that ^{90}Sr specific activity in tissues of macrophytes from Krasnenskaya flood plain increases due to dynamics of radionuclide transformation in soils of water catchment areas and bottom sediments of water bodies. Since after the Chernobyl accident Krasnenskaya flood plain appeared one of the most radionuclide-contaminated areas of the restriction zone, in 1991 – 1995 a complex of flood control dams were constructed here, which changed the hydrological regime of flood plain flows during floods and preventing radioactive substance washing off the soils of contaminated areas (refer to Figure 1). In its turn, this became the reason for intensification of waterlogging and swamping of dam-locked areas. As a result, at the background of general tendencies of ^{90}Sr mobile form increase in the soils of water catchment areas and bottom sediments of the restriction zone water bodies; in swamped soils of Krasnenskaya flood plain fulvic and humic acid concentration increases that decreases pH value in water and intensifies a denuding of water-soluble forms of ^{90}Sr forming soluble complexes with fulvic acids. Hence, an increase of concentrations of mobile forms of the radionuclide and their inclusion to biotic turnover of aquatic ecosystems are observed. This also confirms the increase of ^{90}Sr specific activity, observed in recent years in Krasnenskaya flood plain lakes at the background of stabilization of this index for ^{137}Cs.

The tendency of the root contamination of plant tissues by ^{90}Sr was also observed for terrestrial plants in the restriction zone [12 – 14]. Currently, some authors [15] suggest that ^{90}Sr mobility in soils of the restriction zone is maximal, and this will last during the nearest decade. Thereafter, the rate of radionuclide decomposition will exceed the rate of its mobilization.

Mollusks

Freshwater mollusks are frequently considered as the species-indicators of radionuclide contamination of water objects. These invertebrates accumulate practically all radionuclide detectable in water, and owing to high biomass mollusks play very important role in bioaccumulation and radionuclide redistribution processes in freshwater ecosystems. The data on ^{137}Cs and ^{90}Sr concentration in freshwater mollusks from water bodies of Chernobyl NPP restriction zone are shown in Table 5.

Table 5. Specific activity of radionuclides in freshwater mollusks from the restriction zone of Chernobyl NPP in 1998 – 2003, Bq/kg

Water object	^{90}Sr			^{137}Cs		
	max	min	average	max	min	average
Glubokoe lake	170,300	39,770	61,830	27,150	3,160	9,070
Dalekoe-1 lake	62,830	17,120	32,310	2,410	847	1,523
Azbuchin lake	87,670	34,010	51,120	4,750	2,704	3,620
Cooling pond	2,500	1,600	2,133	2,900	830	1,450
Yanovsky crawl	13,350	7,720	10,350	590	430	510
Uzh river (Cherevach vil.)	216	98	177	30	14	25
Pripyat' river (Chernobyl)	101	77	86	61	31	35

Maximal *CF* values for both ^{90}Sr and ^{137}Cs characterize bivalve mollusks – zebra mussel (*Dreissensia polymorpha* Pall.) and *Unio* sp., which are the most active filtration organisms. Maximal *CF* values of ^{90}Sr exceeding 1,100 were observed for zebra mussel and of ^{137}Cs for *Unio* sp. tissues – about 500. Much lower *CF* of radionuclides were determined for gastropods: *Lymnaea stagnalis* L., *Stagnacolia palustris* L., ram's-horn snail (*Planorbarius corneus* L.), and viviparous pond-snail (*Viviparus vivparus* L.). Minimal *CF* of ^{90}Sr and ^{137}Cs were determined for *Lymnaea stagnalis*: 440 and 137, respectively [16].

The average specific activities of transuranic ^{238}Pu and $^{239+240}$Pu elements in tissues of mollusks inhabiting in Glubokoe and Dalekoe-1 lakes were the following: minima were observed for *Lymnaea stagnalis* – 0.1 and 0.2 Bq/kg in Dalekoe-1 lake and 2.7 and 6.4 Bq/kg in Glubokoe lake, respectively. Higher concentration was determined for *Lymnaea stagnalis* from Glubokoe lake: 14.0 and 36.0 Bq/kg, respectively. Among gastropods of Krasnenskaya flood plain lakes, maximal activities were detected for ram's-horn snail: 0.92 and 2.1 Bq/kg in Dalekoe-1 lake and 24.7 and 53.0 Bq/kg in Glubokoe lake, respectively. Zebra mussel from cooling pond of Chernobyl NPP contained ^{238}Pu and $^{239+240}$Pu in amounts of 2.8 and 6.2 Bq/kg, respectively.

Table 6. Specific activity of radionuclides in fish from the restriction zone of Chernobyl NPP for the period of 1997 – 2004, Bq/kg

Water object	^{90}Sr			^{137}Cs		
	max	min	average	max	min	average
Glubokoe lake	6,800	660	3,011	14,000	1,200	8,250
Dalekoe-1 lake	13,060	410	3,129	27,020	16,110	17,500
Cooling pond	1,520	18	246	13,260	1,100	3,950
Uzh river	24	1	8	290	3	63
Pripyat' river	87	1	18	800	6	113

The specific activity of ^{241}Am in *Lymnaea stagnalis* tissues was minimal, determined within the range of 4.6 – 30.0 (15.2) Bq/kg in Dalekoe-1 lake and within the range of 6.3 – 51.0 (26.5) Bq/kg in Glubokoe lake, respectively. For *Lymnaea stagnalis* from Glubokoe lake concentrations about 75.0 Bq/kg were detected. Maximal values were determined for ram's-horn snail: 18.0 – 29.0 (23.5) Bq/kg in Dalekoe-1 lake and 80.0 – 310.0 (170.0) Bq/kg in Glubokoe lake, respectively. The concentration of ^{241}Am in zebra mussel from the cooling pond of Chernobyl NPP equaled 7.9 Bq/kg.

Fishes

For radioecological investigations of aquatic ecosystems fishes occupying the upper levels of freshwater biocoenosis trophic pyramid and representing one of man's feeding objects are of certain interest. The concentrations of ^{90}Sr and ^{137}Cs radionuclides in fishes of various water bodies in the restriction zone are shown in Table 6.

The concentrations of transuranic ^{238}Pu, $^{239+240}$Pu and ^{241}Am elements were measured in fishes from Glubokoe and Dalekoe-1 lakes. The specific activity of ^{238}Pu in fish tissues equaled 0.4 – 0.5 (0.4) Bq/kg with *CF* equal 72 – 98 (83), $^{239+240}$Pu – 0.7 – 0.9 (0.8) Bq/kg with *CF* equal 68 – 87 (75), and ^{241}Am – 2.2 – 10.0 (6.2) Bq/kg with *CF* equal 367 – 1,667 (1.028).

In all cases, the content of ^{90}Sr and ^{137}Cs radionuclides in fishes from the lakes of Pripyat' river left-bank flood plain far exceeded maximum permissible concentrations (MPC) pursuant to the fish industry standards of Ukraine [17]: on average, 90-fold (at MPC equal 35 Bq/kg) by ^{90}Sr and 120-fold (at MPC equal 150 Bq/kg) by ^{137}Cs, respectively. Maximal recorded values exceeded MPC by 370 and 180 times, respectively. The content of ^{90}Sr in fished inhabiting in the cooling pond also exceeded MPC, by 7 time, on average, maximally reaching 40 times. In all tests, ^{137}Cs concentration also exceeded MPC by 30 times, on average, and by 90 times as a maximum.

Despite the fact that the average concentrations of radionuclides in fishes of Pripyat' river did not exceed MPC, the cases of ^{137}Cs concentration elevated above MPC during the period of tests were observed for 20% of all caught species. Maximal ^{137}Cs concentrations were 5-fold above MPC. In Uzh river ^{137}Cs concentrations exceeding MPC were detected in singular cases.

Table 7. Total concentration and ratio of radionuclides in the main components of the ecosystem of Dalekoe-1 lake

Object	^{90}Sr		^{137}Cs		$^{238+239+240}$Pu, ^{241}Am	
	MBq	%	MBq	%	MBq	%
Bottom sediments	37,000	95.35	51,800	99.11	1,100	99.90
Water	1,650	4.25	236	0.45	0.27	0.03
Seston	58	0.15	155	0.30	–*	–
Biota	96	0.25	73	0.14	0.81	0.07

* – no measurements were performed

Since the requires selection of even-aged individuals was not obtained, reliable differences in concentrations of radionuclides within groups of predatory and nonpredatory fishes can be hardly determined and compared with ecological specialization of one species or another, because even nonpredatory species, characterized by similar trophic orientation, possess *CF* for ^{90}Sr and ^{137}Cs varying within broad ranges. Nevertheless, the averaged data indicate elevated ^{137}Cs concentration for fishes of high trophic levels, whereas MPC exceeding by ^{137}Cs concentrations in the restriction zone were mostly observed for predatory fish species.

Table 8. Total concentration and ratio of radionuclides in the main components of the ecosystem of Glubokoe lake

Object	^{90}Sr		^{137}Cs		$^{238+239+240}Pu$, ^{241}Am	
	MBq	%	MBq	%	MBq	%
Bottom sediments	444,000	89.02	962,000	98.64	25,900	99.80
Water	50,900	10.21	6,200	0.64	10	0.04
Seston	800	0.16	2,471	0.25	–*	–
Biota	3,035	0.61	4,598	0.47	42	0.16

* – no measurements were performed

Radionuclide Distribution in Basic Components of Lake Ecosystems

Distribution and further migration of radionuclides in the components of aquatic ecosystems is defined by biogeochemical repeating pattern of substance motion in the nature. Hence, besides predominant sedimentation and sorption processes of radionuclide depositing to bottom sediments and precipitation in suspensions, their accumulation in living organisms and subsequent invoking in the biotic turnover is also urgent.

The studies of ^{90}Sr, ^{137}Cs, ^{238}Pu, $^{239+240}Pu$ and ^{241}Am distributions in basic biotic and abiotic components of Dalekoe-1 and Glubokoe lakes ecosystems, located at dam-fenced site of Krasnenskaya flood plain, indicated depositing of the main quantity of radionuclides to bottom sediments. Hence, concentrations of radionuclides equaled as follows: by ^{90}Sr – 89 and 95% for Glubokoe and Dalekoe-1 lakes, respectively; by ^{137}Cs – 99%; by transuranic $^{238+239+240}Pu$ and ^{241}Am elements – almost 100% of total content in ecosystems (Tables 7 and 8). The difference in ^{90}Sr contribution to bottom sediments of the lakes and contents of other radionuclides is stipulated by elevated migration activity of ^{90}Sr compared with ^{137}Cs and transuranic elements. This is the determining reason for higher concentration of dissolved ^{90}Sr in waters of both lakes (10.2 and 4.3%) compared with ^{137}Cs (0.6 and 0.5%) and transuranic elements (0.04 and 0.03%) and, vice versa, lower ^{137}Cs concentration in seston (0.16 and 0.15%) compared with ^{137}Cs (0. 25 and 0.30%), respectively.

The content of transuranic elements in biotic components of ecosystems was minimal 0.07 and 0.16%, respectively; for ^{137}Cs it equals 0.14 and 0.47%, and for ^{90}Sr – 0.25 and 0.61%, respectively. Maximal values were determined for Glubokoe lake, where dissolved ^{90}Sr concentration in water was also higher. As suggested, such differences in ^{90}Sr distribution in ecosystems of lakes is, firstly, related to high biomass of the higher aquatic vegetation, intensively overgrowing Glubokoe lake (covering about a half of water plane). It is known that as growths of macrophytes increase and become denser, the oxygen regime becomes worse, significant quantities of organic substances and biogenic elements are accumulate, and pH of the water medium decreases. As pH decreases, radionuclides are desorbed more actively and transit into the dissolved state, ^{90}Sr, first of all [13, 14]. This is also confirmed by lower average pH in Glubokoe lake (8.2) compared with Dalekoe-1 lake

(8.8) and the ratio between specific activity of ^{90}Sr and ^{137}Cs in water equal, on average, 12.5 in Glubokoe lake and 7.3 in Dalekoe-1 lake, respectively. Contamination densities of bottom sediments by different radionuclides are also typical: the average specific activities of transuranic elements in both lakes were practically equal, whereas the ratio between average concentrations of ^{137}Cs and ^{90}Sr in Dalekoe-1 and Glubokoe lakes equaled 1.3 and 2.2, respectively. Therefore, affecting hydrochemical regime of water bodies, the intensity of closed lake overgrowth by associations of higher aquatic plants may also change the type of distribution of radionuclides in the components of ecosystems. It is also obvious that a part of radionuclides transit to water from biomass of dead plants and return to the biotic turnover of the water body.

Table 9. Total concentration of radionuclides in basic biotic components of Dalekoe-1 lake, kBq

Hydrobionts	^{90}Sr	^{137}Cs	$^{238+239+240}$Pu	^{241}Am
Macrozoobenthos:				
Oligochaetes	48	2,128	–*	–
Midges	3	1,637	–	–
Bivalve mollusks	85,160	1,238	284	320
Total	85,211	5,003	284	320
Gastropods:				
Lymnaea stagnalis	860	12	0.04	0.5
ram's-horn snail	476	8	0.03	0.2
Total	1,336	20	0.07	0.7
Macrophytes:				
yellow water-lily	800	3,600	1	2
hornwort	3,400	12,900	38	24
common reed and sedge	4,100	41,900	79	58
Total	8,300	58,400	118	84
Fishes	1,407	5,573	0.7	2
TOTAL	96,254	68,996	403	407

* – no measurements were performed

Total concentrations of radionuclides in the main biotic components of the lakes under study are shown in Tables 9 and 10. In zoobenthos associations of Dalekoe-1 lake the overwhelming majority of ^{90}Sr (99%) is concentrated in bivalve mollusks. Total ^{90}Sr concentration in these mollusks equals about 89% of its content in the lake biota. About 8.5% is accumulated in higher hydrophytes, 1.5% – in fishes, about 1% – in gastropods, and less than 1% – in representatives of "soft" zoobenthos (oligochaetes and midges). Similar situation is observed for distribution of transuranic elements. However, in this case, quantities accumulated in bivalve mollusks are decreased (down to 70 – 80%), as well as these accumulated in gastropods and fishes (to fractions of percent), whereas the value of macrophytes increases to 21 – 29% (Figure 4).

In the benthos of bottom invertebrates the part of ^{137}Cs accumulated in bivalve mollusks is 25% of its total amount or lower. This relates to high *CF* values for this radionuclide in "soft" zoobenthos representatives, in which, despite lower biomass compared with bivalve mollusks (10-fold lower), ^{137}Cs content gives about 65% of its total quantity. About 85% of ^{137}Cs present in the lake biota is accumulated in high aquatic plants, 7 and 8% – in zoobenthos and fish, respectively, and less than 0.1% – in gastropods.

Table 10. Total concentration of radionuclides in basic biotic components of Glubokoe lake, MBq

Hydrobionts	^{90}Sr	^{137}Cs	$^{238+239+240}Pu$	^{241}Am
Gastropods:				
Lymnaea stagnalis	155	12	0.04	0.29
Stagnicolia polustris	25	3	0.04	0.07
ram's-horn snail	115	11	0.17	0.37
Total	295	26	0.25	0.73
Macrophytes:				
yellow water lily and white water	85	178	0.1	0.2
lily	656	1,473	2.5	2.8
water-soldier	108	477	0.3	0.5
hornwort	454	840	12.1	8.2
lesser reedmace	26	28	0.1	0.1
broad leaved pondweed				
joint associations of common reed,	260	1,457	3.4	2.8
sedge, reedmace and bur-reed	1,589	4,453	18.5	14.6
Total				
Macrozoobenthos	1,126	80	3	5
Fishes	25	37	0.01	0.03
TOTAL	3,035	4,596	22	20

In Dalekoe-1 lake biota higher aquatic plants occupying less than 5% of the water plane dominate exclusively in ^{137}Cs distribution, whereas in Glubokoe lake comparative biomass of macrophytes is so high that they accumulate about 90% of all radionuclides present in the biotic component of the lake (Figure 5). The exception is ^{90}Sr, for which the contribution of aquatic plants is slightly more than 50%, and the rest of it is mostly concentrated in gastropod and bivalve mollusk valves.

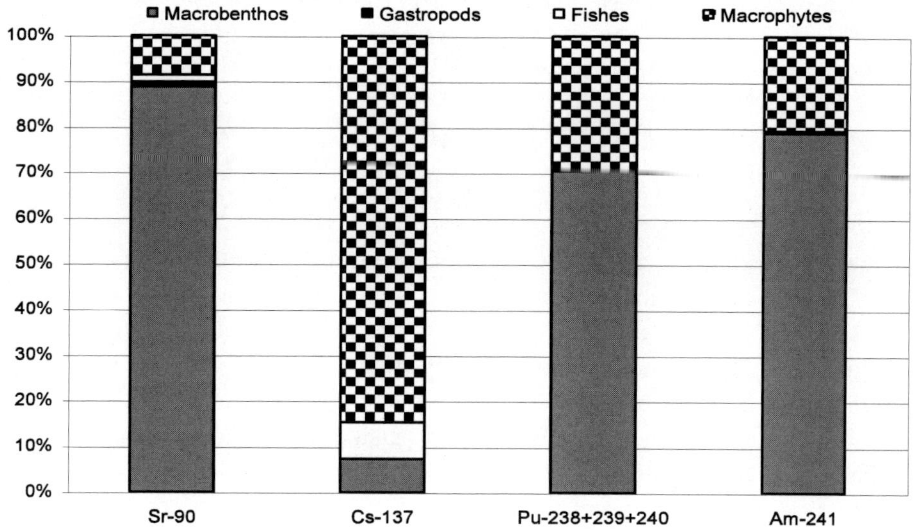

Figure 4. Distribution of radionuclides by main groups of hydrobionts in the biotic component of Dalekoe-1 lake ecosystem

The main factors determining the contribution of various ecological groups of hydrobionts to total content of radionuclides in the biotic component of lake ecosystems are

CF and aquatic organism biomass. Moreover, *CF* values may be so significant that sometimes the biomass factor is displayed in the only cases, when the difference in this parameter between different groups of hydrobionts reaches two orders of magnitude.

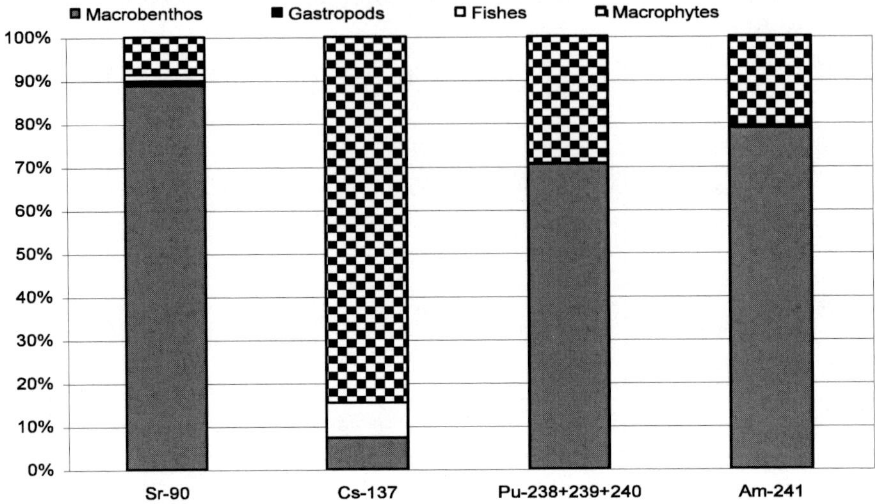

Figure 5. Distribution of radionuclides by main groups of hydrobionts in the biotic component of Glubokoe lake ecosystem

The basic data used for estimation of radionuclides distribution in the main components of lake ecosystems were obtained in 1998, when specific activity of ^{90}Sr in higher aquatic plant tissues did not significantly increase. In this relation, it can be suggested that, currently, ^{90}Sr reserves in the vegetation of tested lakes are increasing. This tendency will be preserved with intensifying swamping processes on the territories under study and corresponding overgrowth of water bodies with macrophytes.

Conclusion

Despite the years passed since the Chernobyl accident, the restriction zone area is still the open source of radionuclide contamination with a complex distribution structure in different landscapes and characterized by dynamic character of the state of radioactive substance forms, which affect their migration and redistribution by components of ecosystems. The basic questions of radioactive safety of the restriction zone concern radionuclides wash-off with surface drainage water to river network, their export outside the restriction zone and affection of the water quality in Dnepr River.

Self purification of closed water bodies in the Chernobyl NPP restriction zone is extremely slow process. Therefore, ecosystems of the majority of lakes, dead channels and crawls possess high level of radionuclide contamination of all the components. The main quantity of radionuclides in aquatic landscapes of the restriction zone are deposited to bottom sediment of closed water bodies, hence, the distribution of radioactive substances in biotic and abiotic components of hydrobiocoenoses is defined by biogeochemical regularities and transformations of radioactive substances in bottom sediments of water bodies and in soils of adjacent territories.

The content of radionuclides in tissues of higher aquatic plants from the restriction zone is characterized by high species specificity. Maximal ^{137}Cs concentrations were observed for aerial-aquatic species and minimal – for Nymphaeaceae family. Maximal ^{90}Sr concentrations were determined for Potamogetonaceae family and minimal – for Cyperaceae and Nymphaeaceae families. Maximal *CF* values of transuranic elements were observed for lesser reedmace (*Typha angustifolia* L.). On average, the values of this parameter are 6-fold higher than *CF* for the rest studied species of plants. This allows for considering lesser reedmace as a hydrophyte specifically accumulating transuranic elements in conditions of water bodies in the Chernobyl NPP restriction zone.

The species specificity of ^{90}Sr and transuranic elements concentrating by freshwater mollusks is, firstly, determined by specific weight of the valve, whereas for ^{137}Cs – by the features of functional ecology and the type of feeding. Among hydrobionts, mollusks dominate in ^{90}Sr biosedimentation processes.

The average values of ^{90}Sr and ^{137}Cs concentration in fishes from rivers of the restriction zone does not exceed MPC pursuant to the fish product standards of Ukraine. However, in the period of 1997 – 2004, ^{137}Cs concentration exceeded MPC in 20% of fish species caught in Pripyat' river, whereas for Uzh river these were singular cases. For ^{90}Sr, MPC exceeding was not observed at all. All fishes sampled in closed or low flowage water bodies showed high concentration of radionuclides. The maximal values were observed in Krasnenskaya flood plain lakes, where average specific activities of radionuclides exceeded MPC by two orders of magnitude. Maximal specific activities of radionuclides in all water bodies of the restriction zone are mostly observed for predatory species.

The construction of a complex of flood retarding dams and degradation of existing melioration systems at the site of the left-bank flood plain of Pripyat' river implied a change of hydrological regime and the character of water object overgrowth. The absence of flowage in water bodies, stagnation effects during spring flooding and seasonal runoffs intensified waterlogging and swamping of dam-fenced territories. As a result, at the background of general tendencies of ^{90}Sr mobile forms increase in the soils of water catchment areas and bottom sediments of the restriction zone water bodies located at dam-fenced sites, as well as increasing intensity of this radionuclide concentrating by higher aquatic plants and, obviously, other autotrophic organisms. For some species of macrophytes, ^{90}Sr concentration compared with the early 1990ies has increased by more than an order of magnitude and exceeds the specific activity of ^{137}Cs. Obviously the specific activity of ^{90}Sr will also increase at higher trophic levels, however, at present, such dynamics was not reliably detected. It is suggested that for radioecological monitoring of aquatic ecosystems in the restriction zone, higher aquatic plants possessing high *CF* values of radionuclides are the most sensitive test-objects for recording increasing specific activity of mobile forms of radionuclides in the water of test site water bodies.

The results of tests performed confirm the tendency to further deterioration of the radiation situation in aquatic ecosystems of the restriction zone. Swamping of contaminated territories leads to acceleration of ^{90}Sr deposited form mobilization processes and their migration and redistribution in closed aquatic landscapes. Hence, an original "depot" of mobile forms of radioactive substances is formed, which in high-flood periods may become a source of increasing ^{90}Sr drainage to Pripyat' river and then outside the restriction zone. In this connection, the necessity to implement hydraulic engineering procedures preventing underflooding of territories with high densities of radionuclide contamination; optimization and enhancement of radioecological monitoring system, and further development of investigations of radionuclide behavior in aquatic ecosystems of the restriction zone – the important components in the complex of measures related to forecasting and minimization of the Chernobyl disaster consequences.

REFERENCES

[1] Aleksandrova V.D., *Flora Classification*, Leningrad, Nauka, 1969, 275 p. (Rus)

[2] Katanskaya V.M., *Higher Aquatic Vegetation of Continental Water Bodies of the USSR*, Leningrad, Nauka, 1981, 1185 p. (Rus)

[3] Kuz'menko M.I., Romanenko V.D., Derevets V.V. *et al.*, *Radionuclides in Aquatic Ecosystems of Ukraine. The Influence of Radionuclide Contamination on Hydrobionts of the Restriction Zone*, Kiev: Chernoibylinterinform, 2001, 318 p. (Ukr)

[4] Kulikov N.V. and Molchanova I.V., *Continental Radioecology: Soil and Freshwater Ecosystems*, Moscow, Nauka, 1975, 185 p. (Rus)

[5] Romanenko V.D., Kuz'menko M.I., Evtushenko N.Yu. *et al.*, *Radioactive and Chemical Contamination of Dnepr River and Its Water Reservoirs After the Accident at the Chernobyl NPP*, Kiev, Naukova Dumka, 1992, 194 p. (Rus)

[6] Derevets V.V., Ivanov Yu.P., Kazakov S.V. *et al.*, *Bulletin of Ecological State of the Restriction Zone and the Zone of Compulsory (Mandatory) Evacuation*, 1999, No. 13, pp. 9 – 19. (Ukr)

[7] Kashparov V.A., Khomutin Yu.V., Glukhovsky A.S. *et al.*, *Bulletin of Ecological State of the Restriction Zone and the Zone of Compulsory (Mandatory) Evacuation*, 2003, No. 1 (21), pp. 67 – 74. (Ukr)

[8] Kanivets V.V. and Voitsekhovich O.V., *Thes. Rep. Intern. Conf. "Radioactivity at Nuclear Explosions and Accidents"*, 2000, Moscow – Saint-Petersburg, Hydrometeoizdat, 2000, p. 128. (Rus)

[9] Gudkov D.I., Zub L.N., Savitsky A.L. *et al.*, *Gidrobiol. Zh.*, 2001, vol. 37(6), pp. 64 – 81. (Rus)

[10] Lainerte M.P. and Seisuma Z.K., *Higher Aquatic and Littoral Aquatic Plants, Thes. Rep. 1st All-Union Conference*, 7 – 9 September, 1977, Borok, pp. 117 – 119. (Rus)

[11] Kaglyan A.E., *The Features of ^{90}Sr and ^{137}Cs Accumulation by Hydrobionts and Abiotic Components of Water Bodies in the Restriction Zone of Chernobyl NPP*, Candidate Dissertation Thesis on Biology, Kiev, 2003, 24 p. (Ukr)

[12] Kashparov V.A., *Bulletin of Ecological State of the Restriction Zone and the Zone of Compulsory (Mandatory) Evacuation*, 1998, No. 12, pp. 67 – 74. (Ukr)

[13] Kashparov V.A., *Formation and Dynamics of Environmental Radioactive Contamination During the Accident at the Chernobyl NPP and After the Accident*, Ed. V.G. Barjakhtar, Kiev, Naukova Dumka, 2001, pp. 11 – 46. (Ukr)

[14] Ivanov Yu.A., 'Dynamics of radionuclide redistribution in soils and vegetation', In Book: *Chernobyl – the Restriction Zone*, Ed. V.G. Barjakhtar, Kiev, Naukova Dumka, 2001, pp. 47 – 76. (Ukr)

[15] Sobotovich E.V., Bondarenko G.N., Kononenko L.V. *et al.*, *Geochemistry of Production Induced Radionuclides*, Kiev, Naukova Dumka, 2002, 332 p. (Ukr)

[16] Gudkov D.I., Derevets V.V., Kuz'menko M.I., and Nazarov A.B., *Radiats. Biol. Radioekol.*, 2001, vol. 41(3), pp. 326 – 330. (Rus)

[17] *Maximum Permissible Concentrations of* ^{137}*Cs and* ^{90}*Sr Radionuclides in Foods and Drinking Water (MPC-97)*, Kiev, Ministry of Health Protection of Ukraine; Committee on Questions of Hygienic Regulation; NKRZU, 1997, 38 p. (Ukr)

In: 20 Years after the Chernobyl Accident
Editors: E. Burlakova and V. Naidich, pp. 287-302
ISBN 1-60021-249-2

Chapter 16

THE ESTIMATION OF CHERNOBYL ACCIDENT INFLUENCE ON THE NATURAL POPULATIONS OF THE BLACK SEA BASIN HYDROBIONTS BY CYTOGENETIC CRITERIA

V. G. Tsytsugina and G. G. Polikarpov
A.O. Kovalevsky Institute of Biology of Southern Seas,
National Academy of Sciences of Ukraine, Sevastopol, Ukraine

ABSTRACT

The results of cytogenetic studies of marine and freshwater hydrobionts in the region affected by the Chernobyl accident, including the 30-kilometer restriction zone of Chernobyl nuclear power plant (NPP), Danube, Dnieper, Black Sea and Aegean Sea are generalized. The data on the level of chromosome mutagenesis in somatic and sexual cells of worms, mollusks, echinoderms, crustaceans, and fishes are presented. The reproductive strategy of Naididae (Oligochaeta) family in water bodies of the 10-kilometer restriction zone of the accident is considered. Basing on the suggested methodology of identification of radiation, chemical and combined impacts, a contribution of radioactive and chemical contamination to the damage of hydrobionts in the 30-kilometer restriction zone of the accident and beyond it was determined.

Keywords: hydrobionts, Chernobyl accident, chromosome aberrations, reproduction, impact identification

INTRODUCTION

Identification of anthropogenic radioactive and chemical contamination impact on natural populations of the living organisms is the urgent but not yet resolved problem of radioecology and toxicoecology. Since 1986, after the Chernobyl accident, the populations of some species of aquatic plants (algae and flowering), invertebrates (worms, mollusks, crustaceans, and insects), fishes, amphibians, and aquatic-boggy complex birds were monitored in the 30-kilometer restriction zone and neighboring areas. Multiple works, analyzed in our reviews [1, 2], indicated damaging effects either at the organism level (qualitative changes in blood corpuscles, biochemical disorders, anomaly gonads, and morphological deformities) or the population level (depression and reduction of specific expansion rate of populations, spermatogenesis disorder, laying anomalies, fry lethality increase, and genetic disorders). Some works indicate changes in the community compositions and decreasing species diversity. However, in other studies not negative effects were observed.

The contradictory results and difficulty of their interpretation are explained by the absence of dosimetric data in the majority of works, as well as the information about concurrent factors on non-radiation origin, chemical pollution of water bodies, in particular, which in most urgent for the Chernobyl zone due to deactivation measures, performed in it.

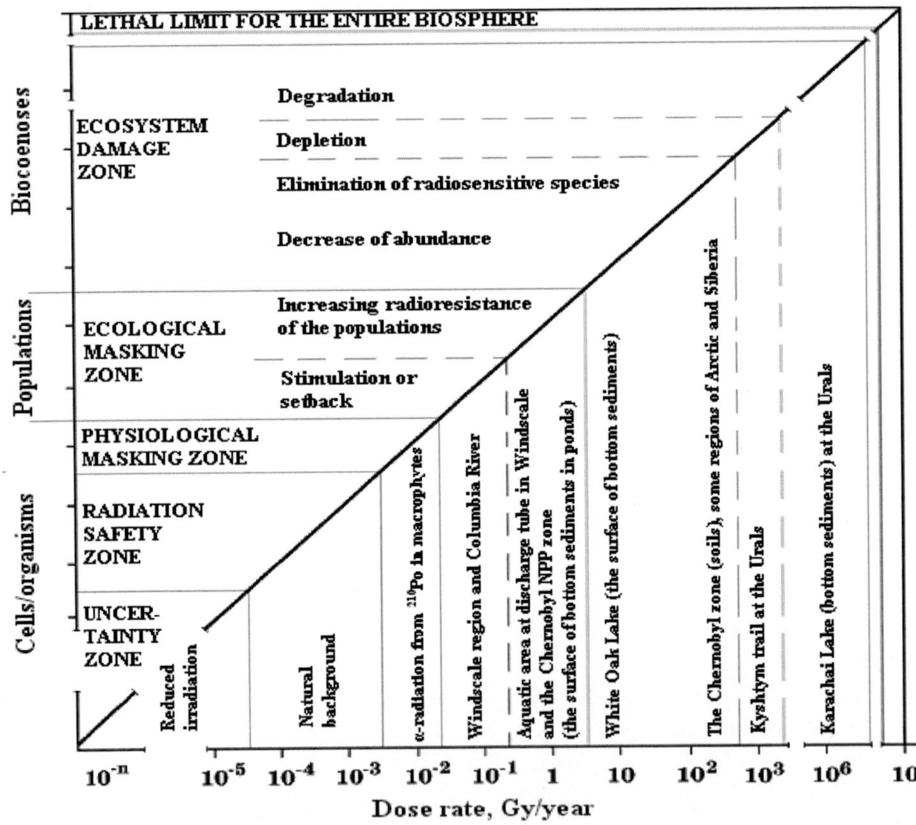

Figure 1. The ionizing radiation dose rate zones and their action on the biosphere [3 – 6]

It has been shown [1, 2] that the dose rates of ionizing radiation, mentioned in the publications under discussion devoted to the Chernobyl accident, generally concern the ecological masking (Figure 1) [3 – 6] and, therefore, the radiation effects may be effaced or modified by the effects of other environmental factors, chemical pollution, in particular.

The combined action of low dose ionizing radiation with other natural or anthropogenic factors, first of all, chemical pollutants of the environment, may cause the additive property, intensification or reduction of the effect with respect to activity and character of interacting agents. Therefore, the radiation effects at low rate of absorbed doses under natural conditions may hardly be determined.

The Level of Chromosome Mutagenesis in Natural Populations of Hydrobionts

Our investigations in the region affected by the Chernobyl accident included the 30-kilometer restriction zone of the Chernobyl accident, Danube, Dnieper, Black Sea and Aegean Sea.

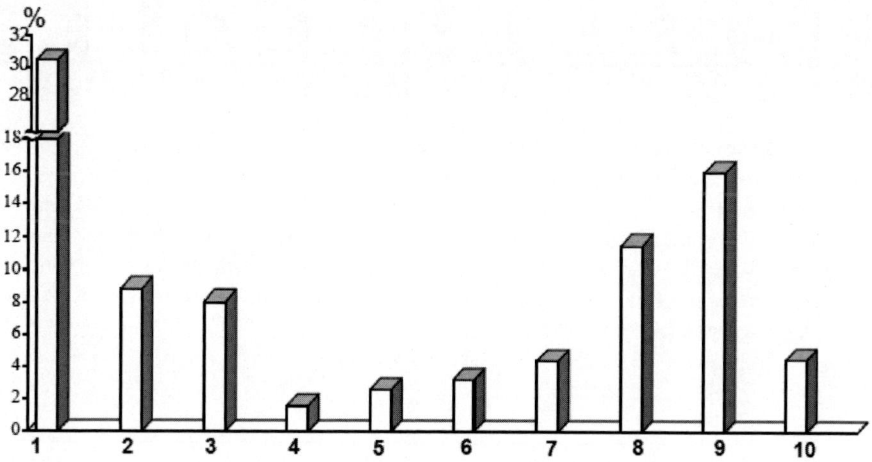

Figure 2. The frequency of cells with chromosome aberrations (%) in hydrobionts from natural populations of the Black Sea and the Aegean Sea (1986 – 1999) [9, 10, 12, 13]; Black Sea: 1 - *Rutilus rutilus* (1986); 2 - *Monodacna caspia* (1986); 3 - *Amphiura stepanovi* (1986); 4 - *Sprattus sprattus* (1987); 5 - *Scophthalmus maeoticus* (1989); 6 - *Trachurus trachurus* (1989); 7 - *Engraulis encrasicholus* (1999); 8 - *Mullus barbatus* (1999); Aegean Sea: 9 - *Plagiostomum aff. girardi* (1987); 10 - *Grubea clavata* (1999)

The rates of absorbed doses in water bodies, located in the 30-kilometer restriction zone, were calculated by radiometric results and estimated with the help of thermoluminescent $CaSO_4$ (Dy) detectors, synthesized by the United Institute for Nuclear Investigations (Dubna, Moscow Oblast) [7], which detected absorbed doses of ionizing radiation on the surface of bottom sediments.

The absorbed dose rates at the sampling moment in the studied water bodies were the following, mGy/day:

Yanov settlement	0.34
Kopachi village	0.24
Pripyat' River crawl	0.048
Nearby Uzh River	0.017
Nearby Strakholesie village	0.0024

According to the data obtained [8], the absorbed dose rate exceeded the natural background: by 300 times in the water cooling pond of Chernobyl NPP; by 5 – 100 times in the Kiev reservoir, and was comparable to the background in Kremenchyg and Kakhovskoe reservoirs.

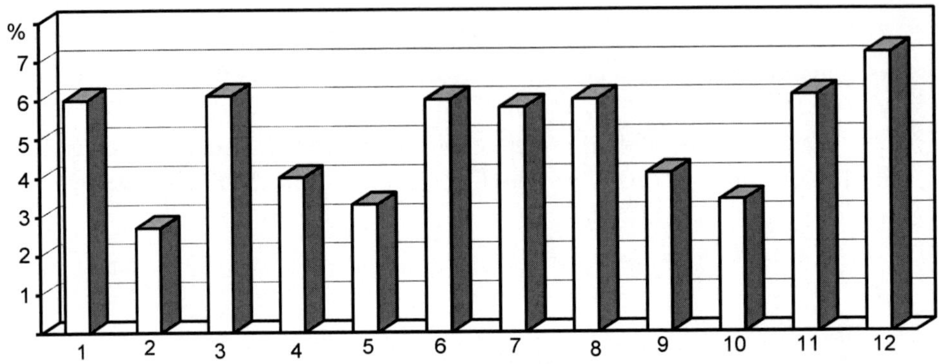

Figure 3. The frequency of cells with chromosome aberrations (%) in hydrobionts of Dnieper River and Danube River (1987 – 1992) [10, 13]; Dnieper River: 1 - Blicca bjoerkna (Kiev reservoir, 1987); 2 - Abramis brama (Kiev reservoir, 1998); 3 - Stizostedium lucioperca (Kiev reservoir, 1998); 4 - Rutilus rutilus (Kiev reservoir, 1998); 5 - Pontogammarus crassus (Kiev reservoir, 1992); 6 - Pontogammarus robustoides (Kremenchug reservoir, 1992); 7 - Pontogammarus crassus (Kremenchug reservoir, 1992); 8 - Pontogammarus robustoides (Kakhovka reservoir, 1992); 9 - Pontogammarus crassus (Kakhovka reservoir, 1992); 10 - Stylaria lacustris (Kakhovka reservoir, 1991); Danube River: 11 - Nais bretsheri (1988); 12 - Nais barbata (1988)

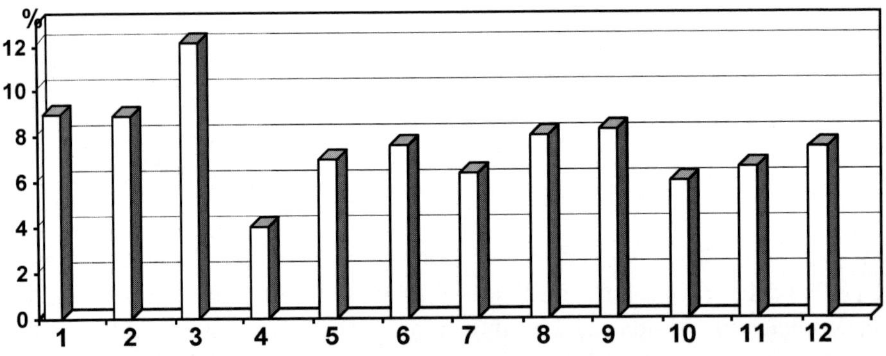

Figure 4. The frequency of cells with chromosome aberrations (%) in hydrobionts from the water bodies, located in the 10-kilometer restriction zone of the Chernobyl accident (1991 – 1996) [10 – 15]; 1 - *Gammarus lacustris* (Kopachi village, 1991); 2 - *Stylaria lacustris* (Kopachi village, 1991); 3 - *Stylaria lacustris* (Pripyat' River, 1991); 4 - *Stylaria lacustris* (Glubokoe Lake, 1993); 5 - *Stylaria lacustris* (Krasnensky Starik, 1993); 6 - *Stylaria lacustris* (water cooling pond, 1993); 7- *Dero obtusa* (Yanov settlement, 1995); 8 - *Nais pseudobtusa* (Yanov settlement, 1995); 9 - *Nais pardalis* (Yanov settlement, 1995); 10 - *Dero obtusa* (Yanov settlement, 1996); 11 - *Nais pseudobtusa* (Yanov settlement, 1996); 12 - *Nais pardalis* (Yanov settlement, 1996)

We have studied the frequency of chromosome aberrations in cells of mollusks, crustaceans and fishes at the early stages of ontogenesis (embryo, fry) in somatic cells of paratomically dividing oligochaetes and in sexual cells of echinoderms and nemerteans and polychaetes.

The material was fixed by a mixture of ethyl alcohol and glacial acetic acid, dyed by 1% acetoorcein, and squash preparations were made. The analysis of chromosome aberrations in somatic cells were performed at the anaphase-telophase mitosis stage and in developing sexual cells – at the anaphase-telophase I meiosis stage.

Figures 2 – 4 present data on the frequency of chromosome aberrations in cells of hydrobionts from natural populations. As observed from the Figures, the highest level of chromosome mutagenesis was observed in the 10-kilometer restriction zone of the accident and in some hydrobionts in the Black Sea and the Aegean Sea in 1986 – 1987 (immediately after the accident).

The Reproductive Strategy of Naididae (Oligochaeta) Family Worms in Water Bodies from the 10-Km Restriction Zone of the Chernobyl Accident

As radioactive contamination affects the natural biocoenoses, the most sensitive species may be eliminated and the species diversity is reduced. Therefore, the forecasting of environmental consequences of radioactive contamination requires the study of a comparative response of species population possessing different ecological valence to the damaging impact.

We have compared the population effects for three abundant species of Oligochaetae class dwelling in the water body located in the nearest zone of Chernobyl accident at Yanov village and the control water body nearby Uzh River.

The study was performed in 1995 and 1996. The object for the investigation was Naididae family oligochaetes: *Dero obtusa*, *Nais pseudobtusa* and *Nais pardalis*.

Tables 1 and 2 present data on hydrochemical parameters and chemical pollution of water bodies. Clearly, there are no significant differences in these parameters between the water bodies. Table 3 shows data on ^{90}Sr concentration in bottom sediments and calculated and measured dose rates of absorbed ionizing radiation at their surface, which were found 20 times different for various water bodies.

Table 1. Hydrochemical characterization of water bodies in 1995 – 1996 [14]

Parameter	Water body	
	Contaminated	Control
pH	6.5	6.5
NO_2^-, μg/l	2.3 ± 2.3	3.0 ± 2.1
NO_3^-, μg/l	5.0 ± 2.9	5.0 ± 3.6
NH_4^+, μg/l	4.0 ± 2.1	5.0 ± 2.5
PO_4^{3-}, μg/l	35.5 ± 15.9	43.3 ± 11.7
SiO_2, mg/l	4.4 ± 1.5	7.6 ± 3.4
ΣCO_2, mg/l	22.8 ± 0.8	24.0 ± 1.0
Ca^{2+}, mg/l	2.0 ± 0.3	2.0 ± 0.3
Mg^{2+}, mg/l	0.30 ± 0.02	0.30 ± 0.03

Table 2. Heavy metal and organochlorine compound concentrations in bottom sediments and aquatic vegetation in water bodies in 1995 – 1996, mg per kg of dry mass

Contaminant	Contaminated water body		Control water body	
	Bottom sediments	Aquatic vegetation	Bottom sediments	Aquatic vegetation
Pb	0.7 ± 0.7	1.5 ± 0.3	0.6 ± 0.6	1.0 ± 0.4
Cd	0	0.4 ± 0.3	0	1.1 ± 0.8
Cu	2.2 ± 1.2	3.1 ± 0.3	1.3 ± 0.9	1.6 ± 0.4
Zn	1.9 ± 1.5	41.7 ± 4.3	2.8 ± 2.8	23.8 ± 8.9
Fe	967 ± 394	2630 ± 988	1212 ± 851	994 ± 82
Mn	5.0 ± 3.9	894 ± 440	7.9 ± 4.0	692 ± 183
Cr	9.5 ± 2.6	0.9 ± 0.3	6.6 ± 1.6	1.6 ± 0.5
Ni	1.5 ± 0.9	3.8 ± 0.5	1.9 ± 1.9	3.7 ± 1.2
Co	1.6 ± 1.6	1.3 ± 1.3	0	0.3 ± 0.3
Hg	0.05 ± 0.01	0.040 ± 0.002	0.03 ± 0.01	0.06 ± 0.01
α+γ-hexachlorocyclohexane	0.003 ± 0.001	0.020 ± 0.004	0.002 ± 0.001	0.03 ± 0.01
DDT + metabolites	0.017 ± 0.006	0.17 ± 0.10	0.010 ± 0.006	0.12 ± 0.02
Polychlorobiphenyls	0.07 ± 0.02	0.919 ± 0.503	0.06 ± 0.01	0.799 ± 0.014

Table 3. The concentration of ^{90}Sr in bottom sediments (0 – 5 cm thick layer) in 1995 – 1996 and calculated (β-irradiation from ^{90}Sr + ^{90}Y) and TLD-measured dose rates [14, 15]

Water body	^{90}Sr concentration, Bq/kg dry mass	Dose rate, μGy/h		
		β	γ	β + γ
Contaminated	5872 ± 3345	5.0 ± 2.9	9.0 ± 1.9	14
Control	105 ± 11	0.09 ± 0.04	0.65 ± 0.12	0.7

Three species of worms mentioned are dwellers of the tangle biotope, and in contrast with other species, *D. obtusa* has gills and build soft tubes from the silt. It is known that Naididae family oligochaetes use both sexual and asexual (paratomic) reproduction. According to data from the literature [16], species with developed sexual apparatus are infrequently observed, whereas asexually reproducing worms are predominant. Therefore, the intensity of paratomic division is one of the most important indices defining the population production. In the zones of paratomic division of various worms, the number of dividing cells varied from 10 to 120. The comparison of *D. obtusa* worm distribution by the number of dividing cells, performed in autumn 1995, indicated more active paratomic division of oligochaetes in contaminated water body compared with the control one ($\chi^2 = 7.27$; $p < 0.01$). In contaminated water body 70% worms had over 30 dividing cells in the paratomic division zones. In the control water body, the amount of such worms was below 30%. In spring 1996, the activity of *D. obtusa* worm paratomic division was lower than in autumn 1995 ($\chi^2 = 3.82$; $p = 0.05$), approaching parameters of the control water body ($\chi^2 = 3.37$; $p > 0.05$). However, it should be noted that although this difference was statistically unreliable, the amount of worms with more than 30 dividing cells in paratomic division zones was 50%, whereas in the control water body – 20% only.

Among other species (*N. pseudobtusa* and *N. pardalis*) from contaminated and control water bodies, neither in 1995, nor in 1996 differences in paratomic division activity were observed ($\chi^2 = 0.57$; $p > 0.05$ and $\chi^2 = 0.16$; $p > 0.05$ for *N. pseudobtusa* and $\chi^2 = 1.64$; $p > 0.05$ and $\chi^2 = 0.83$; $p > 0.05$ for *N. pardalis*).

Table 4. Comparison of the part of species transiting to sexual reproduction with the severity of cytogenetic damages of worms from two water bodies [14, 15]

Time period	Species	Contaminated water body			Control water body		
		Number of worms	Part of worms (%) with chromosome mutagenesis >5%	Part of worms (%) with developing sexual cells	Number of worms	Part of worms (%) with chromosome mutagenesis >5%	Part of worms (%) with developing sexual cells
September, 1995	*D. obtusa*	28	39.3	0.0	19	15.8	0.0
September, 1995	*N. pseudobtusa*	30	60.0	23.3	31	41.9	9.7
September, 1995	*N. pardalis*	32	53.1	21.9	18	38.9	5.6
May, 1996	*D. obtusa*	38	42.1	5.3	15	13.3	0.0
September, 1996	*N. pseudobtusa*	25	36.0	12.0	18	22.2	0.0
September, 1996	*N. pardalis*	32	40.6	3.1	19	10.5	0.0

According to the authors' data [9, 12], the spontaneous level of chromosome mutagenesis in cells of hydrobionts independently of their taxonomic belonging is below 2%. Table 4 shows data on the number of worms with the chromosome mutagenesis level in cells exceeding 5%, i.e. being twice higher than the spontaneous level. This Table also presents the part of worms with developing sexual cells e.g. evolving to the sexual reproduction. The correlation analysis indicates a reliable connection between severity of cytogenetic damaging of worms in the population and the number of species transiting to sexual reproduction (with the correlation coefficient equal 0.81). As shown in Table 4, in autumn 1995 sexual reproduction in *N. pseudobtusa* and *N. pardalis* populations in the contaminated water body was much more active. In 1996, it reduced, but remained higher still than in the control water body. Apparently, unfavorable environmental conditions affect intensification of sexual reproduction, which increases adaptive abilities of the population. It is known that as preserving and fixing valuable heterozygous types and, therefore, providing the highest viability, asexual reproduction limits the possibilities of genetic variability and reduces adaptive potencies of the populations in the changing environment.

Thus, the radiation contaminated water body, in which the absorbed dose rate is at least two orders of magnitude higher than the natural background radiation, demonstrated deviations in the population parameters of worms. Two species (*N. pseudobtusa* and *N. pardalis*) demonstrated intensification of the sexual reproduction at the background of increasing severity of cytogenetic damages in the populations. The third species (*D. obtusa*) demonstrated stimulation of the paratomic division. Apparently, this species is higher stable to the damaging influence. It is also possible that the way of living of this Naididae species in silt tubes, which actively accumulate all kinds of contamination, promotes its higher ecological stability.

Chromosome Aberration Distribution in Cells of Hydrobionts

We have suggested a methodology for identification of radioactive, chemical and combined contamination impact on the natural populations of hydrobionts, which includes several cytogenetic methods and criteria.

Table 5. Distribution of chromosome aberrations by cells of crustaceans and fish embryos (experimental mutagenesis) [10, 13, 18]

Species	Irradiation or chemical mutagen	Dose or concentration	Number of cells with the following number of aberrations				χ^2 fitting criterion with the distribution type:	
			0	1	2	3	Poisson	Geometrical
Crustacean								
Idothea baltica	$^{90}Sr - ^{90}Y$	3.0 Gy	1950	173	5	0	0.83	6.44
Idothea baltica	$^{90}Sr - ^{90}Y$	0.5 Gy	1731	100	3	0	0.003	1.16
Gammarus olivii	$^{90}Sr - ^{90}Y$	3.0 Gy	8114	819	28	0	3.79	32.90
Gammarus olivii	$^{90}Sr - ^{90}Y$	1.5 Gy	1756	113	4	0	0.03	1.27
Gammarus olivii	^{137}Cs	0.5 Gy	611	31	1	0	0.05	0.20
Gammarus olivii	$^{90}Sr - ^{90}Y + ^{137}Cs$	2.0 Gy	1204	93	2	0	0.68	3.62
Gammarus olivii	External γ-radiation	5.0 Gy	2640	371	27	3	2.26	10.87
Gammarus olivii	External γ-radiation	3.0 Gy	740	90	4	0	0.34	3.92
Gammarus olivii	Lead acetate	1.0 mg/l	926	38	5	0	15.14	5.18
Gammarus olivii	Chlorophene	0.2 mg/l	1041	52	5	0	7.98	1.70
Gammarus olivii	^{137}Cs + chlorophene	0.5 Gy + 0.2 mg/l	531	25	2	0	2.64	0.42
Gammarus olivii	$^{90}Sr - ^{90}Y + ^{137}Cs$ + lead acetate + chlorophene	2 Gy + 1 mg/l + 0.2 mg/l	1190	86	2	0	0.39	2.64
Gammarus olivii	External γ-radiation + lead acetate	3 Gy + 1 mg/l	720	94	5	1	1.83	3.60
Fishes								
Scorpaena porcus	External γ-radiation	6.0 Gy	1460	232	11	1	2.56	17.50
Scorpaena porcus	External γ-radiation	1.0 Gy	1541	193	13	0	0.07	4.50
Scorpaena porcus	DDT	0.1 mg/l	807	77	14	2	20.19	3.95

The first method [10, 13, 17, 18] allowing for determination of the contribution of radioactive and chemical contamination into damaging of natural populations of hydrobionts is based on the analysis of chromosome aberration distribution by cells of hydrobionts. The method was designed on the basis of the authors' experimental data on the separate and combined action of ionizing γ- and β-radiation and chemical mutagens (heavy metals and organochlorine compounds) on crustacean and fish embryos, and the results of the field tests before and after the Chernobyl accident. The second method based on the analysis of mutagen-equivalent dose effects (by the number of cells with chromosome aberrations) [13, 19 – 21] is a modification of the method prorosed earlier[22, 23].

Table 6. Distribution of chromosome aberrations in cells of hydrobionts (worms, mollusks, erinoderms, crustaceans and fishes) from natural populations [9 – 15]

Species	Habitat	Year	Number of cells with the number of aberrations				χ^2 fitting criterion with the distribution:	
			0	1	2	3	Poisson	geometrical
Engraulis encrasicholus (embryos)	Aegean Sea	1968	989	52	5	0	7.12	1.36
Mysidacea gen. sp. (embryos)	Atlantic Ocean	1978	247	28	3	1	2.12	0.03
Harpacticoida gen. sp. (embryos)	Black Sea	1982	408	33	6	0	9.49	2.43
Anchylomera Brossevilei (embryos)	Mediterranean Sea	1985	1293	60	4	0	3.94	0.40
Amphiura stepanovi (sexual cells)	Black Sea	1986	396	32	5	0	1.73	0.06
Monodacna caspia (fry)	Black Sea	1986	408	34	2	1	1.21	0.05
Rutilus rutilus (embryos)	Black Sea	1986	1760	589	95	0	1.31	50.47
Plagiostomum aff. girardi (sexual cells)	Aegean Sea	1987	331	62	7	0	0.17	1.62
Naididae gen. sp. (somatic cells)	Danube River	1988	399	22	2	0	2.04	0.43
Pontogammarus crassus (embryos)	Dnieper River, Kakhovka reservoir	1992	1209	41	4	0	9.70	2.71
Pontogammarus robustoides (embryos)	Dnieper River, Kakhovka reservoir	1992	1317	60	7	0	15.44	4.55
Pontogammarus robustoides (embryos)	Dnieper River, Kremenchug reservoir	1992	1127	69	5	1	4.91	0.38
Stylaria lacustris (somatic cells)	Dnieper River, Kremenchug reservoir	1992	607	32	3	0	3.69	0.74
Grubea clavata (sexual cells)	Aegean Sea	1999	954	42	4	0	7.20	1.79
Engraulis encrasicholus (embryos)	Black Sea	1999	628	30	3	0	5.40	1.20
Mullus barbatus (embryos)	Black Sea	1999	411	42	6	1	5.96	0.62
Stylaria lacustris (somatic cells)	Water body in Strakholesie village (at the border of the 30-km restriction zone)	1991	830	49	8	0	17.48	5.58

Table 6 - Continued

Water bodies in the 10-km restriction zone of the Chernobyl accident								
Gammarus lacustris (embryos)	Kopachi village	1991	1339	111	9	0	3.60	0.23
Stylaria lacustris (somatic cells)	Kopachi village	1991	1946	166	11	0	1.74	0.67
Stylaria lacustris (somatic cells)	Pripyat' River	1991	550	78	9	0	1.53	0.47
Stylaria lacustris (somatic cells)	Chernobyl NPP water cooling pond	1993	392	34	1	0	0.14	1.18
Stylaria lacustris (somatic cells)	Krasnensky Staric	1993	540	17	1	0	1.74	0.10
Stylaria lacustris (somatic cells)	Glubokoe Lake	1993	967	40	0	0	0.87	1.68
Nais pseudobtusa (somatic cells)	Yanov village	1995 - 1996	774	57	1	0	0.50	2.34
Nais pardalis (somatic cells)	Yanov village	1995 - 1996	906	74	6	1	3.37	0.03
Dero obtusa (somatic cells)	Yanov village	1995 - 1996	1019	67	2	0	0.02	1.21

The laboratory and field cytogenetic studies of hydrobionts populations (Tables 5 and 6) helped in setting the regularities of by-cell distribution of chromosome aberrations [9 – 15, 17, 18]:

Mutagen	**Distribution of chromosome aberrations in cells**
R (exclusively) and (R' + X)	Generally corresponds to the Poisson theoretical distribution: $P(n) = m^n \times e^{-m}/n!$
X (exclusively) and (X + R) (X' + R)	Generally corresponds to theoretical geometrical arrangement: $P(n) = (1 - q)q^n$,

where R is the radiation impact; X represents chemical mutagens; R' and X' represent greater effectiveness of radiation and chemical mutagens at the combined action; m is the mean number of aberrations per cell; n is the number of aberrations in the cell; $q = m/(m+1)$ is the part of aberrant cells.

Our investigations indicated that before the Chernobyl accident all studied hydrobionts, independently of their taxonomic belonging, showed the by-cell distribution of chromosome aberrations being mostly geometrical (refer to Table 6), i.e. obviously the chromosome aberrations were induced by chemical pollution. After the Chernobyl accident, the highest adequacy of by-cell distribution of the chromosome aberrations to the Poisson distribution was observed exclusively for hydrobionts from biotopes with higher radioactive contamination level, in the 10-kilometer restriction zone, in particular, as well as for two species of hydrobionts (*Rutilus rutilus*, *Plagiostomum aff. girardi*), sampled in the Black Sea and the Aegean Sea in the initial post-accident years. Spawn laying of *Rutilus rutilus* was sampled in June, 1986 in Dnieper-Bug lagoon, and *Plagiostomum aff. girardi* worms were captured in December, 1987, in the Aegean Sea on a floating *Sargassum sp.* branch. In both cases, as combined with high level of the chromosome mutagenesis (30.5 and 16.0%, respectively), the Poisson distribution induces a conclusion about the impact of a strong ionizing radiation source, apparently, the "hot" particles of the Chernobyl origin. As for the roach embryos, this suggestion was confirmed by γ-spectroscopic measurements. In bottom sediments, where the spawn laying was found, ^{103}Ru "hot" particle was detected [24]. So far as concerns *Plagiostomum aff. girardi* worms, such measurements were not performed; however, according to the data from Greece [25], obtained in 1986 – 1987, "hot" particles were detected in Greece. For other hydrobionts in some water bodies of the 30-kilometer zone and outside it, the by-cell distribution of chromosome aberrations is mostly geometrical. Apparently, in these cases, the chromosome aberrations have been induced either by chemical pollution or combined action of radioactive contamination and chemical pollution, the latter being predominant in the damage induced.

The Comparison of Effects in Mutagenic Equivalent Doses

To estimate the influence of ionizing radiation at the background of high chemical pollution, i.e. in the cases, when the by-cell distribution of chromosome aberrations in hydrobionts is most close to geometrical distribution, we have used the method of the effect comparison in mutagenic-equivalent doses [22, 23] after its adjustment to out data [13, 19 – 21]. This method allows for comparing efficiency of mutagens possessing different action mechanism. Hence, the dose is measured in units of a single, well-determined index, and all other effects are adjusted to this standard. Such method gives an opportunity to compare spectra of genetic effects, induced by different mutagens. The genetic efficiency of ionizing radiation (of relatively low doses) and some chemical mutagens (heavy metals and organochlorine compounds) were compared in the experiments with crustacean and fish embryos. The mutagen-equivalent dose was estimated by the quantity of cells with chromosome aberrations (%). This criterion was taken for as a standard. For the additional index the mean number of aberrations per aberrant cell was taken. Figure 5 shows that separate action of ionizing radiation and chemical mutagens in mutagen-equivalent doses, which provide practically equal number of cells with chromosome aberrations, the severity of aberrant cell damages (the number of aberrations per aberrant cell) is higher for chemical mutagens. As comparing the action of ionizing radiation and the combined action of radiation and chemical mutagens in mutagen-equivalent doses, in the second case, higher severity of aberrant cell damage is observed, compared with the single radiation impact. As comparing the chemical mutagen effect and combined action of this mutagen and ionizing radiation, in the mutagen-equivalent doses, of special significance is lower severity of the aberrant cell damaging at the combined action.

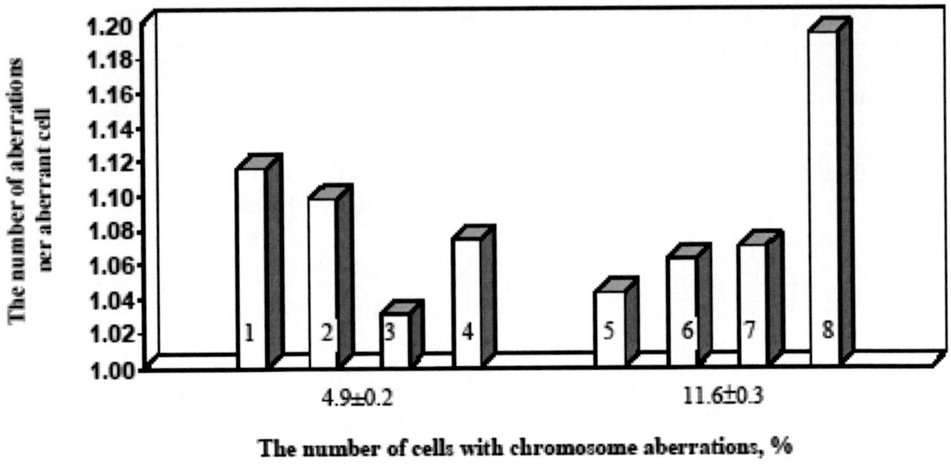

Figure 5. The number of aberrations per aberrant cell in hydrobionts at separate and combined action of radionuclides and chemical mutagens in mutagenic-equivalent doses [20]; 1 - *Gammarus olivii*, $Pb^{(2+)}$ (1.9 μmol/l); 2 - *Gammarus olivii*, chlorophene (0.4 μmol/l); 3 - *Gammarus olivii*, ^{137}Cs (0.5 Gy); 4 - *Gammarus olivii*, ^{137}Cs (0.5 Gy) + chlorophene (0.4 μmol/l); 5 - *Gammarus olivii*, external γ-irradiation (3 Gy); 6 - *Scorpaena porcus*, external γ-irradiation (1 Gy); 7 - *Gammarus olivii*, external γ-irradiation (3 Gy) + $Pb^{(2+)}$ (1.9 μmol/l); 8 - *Scorpaena porcus*, DDT (0.2 μmol/l)

Thus, the comparison of effects in mutagen-equivalent doses gives an opportunity to judge about higher or lower influence of ionizing radiation at the background of chemical pollution.

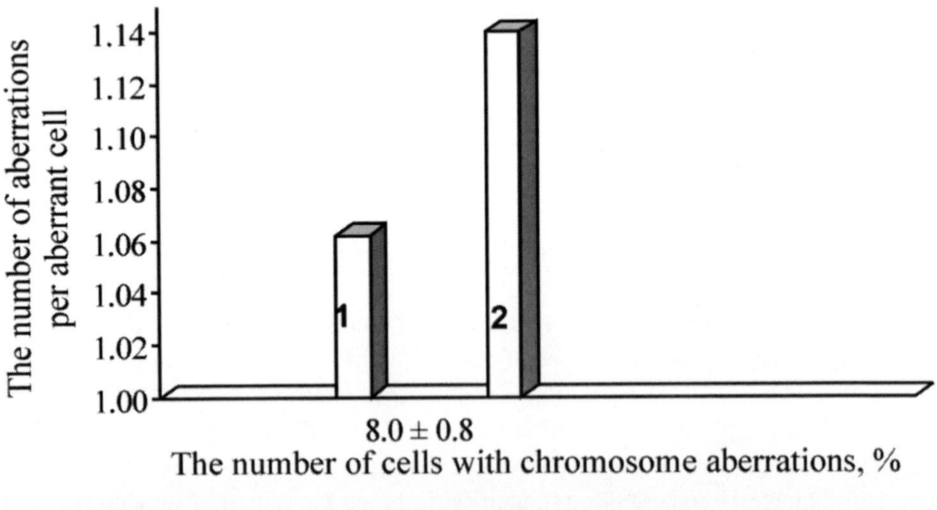

Figure 6. The number of aberrations per aberrant cell of *Stylaria lacustris* from the water bodies, located in Kopachi and Strakholesie villages at equal average quantity of cells with chromosome aberrations [20]: 1 – Kopachi village; Strakholesie village

Previously, we have studied *Stylaria lacustris* (Oligochaeta class, Naididae family) worms from several water bodies located in the 30-kilometer restriction zone of the

Chernobyl accident and outside it [10, 11, 13]. The chromosome aberrations were analyzed in somatic cells of asexually dividing species at the stage of anaphase-telophase mitosis. In different water bodies, the level of chromosome mutagenesis in the worm cells varied within the range of 5 – 12%.

For the comparison of the effects in mutagenic-equivalent doses, we have chosen *S. lacustris* from the water body in Kopachi village (in the 10-kilometer restriction zone of the accident) and the water body in Strakholesie village (at the border of the 30-kilometer restriction zone). The number of cells with chromosome aberrations in worms from two water bodies was approximately equal (about 8%), and by-cell distribution of the chromosome aberrations mostly corresponded to the geometrical distribution (refer to Table 6). Figure 6 presents data on the mean number of chromosome aberrations per aberrant cell in worms from two water bodies. It is obvious that severity of aberrant cell damage in worms from the water body located in Kopachi village is much lower.

Thus, basing on the cytogenetic investigation it may be concluded that in the first water body radioactive contamination effect is predominant in the damage formation. We have compared radiation dose rates and the level and character of chemical pollution of these water bodies [13, 20, 21]. Table 7 shows the results obtained. It is obvious that the dose rate in the water body at Kopachi village is 100 times higher than in the second water body, where it is comparable with the natural background. At the same time, the water body in Strakholesie village demonstrated much higher contamination level by heavy metals (zinc and manganese concentration in bottom sediments is much higher). As shown, dosimetric data and chemical analysis of the contamination structure confirm the conclusion made on the basis of cytogenetic investigation about a reliable influence of ionizing radiation on the worms from the water body at Kopachi village.

Table 7. Radiation dose rates (mGy/day) and concentration of chemical pollutants (mg/kg wet weight) in bottom sediments of two water bodies [20]

Ionizing radiation and chemical pollutants	Water bodies	
	Kopachi village	Strakholesie village
Dose rate	0.24	0.0024
Pb	0.73	0.83
Cd	0.04	0.07
Zn	2.47	26.91
Cu	0.11	1.24
Fe	631.73	464.05
Mn	8.53	52.21
Ni	1.65	0.70
Co	0.55	0.28
Heptachlorocyclohexane	0.0104	0.0046
Polychlorobiphenyls	0.111	0.120

We have also compared severity of the aberrant cell damaging (at identical mean number of cells with chromosome aberrations) for different species of marine and freshwater hydrobionts (worms, crustaceans and fishes) from different biotopes, studied by the authors in different years before and after the Chernobyl accident [10 – 13]. All these hydrobionts possessed the level of chromosome aberrations higher than the spontaneous one, which for different species of hydrobionts [9, 12], independently of their taxonomic belonging, did not exceed 2%. All hydrobionts showed by-cell distribution of chromosome aberrations mostly adequate to the geometrical one (refer to Table 6). Figure 7 indicates that in the first case (at the average quantity of cells with chromosome aberrations (4.3 ± 0.1)%) the average number of aberrations per the aberrant cells is practically identical for different species of hydrobionts from Dnieper River, Black and Aegean Seas. In the second case ((5.9 ± 0.1)% cells with chromosome aberrations), for all studied species of marine and freshwater organisms from Dnieper River, Danube River and Aegean Sea, the severity of the aberrant cell damaging was also almost identical . Apparently, this phenomenon can be explained by global distribution of the same basic chemical pollutants in the hydrosphere (heavy metals and organochlorine compounds).

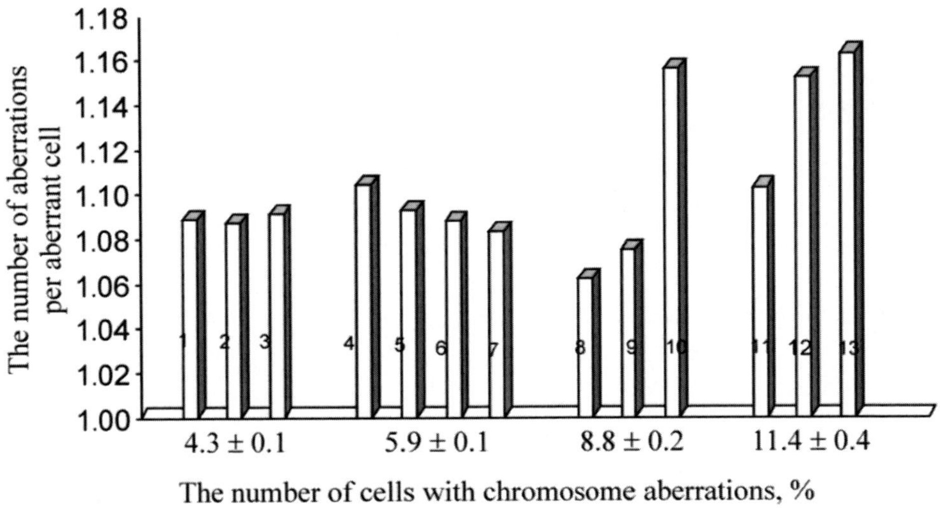

Figure 7. The number of aberrations per aberrant cell in hydrobionts from natural populations at equal mean quantity of cells with aberrations [20]; 1 - *Pontogammarus crassus*, embryos (Dnieper River, Kakhovka reservoir, 1992); 2 - *Grubea clavata*, sexual cells (Aegean Sea, 1999); 3 - *Engraulis encrasicholus*, embryos (Black Sea, 1999); 4 - *Pontogammarus robustoides*, embryos (Dnieper River, Kakhovka reservoir, 1992); 5 - *Pontogammarus robustoides*, embryos (Dnieper River, Kakhovka reservoir, 1992); 6 - *Engraulis encrasicholus*, embryos (Aegean Sea, 1968); 7 - *Naididae gen. sp.*, somatic cells of asexually dividing species (Danube River, 1988); 8 - *Stylaria lacustris*, somatic cells of asexually dividing species (10-km zone of the Chernobyl accident, Kopachi vil., 1991); 9 - *Gammarus lacustris*, embryos (10-km zone of the Chernobyl accident, Kopachi vil., 1991); 10 - *Harpacticoida gen. sp.*, embryos (Black Sea, 1982); 11 - *Stylaria lacustris*, somatic cells of asexually dividing species (10-km zone of the Chernobyl accident, Pripyat' River, 1991); 12 - *Mysidacea gen. sp.*, embryos (Atlantic Ocean, 1976); 13 - Mullus barbatus, embryos (Black Sea, 1999)

In two other cases, the situation is different (Figure 7). They demonstrate (8.8 ± 0.2)% and (11.4 ± 0.4)% cells with chromosome aberrations. Beside Black Sea and Atlantic Ocean hydrobionts, worms and crustaceans from the water bodies located in the 10-km restriction zone of the Chernobyl accident. Figure 7 shows that for hydrobionts from the 10-km restriction zone (Kopachi village, the dose rate 0.24 mGy/day; Pripyat' River, the dose rate 0.048 mGy/day), the average number of aberrations per aberrant cells is much lower than for hydrobionts from the Black Sea and Atlantic Ocean. Obviously, this indicates a considerable effect caused by radioactive contamination on hydrobionts from the water bodies of the accident zone.

CONCLUSION

We suggest that the analysis of the data obtained allows for a conclusion that systemic methods suggested (comparison of cytogenetic effects in mutagen-equivalent doses, analysis of by-cell distribution of chromosome aberrations and severity of aberrant cell damages) may be used for identification of radioactive, chemical and combined effects in natural populations of hydrobionts.

The results obtained indicate a considerable radiation impact on hydrobionts populations observed exclusively in the 30-kilometer restriction zone of the Chernobyl accident.

Various species of hydrobionts from the Black Sea, Dnieper River and Danube River are exposed to general damaging effect of chemical pollution.

REFERENCES

[1] Polikarpov G.G. and Tsytsugina V.G., *Radiats. Biol. Radioecol.*, 1995, vol. 35(4), pp. 536 – 549. (Rus)

[2] Polikarpov G.G. and Tsytsugina V.G., *Radioecology and the Restoration of Radioactive-Contaminated Sites*, Eds. F.F. Luykx and M.J. Frissel, 1996: Kluwer Academic Publishers, pp. 269 - 277.

[3] Polikarpov G.G., *Radiat. Prot. Dosim.*, 1998, vol. 75, pp. 181 - 185.

[4] Polikarpov G.G., *Marine Pollution*, Proc. Symposium held in Monaco, 5-9 October 1998, organized by IAEA, co-sponsored by IOC of UNESCO, UNEP and IMO in co-operation with CIESM. IAEA-TECDOC – 1094, Printed by IAEA in Austria, 1999, pp. 38 - 43.

[5] Polikarpov G.G, "Comparative evaluation of environmental toxicants – health effects of environmental toxicants derived from advanced technologies", *Proc. International Workshop on Comparative Evaluation of Health Effects of Environmental Toxicants Derived from Advanced Technologies*, Chiba, January 28 – 30, 1998. Eds. J. Inaba and Y. Nakamura, Kodansha Scientific, Ltd. Tokyo, 1998, ISBN 4 – 906464 – 06 – 8, pp. 3 – 15.

[6] Polikarpov G.G., Zaitsev Ju.P., and Fuma S., "Equi-dosimetry of deleterious factors at the level of populations and communities of aquatic organisms", *Mor. Ekol. Zh.*, 2004, vol. III(1), pp. 5 – 14.

[7] Kaskanov T.Ya. and Khenninger Yu., "The technology of thermoluminophores $CaSO_4$-4:Dy:Tm synthesis and their dosimetric characteristics", *Soobshch. OIYaI*, Dubna, 1984, pp. 16 – 84 – 635. (Rus)

[8] Kryshev I.I. and Sazykina T.N., *J. Environ Radioactivity*, 1994, vol. 28(1), pp. 91 – 103.

[9] Tsytsugina V.G., *Proc. Seminar on Comparative Assessment of the Environmental Impacts of Radionuclides Released During Three Major Nuclear Accidents: Kyshtym, Windscale, Chernobyl*, Luxembourg, 1 – 5 Oct., 1990, vol. II, pp. 895 – 904.
[10] Polikarpov G.G. and Tsytsugina V.G., *Radiobiologia*, 1993, vol. 32(2), pp. 205 – 212. (Rus)
[11] Polikarpov G.G. and Tsytsugina V.G., *Radiats. Biol. Radioekol.*, 1996, vol. 36(4), pp. 460 – 468. (Rus)
[12] Tsytsugina V.G., *Int. Symp. Marine Pollution*, Monaco, 5 - 9 Oct., 1998, IAEA, Extended Synopses, P. 669 – 670.
[13] Tsytsugina V.G. and Polikarpov G.G., *Readings to the Memory of N.V. Timofeev-Resovsky*, JSBN 966–02–1750–1, Sevastopol, 2000, pp. 70 – 79. (Rus)
[14] Tsytsugina V.G. and Polikarpov G.G., *Radiats. Biol. Radioekol.*, 2000, vol. 40(2), pp. 226 – 230. (Rus)
[15] Tsytsugina V.G. and Polikarpov G.G., *J. Environ. Radioactivity*, 2003, vol. 66, pp. 141 – 154.
[16] Chekanovskaya O.V., *Aquatic Oligochaetes in the Fauna of the USSR*, Moscow – Leningrad, Izd. AN SSSR, 1962, 412 p. (Rus)
[17] Polikarpov G.G. and Tsytsugina V.G., *Newsletter IUR*, 1997, No. 26, pp. 2 – 3.
[18] Tsytsugina V.G., *Radiat. Prot. Dosim.*, 1998, vol. 75(1– 4), pp. 171 – 173.
[19] Polikarpov G.G. and Tsytsugina V.G., *Dopovidi NAN Ukraini*, 1999, vol. 36(6), pp. 199 – 202. (Ukr)
[20] Tsytsugina V.G. and Polikarpov G.G., *Radiats. Biol. Radioekol.*, 2002, vol. 42(5), pp. 569 – 573. (Rus)
[21] Tsytsugina V.G. and Polikarpov G.G., *Gidrobiol. Zh.*, 2004, vol. 40(5), pp. 78 – 89. (Rus)
[22] Auerbah Sh., *The Problems of Mutagenesis*, Ed. N.I. Shapiro, Moscow, Mir, 1978, 463 p. (Rus)
[23] Stadler L.J. and Sprague G.F., *Science*, 1937, vol. 85, pp. 57 – 58.
[24] Tsytsugina V.G., Lazorenko G.E., Radchenko L.A., and Skotnikova O.G., *Proc. I Intern. Conference on "Biological and Radioecological Aspects of the Chernobyl NPP Accident Consequences"*, 11 – 18 Sept., 1990, Zeleny Mys, Kiev, Naukova Dumka, 1991, pp. 114 – 120. (Rus)
[25] Kritidis P., Catsaros N., and Probonas M., *Bergbau – und Indestriemuseum Theuem*, Ostbauern, 1988, Bd. 16/88, pp. 115 – 119.

In: 20 Years after the Chernobyl Accident
Editors: E. Burlakova and V. Naidich, pp. 303-329
ISBN 1-60021-249-2

Chapter 17

BIOCHEMICAL CONSEQUENCES OF RADIOACTIVE CONTAMINATION IN THE EXCLUSION ZONE OF THE CHERNOBYL ACCIDENT IN POPULATIONS OF WILD RODENT

A. G. Kudyasheva*[14]*, L. N. Shishkina*[2*]*,
***N. G. Zagorskaya*[1] *and A. I. Taskaev*[1]**
[1]Institute of Biology, Komi SC UrD RAS, Syktyvkar, Russia
[2]N.M. Emanuel Institute of Biochemical Physics, RAS, Moscow, Russia

ABSTRACT

The results of initial eight years of investigations (1986 – 1993) of the effect of radioactive contamination consequences on the regulation of lipid peroxidation (LPO) and the energy exchange in organs and tissues of Muridae family rodent inhabiting in the 30-kilimeter exclusion zone of the Chernobyl accident are generalized. The regularities of technogenic contamination effect on the dynamics of LPO physicochemical regulation system, generalized indices of the lipid composition and activity of dehydrogenation enzymes in organs of wild rodent possessing different radiosensitivity with respect to the level of radioactive contamination and time of radioactive factor effect are determined. Different sensitivity of LPO regulation parameters and dehydrogenation enzymes, and species of natural Muridae family rodent populations to radioactive contamination of the environment is shown. The analysis of studied biochemical indices and data from the literature testify about formation of qualitatively new subpopulations of wild rodent in the accident zone due to transition of the regulation cellular systems to a new functioning level.

[4] Main addressees: 28 Kommunisticheskaya street, 167982 Syktyvkar, Russia; Institute of Biology, Komi SC UrD RAS; Tel: (7-8212) 43-04-78; E-mail: kud@ib.komisc.ru; [2] 4 Kosygin street, 119991 Moscow, Russia; N.M. Emanuel Institute of Biochemical Physics, RAS; Tel: (7-095) 939-71-86; E-mail: chembio@sky.chph.ras.ru

Keywords: Muridae family rodent, radioactive contamination, the Chernobyl accident, the 30-kilometer exclusion zone, lipid peroxidation, antioxidant activity, composition of lipids, dehydrogenation enzyme activity

The Chernobyl accident has disclosed many complicated, scientifically uninvestigated problems, one of which is the study of damage and adaptation mechanisms of animals to the technogenic radioactive contamination of the habitat. Radioecological conditions, formed after the accident on a considerable territory of Ukraine, Belarus and Russia, provided for the occurrence of biota exposed for a long-term low-dose radiation and made these territories the zones of increased environmental risks. The models for estimating intensity of disorders in ecosystems under critical anthropogenic loads, including radioactive contamination, are usually based on small mammals, which by definition are the species-indicators [1, 2]. Muridae family rodents represent one of the most suitable test-objects for radioecological monitoring due to their tight contact with radionuclides in the soils, abundant population, relative inhabitance, restriction of the individual territories, availability of the food supply basis, etc. [1, 3, 4].

The studies of biological consequences of the Chernobyl accident for the populations of wild rodents analyzed accumulation of radionuclides in organisms of different species of small rodents [5 – 12], changes in the population [5, 9 – 11, 13 – 15], genetic damages [16 – 21], morphological changes in organs and tissues of these animals [7, 22 – 24], and biochemical parameters [25 – 27]. At the level of a population, the radiation impact is combined with the action of other environmental factors, which requires searching for metabolically important indices, which allow for objective estimation of the impact and judge about the change of the whole metabolic system in the species, appeared in the radioactive contamination zone. However, searching for test-parameters is complicated due to some reasons, for example, highly expressed nonlinearity of the "biological effect – dose" dependence in the systems of different complexity and high heterogeneity of the system response to a weak effect [28 – 30], which may also be caused by the effect dependence on the values of corresponding parameters of rodents from control territories [31 – 33].

One of perspective directions in the searching for metabolically important indices is the study of the lipid peroxidation (LPO) proceeding in all types of membranes and playing the very important role in the regulation of cellular metabolism in norm and under the effect of various damaging factors, including ionizing radiation [34 – 37]. Generally high interest to the question of LPO regulation is induced by its ranking in the evolutionary process as one of the earliest regulatory mechanisms. The inverse dependence of LPO intensity on both dose rate and the ionizing radiation dose itself, discovered in 1960ies – 1970ies in experiments both *in vitro* and *in vivo* [35, 38, 39], allowed a suggestion about the important role of oxidative processes in lipids in formation of biological consequences of the Chernobyl accident for the populations of Muridae family rodents from radionuclide-contaminated territories.

Basing on the long-term experience of the studies of wild rodent populations inhabiting in biocoenoses exposed to increased natural radiation levels and the results of investigations, performed in the branch of physicochemical biology by the followers of the Academician N.M. Emanuel, we have implemented a complex analysis of the cell regulation system state in tissues and interrelation between parameters of different units of the LPO regulatory system and the energy supportability of tissues of natural populations of rodents, caught in 1987 – 1993 at the areas with different levels of radioactive contamination in the 30-km exclusion zone of the Chernobyl accident. Generalized below are results of the analysis of biochemical indices for spleen, liver, myocardium, brain, and erythrocytes for more than 1,000 animals of three age groups (immature, mature current year's young and hibernated animals). The state of LPO processes (antioxidant and antiperoxide activity of lipids, the amount of peroxides in lipids, the composition of phospholipids, generalized parameters of the lipid exchange) and the energy exchange (activity of dehydrogenation enzymes – pyruvate, succinate and lactate hydrogenases) were studied. The methods applied are discussed in detail in refs. [31, 32, 40].

DESCRIPTION OF MURIDAE FAMILY RODENT TRAPPING AREAS

The test objects were represented by phytophagous species of rodents: voles (bank vole – *Clethrionomys glareolus* Schreber, common vole – *Microtus arvalis* Pallas, and root vole – *Microtus oeconomus* Pallas) and mice (striped field-mouse – *Apodemus agrarius* Pallas and yellow-necked mouse – *Apodemus flavicollis* Melchior). Rodents of Muridae family were caught during 8 years in the 30-km exclusion zone of the Chernobyl accident at seven areas squared 3 – 5 ha each. Moreover, in 1991 – 1993 rodents of the studied species were also caught in the neighborhood of the Kiev City (Teremki, Romankov, Kicheevo, Lysaya gora, Bezradichi) at the areas with practically normal background radiation. Trapping areas were selected so that they should be presented by several identical biotopes, occupied by the same species of rodents, and differing by the radiation contamination only [5]. The selection of rodent trapping areas and detailed radioecological characterization of the accident zone are presented in refs. [5, 8, 23, 31, 41 – 43]. As indicated by long-term investigations performed in biocoenoses with increased background radioactivity level [44], quantitative estimation of the individual dose for small rodents in natural populations is rather complicated due to active movement of animals by the territory with quite frequent mosaic structure of radioactivity, exposure dose rate and composition of radionuclides. The main contribution into the dose perceived by the rodents in the Chernobyl accident zone may be obtained from the following sources:

- external β- and γ-radiation from radionuclides fallout to the soil and vegetation cover;
- tectorial α- and β-radiation of animals by radionuclides precipitated on the external body surface (hair, skin, etc.);
- peroral and inhalation delivery of radionuclides during respiration, feeding and water.

Hence, it may not be excluded both chemical effect of incorporated heavy radionuclides and permanently acting environmental factors of the non-radiation origin, which may modify biological effects [23, 31, 32, 45]. Therefore, for characterization of the radiation impact on the studied groups of rodents the data on external γ- and β-radiation were used, deduced from the measurement results obtained with the help of radiometric instruments and individual thermoluminescence dosimeters [8]. In spring 1987, by the level of technogenic contamination the areas were divided into three groups: with low external γ-radiation (0.02 – 0.1 mR/h: areas 5 – Stechanka, 6 – Raz'ezzha, 7 – Kupovatoe), moderate level (2 – 20 mR/h: areas 2 – Chistogalovka, 3 – Shepelichi, 4 – Izumrudnoe) and high level (150 – 200 mR/h: area 1 – Yanov). The areas with higher radioactive contamination level were selected in the zone of the western fallout trail from the Chernobyl NPP. Table 1 presents external γ-radiation doses at the trapping areas. The dose rate of external γ-radiation at highly and moderately contaminated areas was stabilized 6 – 7 years after the accident. However, in 1995 it exceeded the natural background radiation by dozen times [46]. The dose rates of external β-radiation on the contaminated areas were always higher than γ-radiation level, on average: by 57 – 60 times in 1986, 35 – 37 times in 1987, 2 – 2.5 times in 1988, 1.2 – 1.7 times in 1989 [31].

Table 1. The ranges of external γ-radiation dose rate variation at the areas, mR/h

Area No.	Sept. 1986	Aug. 1987	Aug. 1988	Aug. 1989
1	300-500	150-200	45-70	10-18
2	40-50	15-20	3.6-6.2	2-4
3	40-50	15-20	4.2-8.3	2.3-4.6
4	4-6	2-3	0.5-0.9	0.3-0.6
5, 6	0.05-0.3	0.02-0.1	0.017	0.01
7	0.05-0.1	0.04-0.07	0.03-0.05	0.04-0.05

The external radiation doses were calculated from the concentration of radionuclides (cesium and strontium) in the organisms of caught rodents. The predominant accumulation of ^{137}Cs in tissues and ^{90}Sr in bones of Muridae family rodents was shown [31]. The highest concentration of radionuclides was detected at the areas with the highest contamination level. Direct correlation between the external background γ-radiation and total concentration of cesium isotopes in the organism of *C. glareolus* is also indicated in ref. [9]. However, the level of radionuclide accumulation significantly depended on the animal habitat conditions. For example, bank voles and field mice [31] that may be explained by their inhabitance under the cover of trees, in which bark the concentration of radionuclides was also high [43].Strieped field mice and yellow-necked mice are much "cleaner", which is due to the environmental conditions of habitat [31]. Higher concentration of cesium radioisotopes in tissues and body of bank voles compared with other Muridae family rodents was also detected in refs. [11, 12].

The species composition and abundance of animals at the areas were defined by the biotope type, the area occupied and the landscape features. We are just to note that the root vole was mostly caught at the areas 4 and 6, strieped field-mouse – at the areas 1, 2, 3 and 5, bank vole – at the areas 2, 3, 5, common vole – at the areas 1 – 3, and yellow-necked vole – at the areas 1, 2, 3 and 7. So far as concerns variations of the relative abundance of rodents caught at all areas of the accident zone, it significantly depended on the species of animals (Figure 1). In autumn 1986, reduction of the total abundance of Muridae family rodents and predominance of immature species among caught animals are induced by the reproduction termination and death of irradiated animals in the initial hours and days after the accident [5, 8]. High total abundance of the rodents in the accident zone in 1987 and its subsequent abrupt decrease is also indicated in refs. [7, 9, 10]. The detailed analysis of accounting for the abundance, maturity structure and dynamics of rodent reproduction at the trapping areas in the initial years after the accident are presented in refs. [5, 8, 10]. As shown by the analysis of the data from Figure 1, in some cases in the 30-km exclusion zone of the Chernobyl accident an obligate phase change in the abundance of animals are observed, which are typical of natural populations inhabiting on the territories with normal background radiation. The long-term periods of the abundance decrease, specifically for the root vole, testify about inhabitance of the rodent populations under unstable conditions typical of the border and impact populations [32, 45].

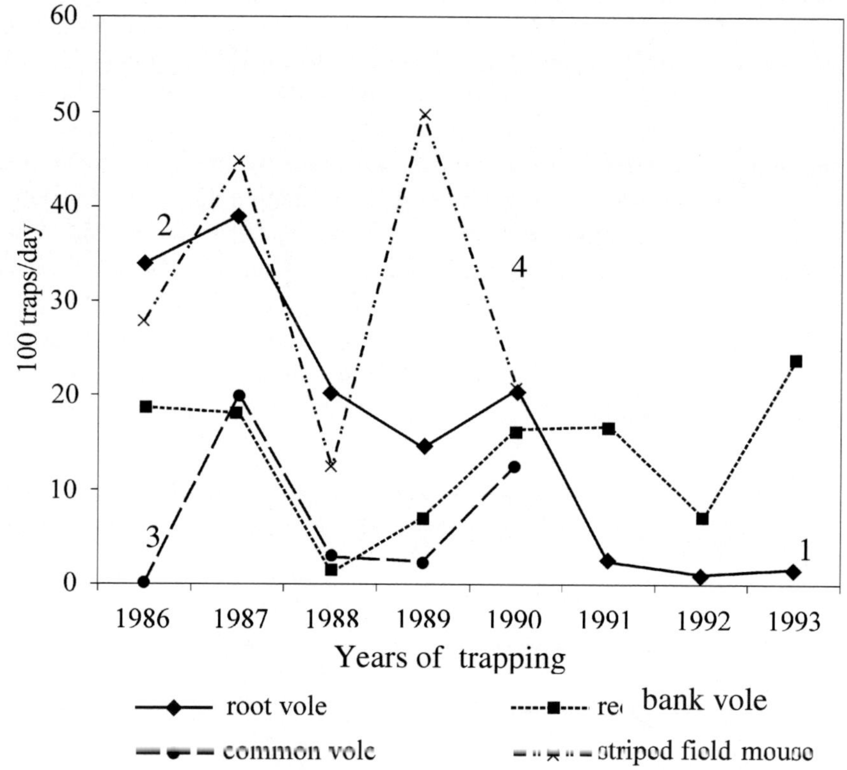

Figure 1. Dynamics of the relative abundance of wild rodents in the 30-km exclusion zone of the Chernobyl accident in 1986 – 1993: 1 – root vole; 2 – bank vole; 3 – common vole; 4 – striped field-mouse

The State of the Lipid Peroxidation in Tissues of Muridae Family Rodents from the Kiev Population

As known, lipids (LH) representing one of the most important structural components of the membrane system of the cell are not only energetic substrates of the cell, but also implement structural and barrier functions, and represent specific regulators of intracellular processes. The rapid development of membranology allowed for justifying the biological importance of free-radical reactions proceeding in different compartments of the cell, cellular metabolism regulation in norm and development of pathologies [34 – 37, 47 – 49]. The change of phospholipids (PL) composition, their regularity and packing in a bilayer play the most important role in the cell adaptation to the environmental conditions [50 – 52]. In the intact organism, physiologically normal level of free-radical processes is provided by different systems of antioxidant (AO) protection: natural antioxidants (AO, [InH]), (tocopherols, ubiquinoones, flavonoids, bilirubin, glutathione, β-carotene, etc.); AO protection enzymes (SOD, catalase, peroxidases, glutathione-dependent peroxidases and transferases, in particular), as well as a broad list of endogenic compounds demonstrating AO activity in complex biological systems (ascorbic acid, urea, pyruvates, carnosine, low-molecular amino acids, etc.) [35, 47, 53 – 58]. The intensity of LPO processes in norm is kept at the stationary level by physicochemical regulation system providing for the structural and functional entity of the membranes [59]. At present, the existence of a physicochemical LPO regulation system at the cellular and organ levels is shown [58], parameters of which are antioxidant (AOA) and antiperoxide (APA) activities and the composition of lipids; the initial quantity of peroxides in lipids ($[ROOH]_0$), stipulated by their nonsaturation degree; the oxidation inhibition extent (OIE) of the model substrate by them. Interrelations between parameters of this system are shown in Figure 2. The functioning of this regulation system indicated the correlation interrelations between the generalized PL composition parameters, which characterize the oxidizability of lipids (the ratio of sums of the more easily and the more poorly oxidizable fractions of phospholipids: ΣEOPL/ΣPOPL) and reflecting the structural state of the membrane system of a cell or an organ (phosphatidyl choline/ phosphatidyl ethanolamine (PC/PE) and [sterols]/[PL] ratios), as well as coordinated changes of AO protection system parameters, localized in various subcellular compartments or tissues [32, 40, 58, 60 – 62].

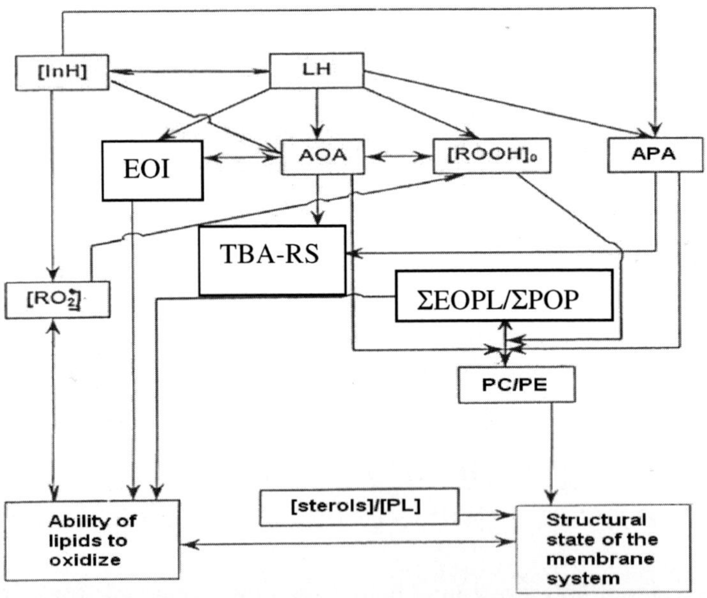

Figure 2. The scheme of interrelations between parameters of physicochemical LPO regulation system at the cell and organ levels

For the population of wild rodent inhabiting in the accident zone, biological consequences of the Chernobyl catastrophe may be estimated only in the case of adequate biological control. The extremely labile parameters of the lipid exchange and scanty literary data on the LPO state in the organs of wild rodents provided for a necessity in experimental determination of physicochemical LPO regulation system in organs and tissues of wild Muridae family rodents, caught at the areas with usual background radiation in the neighborhood of the Kiev City. It is found [31, 63] that the lipid AOA in liver, spleen and brain of rodents vary significantly with respect to their age and species. It should also be noted that the lipid AOA of wild rodent organs is generally lower than corresponding parameters for laboratory animals [31, 35, 63]. In many cases, rodent lipids possess prooxidant properties, i.e. intensify oxidation of methyl oleate. It is shown that the prooxidant activity of lipids is stipulated by both depletion of their AO and high unsaturation level [64, 65]. Obviously, lipids in organs of wild rodents, the inhabitance of which in the natural environment requires adaptation to long-term exposure to low temperatures, are less saturated than lipids of laboratory animal organs. By decreasing AOA value of lipids the organs of tested species of rodents form the following sequence:

liver > brain > spleen.

Similar to laboratory animals, AOA values of lipids in wild rodents change both with respect to the age of animals and the season of their trapping [31]. However, for wild rodents inhabiting in the natural environment or kept in vivarium for a long time much stricter dependence of lipid AOA and, specifically, the ability of lipids to oxidize or their APA on the animal sex is observed compared with the laboratory mice [31 – 33, 35, 63]. Moreover, the content of LPO products (TBA-RS) in the brain and SOD activity in erythrocytes of wild rodent blood is much higher than analogous parameters for laboratory animals [66].

The comparison analysis of the liver lipid AOA of mature animals caught in the neighborhood of the Kiev City in autumn 1991 allowed for forming the following sequence of rodents with respect to reduction of this parameter:

yellow-necked mouse > bank vole > striped field mouse > root vole [63].

In some years, only scale of lipid AOA differences in the liver of rodent of various species. Apparently, this is related to variability of the abundance of the rodent species under investigation, observed in the period of 1991 – 1993 in the neighborhood of the Kiev City, because the phase change in the population cycle is accompanied by the LPO parameter variation in the rodent organs [31 – 33, 63]. Previously observed positive correlation between radiosensitivity of various species and lines of laboratory animals and the AOA lipid level of their organs [35, 67] allows for a suggestion about increasing radiosensitivity of the wild rodents in accordance with lipid AOA decrease in their liver. This is also confirmed by data from the literature on $LD_{50/30}$ values for some tested species [68 – 70].

The qualitative PL composition of spleen, liver and brain of wild rodents of different species inhabiting on the territories with common background radiation is represented by a selection of PL typical of eukaryote cells [31, 50, 71, 72]. However, in some cases wild rodents demonstrated definite differences in the relative concentration of some PL fractions compared with the laboratory animals. For example, high concentration of cardiolipin (CL) and phosphatidic acid (PA), the concentration of which varies from 14 to 20.3% in mature animals of different species, is typical of the spleen PL [31]. It is known that CL and PA participate in calcium ion transport through a membrane, conjugated with activation of phosphatidyl inositol (PI) metabolism [73]. Apparently, this also induces quite high total concentration of PI and phosphatidyl serine (PS) in spleen PL of mature and hibernated rodents of different species from the Kiev population: 22.5, 14.9 and 15.7% for root voles, yellow-necked mice and bank voles, respectively. Brain PL of the wild rodents caught in the neighborhood of the Kiev City in 1991 are characterized by low relative concentration of PL lysoforms, whereas the species differences in general parameters of the brain PL composition for the wild rodents is lower than in spleen and liver of the same animals [31]. However, in accordance with the decrease of the sums of the more easily to the more poorly oxidizable PL fractions ratio in the brain of mature and hibernated animals of the studied species of rodents form the following sequence:

bank voles > yellow-necked mice > root voles = striped field-mice,

in the spleen PL of these animals only a tendency to decrease was observed as follows:

bank voles > yellow-necked mice > root voles.

Hence, the reliable species differences in organ PL of the rodents by the ratio of the main PL fractions (PC/PE) were determined. In relation to a decrease of this parameter, these species form the following sequences:

Brain PL:

yellow-necked mice > striped field-mice = root voles > bank voles;

spleen PL:

root voles > bank voles > yellow-necked mice [31].

The systematic analysis of PL composition of organs of root voles inhabiting on the territories with usual background radiation in Komi Republic has indicated significant changes of the relative concentration of some PL fractions with respect to the population cycle phase [72]. In the phase of the abundance depression of the population a sharp increase of PL lysoforms (LPC) fraction in liver and spleen lipids and the occurrence of their traces in the brain PL, as well as a reduction of the main PL fractions, specifically PE, in all studied tissues was observed [32, 72]. It is obvious that so high relative concentration of lysoforms in spleen (14.3 ± 1.9) and liver (18.6 ± 2.0) PL of the root vole mature males, caught in the neighborhood of the Kiev City in autumn 1991 [31, 40], was explained by currently occurring depression of the population abundance at that period.

Generally, despite the definite differences of the LPO regulation system parameters in tissues of wild rodents compared with these of laboratory animals, a reverse correlation dependence between the ability of lipids to oxidize (ΣEOPL/ΣPOPL) and the structural state of the membrane system of organs (PC/PE) of laboratory mice [58, 61, 62] and root voles [32, 40], stipulated by functioning of the LPO membrane system of the cell and organs as the whole. Moreover, the presence of relation between AOA of liver lipids and resistance to acute radiation impact of both laboratory animals and wild rodents allows a suggestion that the AO protection system is quite stable and capable of preserving the species specificity of biophysical and biochemical parameters in cells and organs for LPO regulating in tissues of animals from natural populations. High variability of AOA lipids and relatively low values of this parameter in liver due to high nonsaturation of lipids, which is also testified by higher ratios between sums of the more easily and the more poorly oxidizable fractions in liver PL of Muridae family rodents of the Kiev population compared with the laboratory mice [31, 62, 71, 74], obviously, is the necessary condition for supporting homeostasis and normal functioning of the regulatory systems in organs of wild rodents in the inhabitance.

The State of the Lipid Peroxidation and the Energy Exchange in Tissues of Wild Rodents from the Chernobyl Accident Zone

A complex study of the cellular regulation system state performed in organs of Muridae family rodents from natural populations in different areas of the Chernobyl accident zone during 8 years suggested that chronic low-intensive low dose radiation induces a complex of disorders in LPO and dehydrogenation regulation in the organism. The choice of dehydrogenase activity as a functional test is reasoned as follows. First of all, succinate dehydrogenase (SDH) and pyruvate dehydrogenase (PDH) are the membrane-bound enzymes and, consequently, changes in the structure and physical state of membranes effect on the activity of these enzymes; some part of lactate dehydrogenase (LDH) is also membrane-bound in the organism, the size of the membrane-bound fraction in different organs being different [75]. Secondly, all dehydrogenases are SH-containing enzymes. Thirdly, among the cellular systems related to the Krebs cycle, multienzyme complexes PDH and SDH are the most sensitive to radiation [76].

The investigation results give an opportunity to track the main stages of biochemical consequence formation in the regulation cellular systems of several generations of rodents inhabiting in the Chernobyl accident zone. Despite the seeming physiological happiness, in tissues of studied animals the changes of oxidative and energy process parameters were observed.

High total abundance of the animals in the accident zone, observed in 1987, allowed for a comparison analysis of parameters of physicochemical LPO regulation system for all studied species of rodents. First of all, a definite similarity of the technogenic contamination impact on lipid AOA in liver, the composition of lipids and dehydrogenation activity, and the influence of acute radiation on the change of biophysical parameters in tissues of laboratory animals should be noted. For example, liver lipid depletion by antioxidants proceeds in accordance with the natural radioresistance of a species and radiosensitivity of an organ [31, 71]: in liver of animals from the areas 5 and 6 with the lowest contamination level the lowest lipid AOA values were observed for hibernated males and immature and mature females of the most radiosensitive root voles. Moreover, the lipid AOA values in liver of root vole mature males, hibernated bank voles and mature striped field-mice are lower the these parameters of rodents of the same species, caught in autumn 1991 in the neighborhood of the

Kiev City [31, 63]. The level of lipid AOA in spleen of all wild rodent species from the accident zone is always lower than in liver and brain lipids [31].

Changes in the lipid composition of organs are characterized by an increase of the relative amount of PL lysoforms and the ratio PC/PE, reduction of PL concentration in the composition of total lipids in liver, brain and spleen, increasing concentration of free fatty acids in lipids of erythrocytes [31, 40, 71, 74, 77]. Table 2 presents the liver PL composition for hibernated rodents, trapped the accident zone in autumn 1987 at the areas possessing moderate and low contamination levels, and for root voles from the Kiev population. Despite the observation of the depression phase in autumn 1991 and a peak of the rodent abundance in 1987 (Figure 1), the part of lysoforms in the PL composition is by 1.6 – 1.8 times and the ratio PC/PE is 6.8 – 7.3 times higher, and the PL content in the composition of total lipids is 2.7 – 5.4 times lower in the liver of root vole males hibernated at different areas of the accident zone compared with the analogous parameters of the Kiev population animals. In the liver PL of hibernated red-backed voles from a low contaminated area the concentration of radionuclides in tissues of which is higher compared with striped field-mice, the relative concentration of lysoforms (Table 2) and the ratio PC/PE (4.24 ± 0.21) reliably exceed analogous parameters in the liver PL of striped field-mice (PC/PE = 2.74 ± 0.22) caught at moderately contaminated area. The more radioresistant hibernated red-backed voles, caught at the areas 5 and 6 in spring 1991, demonstrated the part of lysoforms in the liver PL 2.9 – 6.5 times lower and the part of PE 1.8 – 3.3 times, and PL concentration among total lipids 1.3 – 2.35 times higher compared with the analogous parameters for the higher radiosensitive root voles [31].

Table 2. The composition of liver phospholipids of hibernated Muridae family rodents of different areas, caught in the neighborhood of the Kiev City and the Chernobyl accident zone

PL fraction, %	Kiev neighborhood, root voles, males, (2 ind.), autumn 1991	Area 6, root voles, males (3 ind.), autumn 1987	Area 4, root voles, males (11ind.), autumn 1987	Area 4, striped field-mice, males and females (6 ind.), autumn 1987	Areas 5 and 6, red-acked voles, males and females (5 ind.), autumn 1987
LPC	18.6 ± 2.0	30.35 ± 1.0	34.2 ± 3.05	20.80 ± 2.65	37.5 ± 0.85
SM	5.50 ± 0.60	11.9 ± 1.9	14.9 ± 1.8	19.0 ± 0.85	9.05 ± 0.20
PC	24.85 ± 1.20	38.45 ± 2.6	29.5 ± 0.95	26.85 ± 1.55	32.2 ± 3.0
PI + PS	14.95 ± 2.20	6.40 ± 0.85	10.0 ± 1.05	10.05 ± 0.50	6.00 ± 0.23
PE	20.05 ± 1.55	4.50 ± 1.65	3.60 ± 0.70	10.05 ± 1.65	7.60 ± 0.42
CL + PA	16.05 ± 2.0	8.35 ± 1.45	7.75 ± 1.45	13.25 ± 1.30	7.6 ± 2.6
PL (%) among total lipids	46.8 ± 2.4	17.1 ±0.5	8.65 ± 1.75	14.45 ± 0.85	5.35 ± 0.30

Both reduction of SDH, PDH and LDH activity in tissues of Muridae family rodents, caught in the accident zone, compared with the analogous values of activity of these enzymes typical of root voles from the Komi population, and noncoordination of dehydrogenation processes were determined [31, 32, 71, 77]. Three years after the accident activity of studied enzymes is 1.5 – 2 times lower than in 1987.

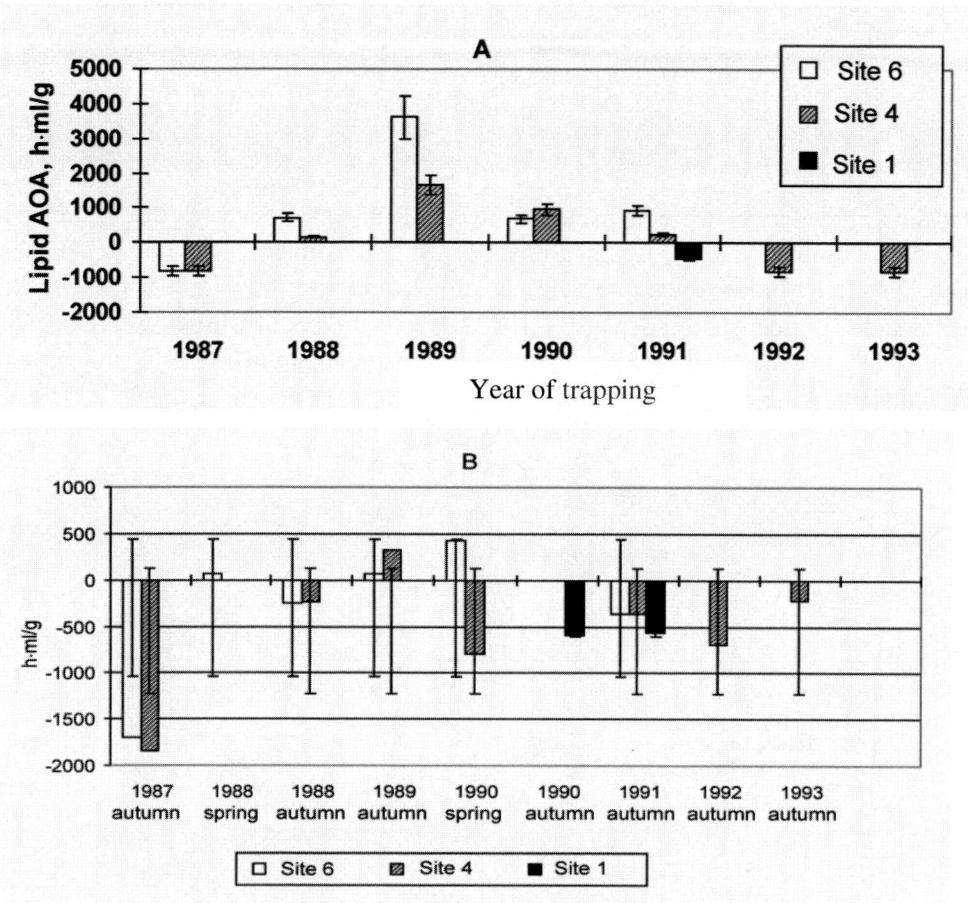

Figure 3. Dynamics of lipid AOA in liver (A) and brain (B) of immature root voles, caught in the Chernobyl accident zone at the areas with low (site 6), moderate (site 4) and high (site 1) contamination

Analysis of the data obtained allowed for a conclusion that the values of studied parameters were stipulated by the level of background γ-radiation at the trapping area, species, sex and age of caught animals, as well as showed an extreme change with respect to the capture season and time passed after the accident [31, 74]. For example, Figure 3 presents dynamics of liver and brain lipid AOA values of immature root voles, caught in the accident zone at the areas with low, moderate and high level of the background γ-radiation. In 1989, a considerable increase of lipid AOA in these organs was detected. However, times of occurrence of the lipid AOA maxima in spleen of the studied species of rodents do not correspond to times of their occurrence in other organs [31]. Weak AO properties were observed for the spleen lipids of striped field-mice, caught at the areas 2 and 3 in the summer 1988, and yellow-necked mice, caught in the autumn 1990 at the area 1. Extremely high AOA was shown by spleen lipids of the rodents, caught in autumn 1990, specifically striped field-mice and common voles from the areas 1 and 2, and root voles from the area 6. Low AOA of lipids was detected exclusively for spleen of root voles, caught at a highly contaminated area. The prooxidant activity of spleen lipids of the rodents from the accident zone also corresponds to high concentration of peroxides in them [31]. For example, peroxide concentration in the spleen lipids of root voles, caught in autumn 1991 at the area 4, was almost 20 times higher than the analogous parameter for the animals from the Kiev population. Only in autumn 1990, a weak APA was detected in spleen lipids of root voles, caught at the areas 4 and 6, striped field-mice from the area 2 and common voles from the area 1. It is obvious that the ability to decompose peroxides is one of the reasons for high lipid AOA of spleen in these groups of rodents. Considerable differences in dynamics of both generalized parameters of PL composition in the rodent organs with respect to the species, age and sex of animals and the trapping area (which is illustrated in Figures 4 and 5 on the example of the liver lipids of the root vole), and the quantity of total lipids per 1 g of liver in root voles in the period of 1987 – 1991 [31, 74]. Hence, higher sensitivity of the hepatocyte lipid composition to the effect of technogenic radioactive contamination compared with other tissues under study.

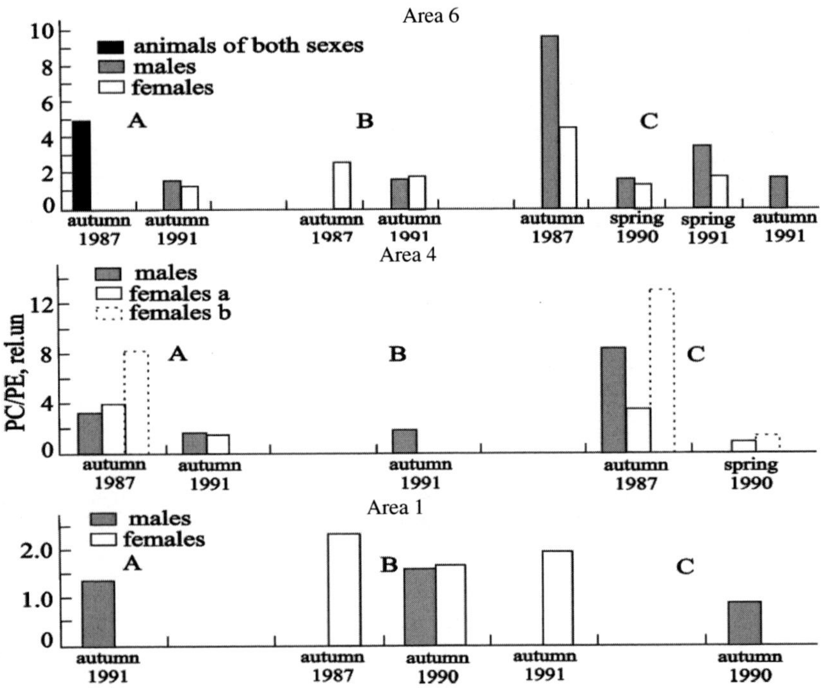

Figure 4. The change of PC/PE ratio in liver phospholipids of root voles, caught in the Chernobyl accident zone during 1987 – 1991: A – immature young of the year; B – mature young of the year; C – hibernated animals; a, b – groups of females

The absence of normalization of AO status parameters, the composition of lipids and activity of dehydrogenation enzymes, and their different sensitivity to long-term low-intensity radiation impact provide for the preservation of disorders in the functioning of physicochemical LPO regulation system and dehydrogenation processes in the organs of rodents during far remote period after the accident [15, 31] and significantly affect the formation of biological consequences of radioactive contamination of the environment. Considerable changes in the PL composition, the nonsaturation degree and provision of AO lipids promote an occurrence of a complex of structural and functional disorders in spleen of root voles and the multiplicity of types of degenerative changes in the liver parenchyma of the wild rodents from the accident zone [22, 23, 77 – 79].

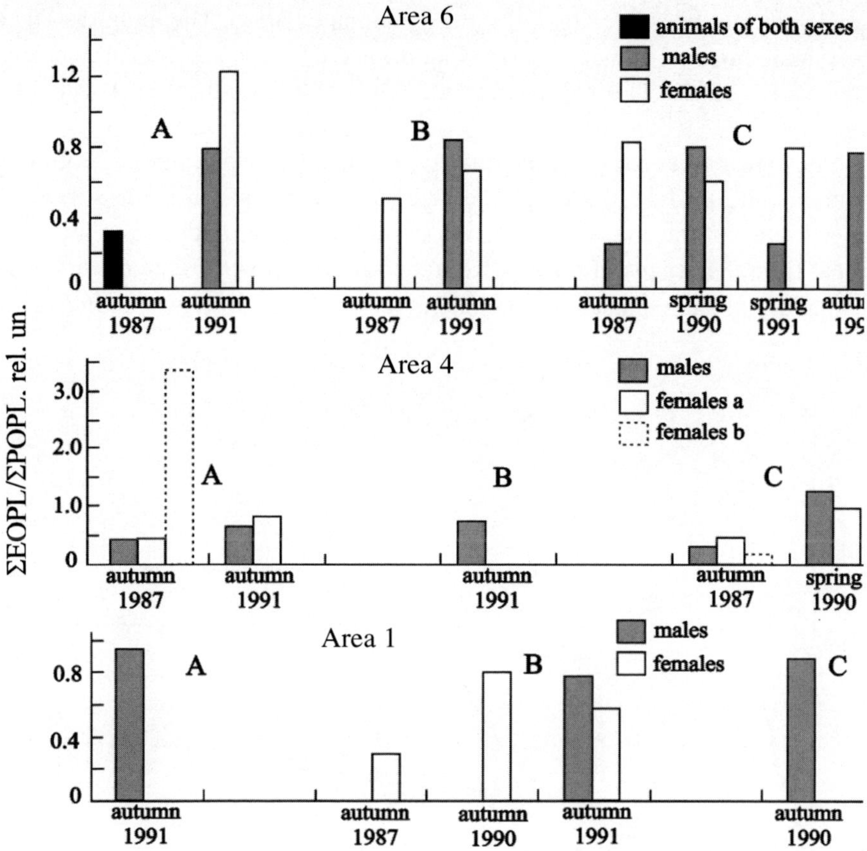

Figure 5. The relation between sums of the more easily to more poorly oxidizable fractions in liver phospholipids of root voles, caught in the Chernobyl accident zone during 1987 – 1991: A – immature young of the year; B – mature young of the year; C –hibernated animals; a, b – groups of females

The long-term low dose rate radiation impact also indicates the features of changes in the studied parameters in wild rodent tissues compared with the analogous parameters of animals inhabiting outside the accident zone. To our point of view, the following detected phenomena are of the greatest importance. Firstly, in many cases, a change in the order of organs ranking by lipid AOA values and dehydrogenation enzyme activity compared with the analogous sequences for the rodents, caught at the control areas, were found. For example, the organs of hibernated striped field-mice, caught at the areas 4 and 7, and bank voles, caught at the areas 5 and 6, are ranked by decreasing values of lipid AOA as follows:

brain >> liver = spleen.

Moreover, the lipid AOA values in brain were higher for the rodents, caught in 1987 at the areas with increased radioactive contamination level [31, 80]. This parameter changes differently and in broad ranges for different species of rodents, though brain is related to organs with the most stable biophysical and biochemical characteristics [50]. For example, in 1987, the value of brain lipid AOA increases with the dose rate of γ-radiation at the trapping areas by 1.5 times for striped field-mice and bank voles, and 1.85 times for yellow-necked mice compared with the rodents, caught at low contaminated areas. The highest LDH activity was also observed in brain of immature striped field-mice from the areas 5 and 6 and common voles from the area 1, caught in 1990, and striped field-mice from the area 1, caught in 1988 [31].

Secondly, in some cases, considerable changes in the quantitative ratio of both minor and basic PL fractions (PC and PE), kinetic characteristics and nonsaturation degree of lipids provide for occurrence of unusual kinetic curves of methyl oleate oxidation in the presence of lipids, extracted from the organs of rodents from the accident zone (Figure 6, curves 4 and 5). The relative quantity of such unusual oxidation curves increases 5 – 7 years after the accident [31].

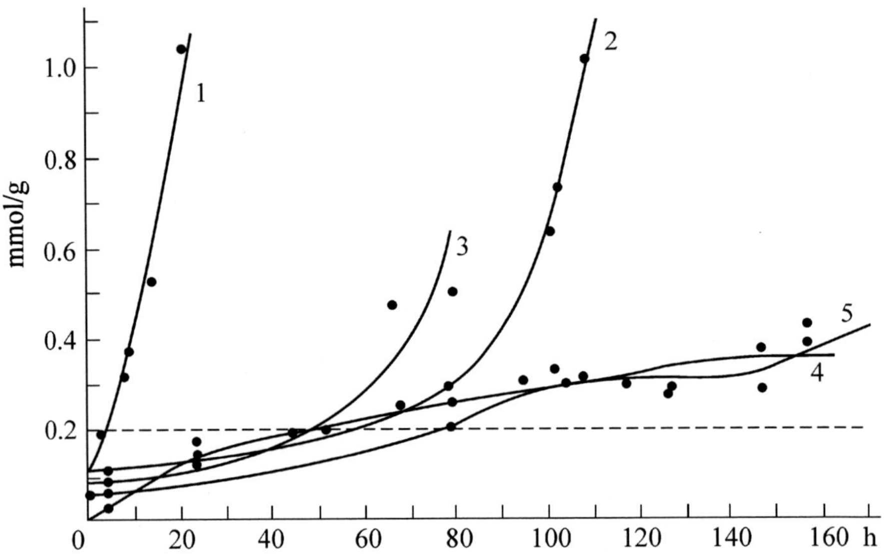

Figure 6. Kinetic curves of peroxide accumulation at thermal autooxidation of methyl oleate (1) and its solutions with liver lipids of rodents, caught in 1987 in the Chernobyl accident zone: 2 – bank voles from the area 3 (15 – 20 mR/h); 3 – common voles from the area 2 (15 – 20 mR/h); 4 – striped field-mice from the area 3 (15 – 20 mR/h); 5 – yellow-necked mice from the area 7 (0.04 – 0.07 mR/h)

Thirdly, a polymodal "biological effect – dose" dependence was determined, displayed in the absence of the monotonous decrease of the lipid AOA level, activity of dehydrogenation enzymes and the scale of the lipid exchange parameter variation with increasing dose rate of the external γ-radiation at the trapping areas of rodents (Figure 7). The most considerable changes of the majority of studied parameters were observed in the tissues of rodents, caught at the areas with the moderate radiation contamination level in the initial years after the accident. In the tissues of animals, caught at the areas with high level of radioactive contamination, these deviations are not always significant. Apparently, the increase of the absorbed dose and the dose rate of γ-radiation at the rodent trapping areas initiate the repair processes, the induction of antioxidant protection enzymes and affect the ratio of both DNA damaging and repair processes and PL biosynthesis and degradation rates [32, 81 – 83]. The absence of monotonous shifts in the organs of animals, caught in the exclusion zone of the Chernobyl accident or exposed to radionuclide-contaminated areas, with respect to the dose rate of external γ-radiation was indicated on both biochemical [25 – 27, 84 – 86] and cytomorphological and genetic levels [16, 18, 20 – 22, 77 – 79, 87]. A complex polymodal character of biochemical changes with respect to the level of radioactive contamination is preserved during several years after the accident. The absence of monotonous character and bimodal dependence of the changes in biophysical and biochemical parameters of cells and the organism at their irradiation by low doses of different intensity were also observed in multiple experimental tests.

Fourthly, it is found that the scale of biochemical parameter changes in the organs of animals from the Chernobyl accident exclusion zone is comparable or even exceeds the change scale of these parameters for the laboratory animals, exposed to sublethal and lethal radiation doses. Obviously, this is the reason for the histomorphological state of the root vole liver from the accident zone reminding the intensively irradiated tissue independently of the season [79], and in 1991 the hemopoietic system state of rodents from the 30-km exclusion zone of the Chernobyl accident being corresponded to the picture of a chronic radiation disease [7].

Fifthly, high heterogeneity of the studied parameters and, specifically, PL composition was found, which was most considerably expressed in spleen and liver of root voles, caught at a area with the moderate level of technogenic contamination [31, 80, 88]. Apparently, heterogeneity of the population of rodents provides for the variability of response to a weak effect, which scale significantly depends on the values of corresponding initial parameters of the animal tissues [31 – 33, 89, 90]. However, it should be taken into account that realization of the damaging factors impacting strictly depends on the physicochemical stage of the agent interaction with the system. Hence, the ionizing radiation impact on the biological systems modifies parameters as follows: inhomogeneous primary distribution of energy between molecules, high local density of the energy, approaching and redistribution of characteristic times of physical, physicochemical and chemical processes, the leading role of highly reactive particles, the concentration of which in the condensed phase is by many orders of magnitude higher than the equilibrium one [91].

The features of chronic impact of the low dose rate radiation at low-dose detected may be stipulated by total action of external and internal radiation of organs by incorporated radionuclides or a synergic effect of β- and γ-radiation on rodents [92]. Hence, the maximal probability of synergic effect of various physical and chemical factors is observed directly in the range of low doses and intensities [93 – 96]. The radiation quality should also be taken into account, because it defines both composition and spatial distribution of active particles, and the quantitative response of the cells and tissues to the impact [91, 97, 98], as well as higher vulnerability of membranes exposed to low-dose low-intensity radiation impact, when damages of the membrane components and functional tests are much more considerable [58, 81, 99 – 102].

The complex analysis of the metabolism regulation system state in tissues of rodents, investigated from the Chernobyl accident zone, allowed for a conclusion that root voles can be used as the test-objects for bioindication of the technogenic radioactive contamination of the natural biocoenoses. Previously [3, 23, 44], this species of rodents has been suggested as a test-object for the estimation of biocoenoses state with increased natural background radiation.

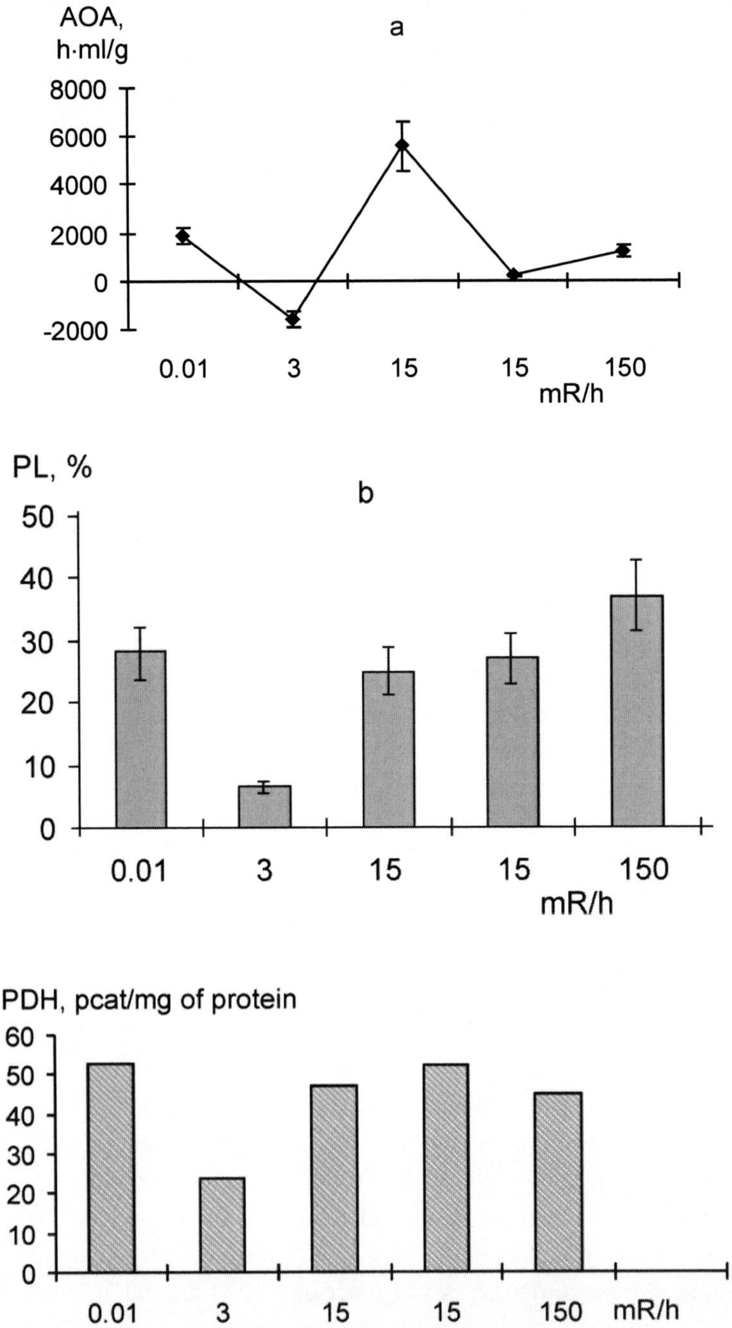

Figure 7. Polymodal character of changes in LPO regulation system parameters and functional tests in the liver of immature striped field-mice from the accident zone with respect to γ-radiation dose rate at the trapping area (1987): a – lipid AOA, h·ml/g; b – content of phospholipids in the composition of total lipids, %; c – pyruvate dehydrogenase activity, pcat/mg of protein

CONCLUSION

The populations of animals represent complex self-regulating systems of the open type, capable of adaptation to anthropogenic contamination of the environment [103, 104]. Such populations are significantly different from those inhabiting on the territories with natural background; however, in this case, the mammals should be selected during 20 – 30 generations [105]. Despite the forecasted ability of the Nature to self preservation and purification via the mechanisms of global redistribution [106], systematic investigations of the Muridae family rodent population state, performed during several years after the Chernobyl accident, indicate formation of qualitatively new subpopulations on the contaminated territories. It is unambiguous that already in the first year after the accident the rodents were examined, which had passed through an original filter of the natural selection and representing the most viable species exposed to permanent radiation impact. The unequal sensitivity to chronic low-intensity radiation and possibility of normalization of the LPO regulation system parameters and dehydrogenation processes in organs and tissues of the rodents induces a break of interrelations between parameters, coordinated in norm. This is confirmed by the comprehensive analysis of the correlations between parameters under study, which allowed for determination of organ and species differences of biological consequences for rodents with respect to both the level of radioactive contamination and the sex and age of the animals [31, 80, 88].

As shown by the laboratory tests, the relation between lipid AOA and their effect as a inhibition factor on the model substrate (methyl oleate) oxidation under the combined action of low-dose X-ray radiation and SAS [61], whereas under the effect of acute radiation, primary disorders in the physicochemical system of LPO regulation is observed in the AOA – lipid composition unit [59]. The lipids in organs of wild rodents have also demonstrated direct correlation between AOA and extent of the oxidation inhibition of methyl oleate by them (Figure 8a), the linear regression coefficient of which decreases with the increase of the radical initiation rate in the system. A linear dependence between these parameters is also preserved in the lipids of organs of the rodents from the accident zone. However, linear regression coefficients (b) e.g. the interrelation scale are stipulated not only by the intensity of oxidative processes, but also by the animal species, characteristics of lipids and time passed after the accident [31, 88]. For example, in 1987 the reliable differences in b values at equal rates of the radical initiaition in the oxidation process were detected for the liver lipids of striped field-mice ($b = 57.23 \pm 0.14$) and root voles ($b = 55.83 \pm 0.06$), whereas the trapping area, age and sex of the animals were practically insignificant for this parameter [31]. However, for the liver lipids of striped field-mice, caught in autumn 1990, the linear regression coefficient of the relation between lipid AOA and their oxidizability decreases with the increasing dose rate of γ-radiation at the trapping areas (Figure 8b) at an abrupt decrease of the correlation coefficient between lipid AOA and EOI for rodents from the area 1 (Figure 8b, line 2: $R = 0.79 \pm 0.19$). Moreover, the values of b for the liver lipids of striped field-mice, caught in the accident zone in 1990, were reliably below the analogous values for rodents from the Kiev population ($b = 8.446 \pm 0.007$, Figure 8a) or the animals, caught in 1987 in the accident zone (Figure 8b, line 1). In 1987, the interrelation between brain lipid EOI and AOA of the rodents was practically independent of the species and age of animals or the level of radioactive contamination. However, in 1990ies a decrease of parameter b value in rodents from highly contaminated areas was observed, the most clearly determined for radiosensitive species (root voles and striped field-mice) [31, 88].

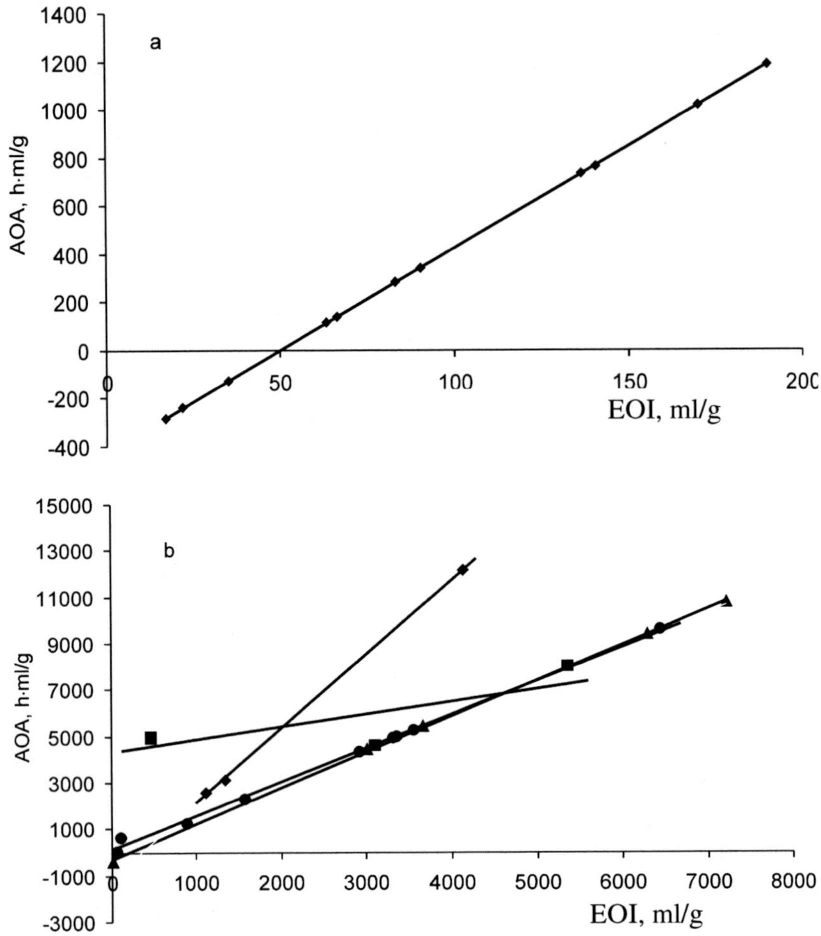

Figure 8. The interrelation between lipid AOA in liver of striped field-mice and the methyl oleate oxidation inhibition extent by lipids: a – immature and mature rodents of both sexes, caught in the Kiev neighborhood in autumn 1991; $W_0 = 3.28\times10^{-10}$ M/s; b – immature animals, caught at the area 3 in autumn 1987 (1), $W_0 = 3.47\times10^{-10}$ M/s; immature and mature animals, caught in autumn 1990 at the areas 1 (2), 5 and 6 (3), 2 and 3 (4), $W_0 = (3.28 \pm 0.11)\times10^{-10}$ M/s

In the first year after the accident, a reliable correlation between AO status parameters (lipid AOA, EOI, APA) and the relative concentration of some PL fractions or generalized indices of PL composition in the liver lipids of root voles, caught at low contaminated areas, is observed for the lipid AOA level and the fraction of PL lysoforms only, whereas for root voles caught at the area 4 – between lipid AOA and PL content among total lipids or the ratio ΣEOPL/ΣPOPL [31]. In 1991, the liver lipids of root voles, sampled from the same sites, demonstrated recovery of the interrelation between AO status parameters and the composition of lipids not by all parameters [31]. Similar situation was observed in the correlation analysis of the brain parameters of these rodents [31, 88]. The absence of interrelated changes in AO properties and the lipid composition in liver of mature striped field-mice, caught in autumn 1993 at the areas with different contamination levels, is also confirmed by data presented in Figure 9. The animals captured simultaneously in the neighborhood of the Kiev City demonstrated a linear correlation between the lipid AOA and PC/PE ratio and the reverse dependence between the lipid AOA and ΣEOPL/ΣPOPL.

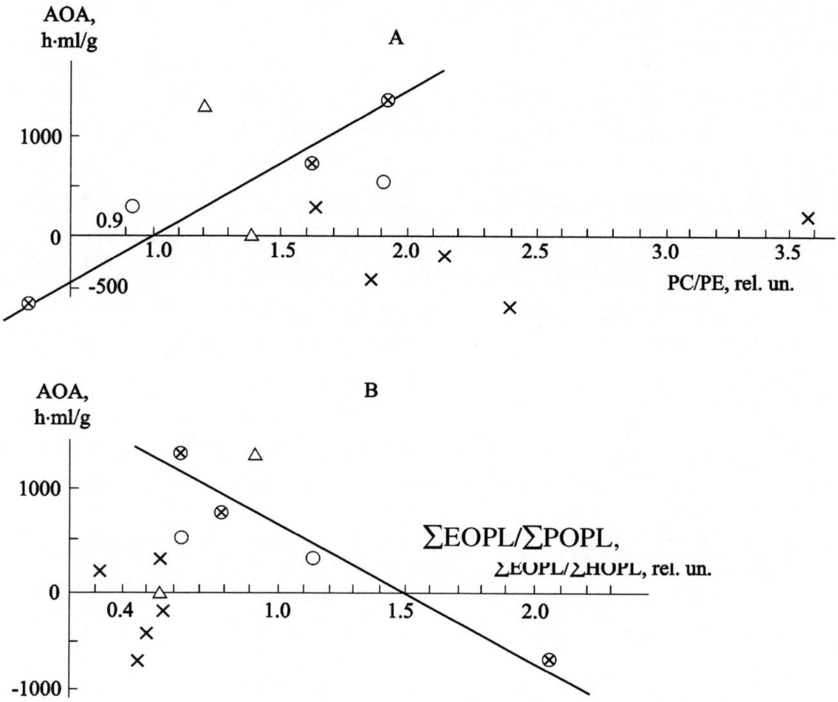

Figure 9. A relation between lipid AOA and PC/PE ratio (A) or the ratio of sums of the more easily and the more poorly oxidizable fractions (B) in liver phospholipids of striped field-mice, caught in autumn 1993 in the Chernobyl accident zone and the Kiev neighborhood

Different sensitivity of the lipid oxidability and the structural state of the membrane system of an organ to the technogenic contamination is also testified about by the change in the scale of interrelation between the ratio of sums of the more easily and more poorly oxidizable fractions and PC/PE ratio in the liver and brain PL of the rodents, caught in the accident zone. In 1987, the reverse correlations between these parameters were observed in the liver PL of hibernated root voles (R = 0.998 ± 0.002; b = –0.031 ± 0.001) [40] and immature males of root voles (R = 0.75 ± 0.18; b = – 0.270 ± 0.084) from the area 4, common voles of both sexes from the area 2 (R = 0.97 ± 0.03; b = –0.133 ± 0.016), and yellow-necked mice of both sexes from the area 7 (R = 0.91 ± 0.10; b = –0.116 ± 0.038). At the same time, hibernated root voles, caught at low contaminated sites, demonstrated a direct correlation between these parameters in the liver PL, detected in 1987 (R = 0.83 ± 0.16; b = 0.15 ± 0.07). Hence, in 1991, the reverse correlation was observed (R = 0.996 ± 0.005; b = –0.31 ± 0.02).In the brain PL of root voles, caught in 1991 at a area with the moderate contamination level, the interrelation between oxidizability of lipids and the structural state of the membrane system of the organ is absent in the presence of reverse correlation between these parameters in animals, caught simultaneously at a area with low contamination level. Generally, the root voles caught at low contaminated areas demonstrated deeper disorders of the lipid exchange regulation in liver, whereas the same disorders in brain were observed for the animals, caught at the area with the moderate radioactive contamination level.

Dehydrogenation imbalance in the tissues of rodents, trapped in the accident zone, also testifies about disorders in the relations between activities of dehydrogenases, tightly associated in norm [31, 80]. Hence, root voles from the accident zone demonstrated much higher significance of both activation and suppression of dehysrogenase activation than the animals from the Komi population, which for decades inhabited on the territory with increased concentration of natural heavy radionuclides [31, 32]. Normalization of the interrelations between lipid AO status and functional tests in liver and cerebrum of root voles from the accident zone was not also observed. Even at equal scale and direction of the lipid AOA change the character of activity change for hydrogenases in the organs of animals depends on both the species of rodents and the trapping area and on the origin of the enzyme and the organ, in which enzymatic activity was analyzed [80]. Hence, four years after the accident biochemical functions in liver higher imbalanced, and coordination between lipid AOA, APA and EOI and dehydrogenation enzyme activity is not fully restored [31, 88].

Table 3. The part of Muridae family rodents from the 30-km zone of the Chernobyl accident with low antioxidant activity of the organ lipids

Species and age of rodents	Part of animals with low lipid AOA	
	Liver	Brain
Bank voles of all ages	Decreases	Increases
Mature and hibernated root voles	Decreases	Increases
Striped field-mice of all ages	Increases	Increases

Table 4. Variability of organ lipid antioxidant activity for Muridae family rodent from the 30-km zone of the Chernobyl accident

Species	Lipid AOA variability	
	liver	brain
Root voles	Increases	Increases
Bank voles 1991, areas 5, 6 1993, areas 1, 2	 20-fold increased decreased	 2-fold decreased increased
Mature striped field-mice	Decreased	Appr. normal
Yellow-necked mice	Antibate changes in liver and brain; the change direction is defined by the year of trapping	

This suggestion is also confirmed by results of the rodent natural population state analysis for the accident zone using biometric methods. It is found that radioactive contamination in the accident zone causes a considerable effect on the type of rodent distribution by liver and brain lipid AOA values [31, 63]. Despite the fact that the type of distribution and variability of the lipid AOA are organ and species specific parameters and depend on the age of animals, the contamination level of the trapping area and the period of the radiation impacting, definite regularities generalized in Tables 3 and 4 are observed. Moreover, for higher radioresistant bank voles, the type of distribution of the animals from the accident zone by brain lipid AOA values is changed in later periods after the accident compared with higher radiosensitive striped field-mice and root voles. The totality of biometric analysis data allows for a conclusion that radioactive contamination in the accident zone induces qualitative changes in the state of the natural populations of Muridae family rodents [63]. This is confirmed by the change of the most abundant species of rodents in the 30-km exclusion zone of the Chernobyl accident, observed in the period of 1986 – 1996 [107], the increasing variability of signs and radioresistance of bank voles in biocoenoses of Mogilev Region, exposed to chronic ionizing radiation [108], and biological changes in the populations of rodents inhabiting on the radionuclide-contaminated territories of Bryansk Region [11, 14, 109 – 111]. The existence of qualitatively different subpopulations of root voles, which for decades inhabited on the territories of Komi Republic, contaminated by natural heavy radionuclides, was experimentally confirmed [32, 44].

High sensitivity of the LPO regulation system parameters stipulating the increasing role of membranes as a coordinator of oxidative reaction regulation in the field of low effects and dehydrogenation processes proceeding in the wild rodent organs to chronic impact of the low-dose low-intensity radiation, as well as a long-term preservation of the changed scale and direction of interrelations between them give an opportunity to use them for testing biological consequences of radioactive contamination of the biota. It should be noted that in the case of radioactive contamination of the environment the depth of the lipid exchange disorders, tissue supply with energy [31, 32, 80, 112] and genetic consequences for natural populations of rodents [16, 20, 21, 32, 113 – 115] were found much more significant than observed in laboratory tests with animal exposure to chronic external γ-radiation. This requires careful extrapolation of the experimental results to forecasting of the biological consequences the technogenic contamination of biota.

Despite the complexity of interpretation of the results obtained, the totality of self data and analysis of data from the literature allow for a conclusion that adaptation of rodents in the 30-km exclusion zone of the Chernobyl accident, stipulated by transition of the cellular regulation systems to a new functioning level, is aimed at the survival of the population and supporting homeostasis under changed radioecological conditions. Obviously, this is the reason for preservation of the biodiversity, abundance and the age structure of the populations of bank vole, which is the most radioresistant species, during initial eight years of inhabitance in the areas of Bryansk Region, contaminated by radionuclides [116], as well as mobilization of compensatory processes in the liver of root voles from the 30-km exclusion zone of the Chernobyl accident 3 – 5 years after [79]. Basing on a supposition that nonlinearity of the dose dependence is the mathematical presentation of the cell ability to choose between "passive" and "active" strategies of adaptation with respect to intensity of the external impact [117], one may suggest that the change of expressivity of interrelation between parameters of physicochemical LPO regulation system is the information signal determining the selection of this strategy.

REFERENCES

[1] Sokolov V.E., Krivolutskyi D.A., and Usachev V.L., *Wild Animals in the Global Radioecological Monitoring*, Moscow, Nauka, 1989, 150 p. (Rus)

[2] Krivolutskyi D.A., *The Soil Fauna in the Ecological Control*, Moscow, Nauka, 1994, 270 p. (Rus)

[3] Maslov V.I., *Radioecological Investigations in Natural Biocoenoses*, Moscow, 1972, pp. 216 – 226. (Rus)

[4] Isaev S.I. and SHilova S.A., *Bioindication of Radioactive Contaminations*, Ed. D.A. Krivolutskyi, Moscow, Nauka, 1999, pp. 36 – 41. (Rus)

[5] Taskaev A.I., Testov B.V., Materiy L.D., and Shevchenko V.A., Environmental and morphophysiological consequences of the accident at the Chernobyl NPP for the populations of Muridae family rodents, *Ser. Nauchnye Doklady Komi Nauch. Tsentr UrO AN* SSSR, Syktyvkar, 1988, 56 p. (Rus)

[6] Mascanzoni D., von Bothmer S., Mattei T., and Cristaldi M., *The Science of the Total Environment*, 1990, vol. 99, pp. 61 – 66.

[7] Krivolutskyi D.A., Martyushov V.Z., and Ryabtsev I.A., *Bioindication of Radioactive Contaminations*, Ed. D.A. Krivolutskyi, Moscow, Nauka, 1999, pp. 106 – 122. (Rus)

[8] Taskaev A.I. and Testov B.V., *Bioindication of Radioactive Contaminations*, Ed. D.A. Krivolutskyi, Moscow, Nauka, 1999, pp. 200 – 205. (Rus)

[9] Rozhdestvenskaya A.S., *Bioindication of Radioactive Contaminations*, Ed. D.A. Krivolutskyi, Moscow, Nauka, 1999, pp. 226 – 231. (Rus)

[10] Frantsevitch L.I., Komissar A.D., Ermakov A.A. *et al.*, *Chernobyl-88. Rep. I All-Union Sci.-Tech. Conf. on the Results of Liquidation of the Chernobyl Accident Consequences*, Chernobyl, 1989, vol. 3, Part 2., pp. 110 – 123. (Rus)

[11] Dolgov V.A., Krylova T.V., Tsiperson V.P. *et al.*, *Biol. Nauki*, 1992, No. 11-12 (347), pp. 127 – 133. (Rus)

[12] Goncharova R.I. and Ryabokon' N.I., *Radiats. Biol. Radioekol.*, 1998, vol. 38(5), pp. 737 – 745. (Rus)

[13] Testov B.V., "The influence of radioactive contamination on the population of Muridae family rodents", *Doctor Dissertation Thesis*, Ekaterinburg, 1993, 35 p. (Rus)

[14] Tsiperson V.P. and Soloviev M.Y., *The Science of the Total Environment*, 1997, vol. 203, pp. 105 – 113.

[15] Kudyasheva A.G., Bashlykova L.A., Shevchenko O.G. *et al.*, "Antropogenno-zminene seredovishche: Riziki dlya zdorov'ya naselennya ta ekologichnikh system", *Rep. Intern. Conf.*, Kiev, 2003, Part III, pp. 21 – 31. (Ukr)

[16] Pomerantseva M.D., Ramaia L.K., Testov B.V. *et al.*, *Radiobiologia*, 1990, vol. 30(4), pp. 441 – 445. (Rus)

[17] Cristaldi M., Arcangelo E.D., Ieradi L.A. *et al.* // *Environ. Pollution*, 1990, vol. 64, pp. 1 – 9.

[18] Bashlykova L.A., *Ecological Consequences of Radioactive Contamination of the Environment*, Syktyvkar, 1991, pp. 58 – 64. (Rus)

[19] Goncharova R.I. and Ryabokon N.I., *Radiat. Prot. Dosim.*, 1995, vol. 62(1/2), pp. 37 – 40.

[20] Rakin A.O. and Bashlykova L.A., *The Effect of Radioactive Contamination on Surface Ecosystems in the Chernobyl Accident Zone (1986 – 1996)*, Syktyvkar, 1996, vol. 1, pp. 113 – 122. (Rus)
[21] Goncharova R.I. and Ryabokon' N.I., *Radiats. Biol. Radioekol.*, 1998, vol. 38(5), pp. 746 – 753. (Rus)
[22] Materiy L.D. and Ermakova O.V., "Histomorphological criteria of estimation of environmental radioactive contamination (The impact in animals in the Chernobyl NPP zone), *Ser. Nauchnye Doklady Komi Nauchn. Tsentr UrO RAN*, Syktyvkar, 1993, No. 312, 24 p. (Rus)
[23] Maslova K.I., Materiy L.D., Ermakova O.V., and Taskaev A.I., *The Atlas of Pathomorphological Changes in Root Voles from Local Radioactive Contamination Foci*, St. Petersburg, Nauka, 1994, 192 p. (Rus)
[24] Materiy L.D., Ermakova O.V., and Taskaev A.I., *Morphofunctional Estimation, of the State of Small Mammal Organisms in Radioecological Investigations (on the Example of Root Voles)*, Syktyvkar, 2003, 164 p. (Rus)
[25] Graevskaya B.M., Zolotareva N.N., and Elfimova S.S., *Thes. Rep. All-Union Radiobiol. Congress*, Pushchino, 1989, vol. 5, pp. 1025 – 1026. (Rus)
[26] Zolotareva N.N., *Thes. Rep. I Intern. Conference "Biological and Radioecological Aspects of the Chernobyl Accident Consequences"*, Moscow, 1990, p. 76. (Rus)
[27] Elfimova S.S., In Coll.: *Thes. Intern. Conference "The Fifteenth Aniversary of the Chernobyl Catastrophe. The Experience of Overcoming"*, Kiev, 2001, pp. 21 – 22. (Rus)
[28] Burlakova E.B., *Vestnik RAN*, 1994, vol. 64(5), pp. 425 – 431. (Rus)
[29] Kolomiytseva I.K., Slozhenikina L.V., Fialkovskaya L.A. *et al.*, *J. Biol. Phys.*, 1999, vol. 25, pp. 325 – 338.
[30] Burlakova E.B., *Ros. Khim. Zh.*, 1999, vol. 43(5), pp. 3 – 11. (Rus)
[31] Kudyasheva A.G., Shishkina L.N., Zagorskaya N.G., and Taskaev A.I., *Biochemical Mechanisms of the Radiation Damage of Muridae Family Rodent Natural Populations*, St. Petersburg, Nauka, 1997, 156 p. (Rus)
[32] Kudyasheva A.G., Shishkina L.N., Shevchenko O.G., Bashlykova L.A., and Zagorskaya N.G., *Biological Effects of Radioactive Contamination in Natural Populations of Muridae Family Rodents*, Ekaterinburg, 2003, 224 p. (Rus)
[33] Shishkina L.N., Kudyasheva A.G., Zagorskaya N.G., and Shevchenko O.G., *Radiats. Biol. Radioekol.*, 2005, vol. 45(4). (Rus)
[34] Vladimirov Yu.A. and Archakov A.I., *Peroxide Oxidation of Lipids in Biological Membranes*, Moscow, Nauka, 1972, 252 p. (Rus)
[35] Burlakova E.B., Alesenko A.V., Molochkina E.M., Pal'mina N.P., and Khrapova N.G., *Bioantioxidants in the Radiation Damage and Cancer Development*, Moscow, Nauka, 1975, 214 p. (Rus)
[36] Baraboi V.A., Orel V.E., and Karnaukh I.M., *Peroxide Oxidation and Radiation*, Kiev, Naukova Dumka, 1991, 256 p. (Rus)
[37] *Membrane Lipid Oxidation*, Ed. C. Vigo-Pelfrey, Boca Raton, Ann Arbor. Boston: CRC Press, 1991, vol. III, 300 p.
[38] Burlakova E.B., Gorban' N.I., Dzantiev B.G. *et al.*, *Izvestiya VUZov. Khimia i Khim. Tekhnol.*, 1961, No. 5, pp. 751 – 754. (Rus)

[39] Petkau A. and Chelack W.S., *Biochim. Biophys. Acta*, 1976, vol. 433(3), pp. 445 – 456.
[40] Kudyashova A.G., Shishkina L.N., Zagorskaya N.G. *et al., Radiats. Biol. Radioekol.*, 2000, vol. 40(3), pp. 327 – 333. (Rus)
[41] "The information about the accident at the Chernobyl NPP and its consequences, prepared for IAEA", *Atom. Energia*, 1986, vol. 61(5), pp. 301 – 320. (Rus)
[42] Meshalkin G.S., Arkhipov N.P., Akhipov A.N. *et al., Chernobyl-88. Rep. I All-Union Sci.-Tech. Conf. on the Results of Liquidation of the Chernobyl Accident Consequences*, Chernobyl, 1989, vol. 5, Part 1, pp. 34 – 42. (Rus)
[43] Kozubov G.M. and Taskaev A.I., *Radiobiological and Radioecological Investigations of Woody Plants*, St. Petersburg, Nauka, 1994, 256 p. (Rus)
[44] Aleksakhin R.M., Arkhipov N.P., Bardukharov R.M. *et al., Natural Heavy Radionuclides in the Biosphere: Migration ND Biological Effect on Populations and Biocoenoses*, Moscow, Nouka, 1999, 368 p. (Rus)
[45] Zagorskaya N.G., Kudyasheva A.G., and Bashlykova L.A., *Radioecological and Biological Consequences of Low Intensity Impacts*, Syktyvkar, 2003, pp. 19 – 34. (Rus)
[46] *Bulletin of the Ecological State of the Restriction Zone in the First Half of 1995*, Issue Ed. N.P. Arkhipov, Chernobyl, 1995, 40 p. (Rus)
[47] Burlakova E.B. and Khrapova N.G., *Uspekhi Khimii*, 1985, vol. 54(9), pp. 1540 – 1558. (Rus)
[48] Frankel E.N., *Prog. Lipid Res.*, 1980, vol. 19, pp. 1 – 22.
[49] *Oxidative Stress*, Ed. H. Sies, L.: Acad. Press, 1985, 507 p.
[50] Kreps E.M., *Lipids of Biological Membranes*, Leningrad, Nauka, 1981, 340 p. (Rus)
[51] Antonov V.F., Smirnova E.Yu., and Shevchenko E.V., *Lipid Membranes in Phase Transformations*, Moscow, Nauka, 1992, 136 p. (Rus)
[52] Ramos J.L., Gallegos M.T., Godoy P. *et al., Annu. Rev. Microbiol.*, 2002, vol. 56, pp. 743 – 768.
[53] Wayner D.D.M., Burton G.W., Ingolg K.U. *et al., Biochim. Biophys. Acta*, 1987, vol. 924, pp. 408 – 419.
[54] Cille J.J. and Joengje H., *Membrane Lipid Oxidation*, Ed. C. Vigo-Pelfrey, Boca Raton, Ann Arbor. Boston: CRC Press, 1991, vol. III, p. 1 – 32.
[55] Kenya M.B., Lukash A.I., and Gus'kov E.P., *Uspekhi Sovrem. Biologii*, 1993, vol. 113(4), pp. 456 – 470. (Rus)
[56] Men'shchikova E.B. and Zenkov N.K., *Uspekhi Sovrem. Biologii*, 1993, vol. 113(4), pp. 442 – 455. (Rus)
[57] Zenkov N.K., Lankin V.Z., and Men'shchikova E.B., *Oxidative Stress: Biochemical and Pathophysiological Aspects*, Moscow MAIK Nauka/Interperiodika, 2001, 340 p. (Rus)
[58] Shishkina L.N., Kushnireva E.V., and Smotryaeva M.A., *Radiats. Biol. Radioekol.*, 2004, vol. 44(3), pp. 289 – 295. (Rus)
[59] Burlakova Ye.B., Pal'mina N.P., and Mal'tseva Ye.L., *Membrane Lipid Oxidation*, Ed. C. Vigo-Pelfrey. Boca Raton. Ann Arbor. Boston: CRC Press, 1991, vol. IV, pp. 209 – 237.
[60] Marcon J.L. and Filho D.W., *Comp. Biochem. Physiol. Part C*, 1999, vol. 123, pp. 257 – 263.

[61] Shishkina L.N., Kushnireva E.V., and Smotryaeva M.A., *Radiats. Biol. Radioekol.*, 2001, vol. 41(3), pp. 301 – 306. (Rus)

[62] Urnyisheva V.V., Gulyaeva O.N., Kushnireva E.V., and Shishkina L.N., *Radiats. Biol. Radioekol.*, 2002, vol. 42(5), pp. 481 – 487. (Rus)

[63] Shishkina L.N., Kudyasheva A.G., Zagorskaya N.G., and Taskaev A.I., *Radiats. Biol. Radioekol.*, 1998, vol. 38(6), pp. 924 – 935. (Rus)

[64] Shishkina L.N., Men'shov V.A., and Brin E.F., *Izv RAN, Ser. Biol.*, 1996, No. 3, pp. 292 – 297. (Rus)

[65] Khrustova N.V. and Shishkina L.N., *Kinetika i Kataliz*, 2004, vol. 45(6), pp. 848 – 858. (Rus)

[66] Shevchenko O.G., Zagorskaya N.G., Kudyasheva A.G., and Shishkina L.N., *Ekologia*, 1999, No. 1, pp. 73 – 76. (Rus)

[67] Burlakova E.B., Ivanenko G.F., and Shishkina L.N, *Izv. AN SSSR, Ser. Biol.*, 1985, No. 4, pp. 588 – 592. (Rus)

[68] Graevskaya B.M., *Radiobiologia*, 1972, vol. 12(3), pp. 323 – 335. (Rus)

[69] Il'enko A.I., Mozhekaite R.B., Nizhnik G.B., and Ryabtsev I.A., *Radiobiologia*, 1977, vol. 17(4), pp. 545 – 549. (Rus)

[70] Maslova K.I., *Radiosensitivity of Plants and Animals of Biocoenoses with Increased Natural Background Radiation*, Syktyvkar, 1988, pp. 5 – 15. (Rus)

[71] Kudyasheva A.G., Shishkina L.N., and Zagorskaya N.G., *Ekologia*, 1994, No. 6, pp. 48 – 53. (Rus)

[72] Shevchenko O.G., Shishkina L.N., and Kudyasheva A.G., *Zh. Evol. Biokhim. Fiziol.*, 2002, vol. 38(2), pp. 131 – 135. (Rus)

[73] Shvets V.I., Stepanov A.I., Krylova V.N., and Gulak P.V., *Myo-inositol and Phosphoinositydes*, Moscow, Nauka, 1987, 248 p. (Rus)

[74] Zagorskaya N.G., Kudyasheva A.G., and Shishkina L.N., *Radioecological Monitoring of Natural Ecosystems*, Syktyvkar, 1993, pp. 45 – 53. (Rus)

[75] Brandt R.B., Laux J.B., Spainhour S.E., and Kline E.S., *Arch. Biochem. Biophys.*, 1987, vol. 259(2), pp. 412 – 422.

[76] Cherkasova L.S., Pikulev A.T., and Taits M.Yu., *Metabolic Shifts in Mitochondria of Irradiated Organism, Related to the Krebs Cycle*, Minsk, Nauka i Tekhnika, 1977, 152 p. (Rus)

[77] Shishkina L.N., Materiy L.D., Kudyasheva A.G. *et al.*, *Radiobiologia*, 1992, vol. 32(1), pp. 19 – 29. (Rus)

[78] Materiy L.D., *The Impact of Radioactive Contamination on the Surface Ecosystems in the Accident Zone at Chernobyl NPP (1986 – 1996)*, Syktyvkar, 1996, vol. 1, pp. 12 – 40. (Rus)

[79] Materiy L.D. and Goncharov M.I., *The Impact of Radioactive Contamination on the Surface Ecosystems in the Accident Zone at Chernobyl NPP (1986 – 1996)*, Syktyvkar, 1996, vol. 1, pp. 41 – 57. (Rus)

[80] Kudyasheva A.G., Shishkina L.N., Zagorskaya N.G., and Taskaev A.I., "The influence of industry-induced contamination on regulatory systems of the cell", *Ser. Nauch. Dokl. Komi Nauch. Tsentr UrO AN SSSR*, 1990, No. 248, 40 p. (Rus)

[81] Smotryaeva M.A., Kruglyakova K.E., Shishkina L.N. *et al.*, *Radiats. Biol. Radioekol.*, 1996, vol. 36(1), pp. 21 – 29. (Rus)

[82] Burlakova E.B., Goloshchapov A.N., Zhizhina G.P., and Konradov A.A., *Radiats. Biol. Radioekol.*, 1999, vol. 39(1), pp. 26 – 34. (Rus)

[83] Kolomiytseva I.K., Markevich L.N., Vakulova L.A., and Kuzin A.M., *Doklady RAN*, 1998, vol. 363(4), pp. 549 – 551. (Rus)

[84] Mazhul' L.M., Volykhina V.E., and Gatsko G.G., *Thes. Rep. I Intern. Conference "Biological and Radioecological Aspects of the Accident Consequences at the Chernobyl Nuclear Power Plant"*, Moscow, 1990, p. 189. (Rus)

[85] Serkiz Ya.I., Pinchuk V.G., Pinchuk L.B. *et al.*, *Radiobiological Aspects of the Chernobyl Accident*, Kiev, Naukova Dumka, 1992, 172 p. (Rus)

[86] *The Chernobyl Caatastrophe*, Ed. V.G. Bar'yakhtar, Kiev, Naukova Dumka, 1995, p. 559. (Rus)

[87] Zainullin V.G., Rakin A.O., and Taskaev A.I., *Radiats. Biol. Radioekol.*, 1994, vol. 34(6), pp. 852 – 857. (Rus)

[88] Shishkina L.N., Kudyasheva A.G., and Zagorskaya N.G., *The Impact of Radioactive Contamination on the Surface Ecosystems in the Accident Zone at Chernobyl NPP (1986 – 1996)*, Syktyvkar, 1996, vol. 1, pp. 77 – 100. (Rus)

[89] Treshchenkova Yu.A. and Burlakova E.B., *Radiats. Biol. Radioekol.*, 1997, vol. 37(1), pp. 3 – 12. (Rus)

[90] Shishkina L.N., Kushnireva E.V., Bespal'ko O.F., and Polyakova N.V., *Radiats. Biol. Radioekol.*, 2000, vol. 40(2), pp. 162 – 167. (Rus)

[91] Bugaenko L.T., *Ros. Khim. Zh.*, 2000, vol. 44(4), pp. 40 – 43. (Rus)

[92] Harwell M. and Hatchinson T., *Consequences of a Nuclear War. The Influence on Ecology and Agriculture*, Moscow, Mir, 1988, 551 p. (Rus)

[93] Petin V.G. and Komarov V.P., *Quantitative Description of Radiosensitivity Modification*, Moscow, Energoatomizdat, 1989, 192 p. (Rus)

[94] Malenchenko A.F., Sushko S.N., Plotnikova S.I., and Savchenko V.K., *Doklady AN BSSR*, 1989, vol. 33(11), pp. 1043 – 1045. (Rus)

[95] Malenchenko A.F., Sushko S.N., and Kuz'mina T.S., *Radiats. Biol. Radioekol.*, 1995, vol. 35(5), pp. 777 – 785. (Rus)

[96] Petin V.G., Zhurakovskaya G.P., Pantyukhina A.G., and Rassokhina A.V., *Radiats. Biol. Radioekol.*, 1999, vol. 39(1), pp. 113 – 126. (Rus)

[97] Ono K., Nagata Y., Akuta K. *et al.*, *Radiat. Res.*, 1990, vol. 123, pp. 345 – 347.

[98] Shishkina L.N. and Burlakova E.B., *Khim. Fizika*, 1996, vol. 15(1), pp. 43 – 53. (Rus)

[99] Koteles G.Y., *Atom. Energy Rev. Vienna*, 1979, vol. 17, pp. 3 – 30.

[100] Grigor'ev Yu.G., Popov V.I., Shafirkin A.V. *et al.*, *Somatic Effects of Chronic γ-Radiation*, Moscow, 1986, 200 p. (Rus)

[101] Geard C.R. and Chen C.Y., *Radiat. Res.*, 1990, vol. 124, pp. 56 – 61.

[102] Shishkina L.N. and Smotryaeva M.A., *Biofizika*, 2000, vol. 45(5), pp. 844 – 852. (Rus)

[103] Luk'yanova L.E. and Luk'yanov O.A., *Uspekhi Sovrem. Biologii*, 1998, vol. 118(6), pp. 693 – 706. (Rus)

[104] Shilova S.A., *Uspekhi Sovrem. Biologii*, 1999, vol. 119(5), pp. 487 – 503. (Rus)

[105] Yablokov A.V., *Thes. Rep. Intern. Conference "Biological Effects of Low Doses of Radiation and Radioactive Contamination of the Environment"*, Syktyvkar, 2001, p. 256. (Rus)

[106] Arkhipov N.P., *Thes. V Intern. Sci. Conference "Chernobyl-96. The Results of 10 Year Work on the Liquidation of the Accident Consequences at the Chernobyl NPP"*, Zeleny Mys, 1996, p. 12 – 13. (Rus)

[107] Buntova E.G., Gashchak S.p., Rudenskaya G.A. *et al.*, *Thes. V Intern. Sci. Conference "Chernobyl-96. The Results of 10 Year Work on the Liquidation of the Accident Consequences at the Chernobyl NPP"*, Zeleny Mys, 1996, p. 347. (Rus)

[108] Il'enko A.I. and Krapivko T.P., *Thes. V Intern. Sci. Conference "Chernobyl-96. The Results of 10 Year Work on the Liquidation of the Accident Consequences at the Chernobyl NPP"*, Zeleny Mys, 1996, p. 348. (Rus)

[109] Adamovich V.L. and Merkushina O.S., *Radiats. Biol. Radioekol.*, 1997, vol. 37(3), pp. 303 – 311. (Rus)

[110] Zolotareva N.N. and Ryabtsev I.A., *Radiats. Biol. Radioekol.*, 1997, vol. 37(3), pp. 438 – 444. (Rus)

[111] Sokolov V.E., Krylova T.V., and Skurat L.N., *Bioindication of Radioactive Contaminations*, Ed. D.A. Krivolutskyi, Moscow, Nauka, 1999, p. 123 – 128. (Rus)

[112] Kudyasheva A.G., Shishkina L.N., Shevchenko O.G. *et al.*, *Radiats. Biol. Radioekol.*, 2004, vol. 44(3), pp. 262 – 268. (Rus)

[113] Shevchenko V.A., Pomerantseva M.D., Ramaiya L.K. *et al.*, *The Science of the Total Environment*, 1992, vol. 112, pp. 45 – 56.

[114] Gileva Z.A., *Ecological-Genetic Monitoring Using Rodents (the Ural Experiment)*, Ekaterinburg, Izd. Ural'skogo Universiteta, 1997, 105 p. (Rus)

[115] Nadzhafova R.S., *Bioindication of Radioactive Contaminations*, Ed. D.A. Krivolutskyi, Moscow, Nauka, 1999, p. 239 – 241. (Rus)

[116] *Consequences of the Chernobyl Catastrophe: The Environment Health*, Ed. V.M. Zakharov and E.Yu. Krysanov, Moscow, Center of Ecokolical Policy of Russia; Moscow Dep. "Biotest" Fund, 1996, 170 p.

[117] Geras'kin S.A., Dikarev V.G., Udalova A.A., and Dikareva N.S., *Radiats. Biol. Radioekol.*, 1999, vol. 39(4), pp. 373 – 383. (Rus)

In: 20 Years after the Chernobyl Accident
Editors: E. Burlakova and V. Naidich, pp. 331-343
ISBN 1-60021-249-2

Chapter 18

TWENTY YEARS AFTER THE CHERNOBYL ACCIDENT – IS IT MUCH OR NOT?

I. I. Pelevina*[1*]*, G. G. Afanasjev*[1]*,
***V. Ya. Gotlieb*[1]* and A. A. Konradov*[2]**
[1] N.N. Semenov Institute of Chemical Physics,
Russian Academy of Sciences, Moscow, Russia
[2] N.M. Emanuel Institute of Biochemical Physics,
Russian Academy of Sciences, Moscow, Russia

ABSTRACT

The results of long-term investigations, implemented in the zones affected by the Chernobyl accident consequences are presented. They were obtained for the cells in tissue culture, laboratory animals (mice), blood lymphocytes of citizens from contaminated areas (children and adults). It is concluded that the presence in contaminated areas induces genomic instability observed in a multiplicity of effects at cytogenetic, cellular and organism levels. It is indicated that in late generations after exposure to radiation, in the accident zone intensified cell death, increased level of micronuclei, giant cells, elevated radiosensitivity, and the absence of the adaptive response are observed. The culture of embryonic fibroblasts from exposed mice demonstrated elevation of the number of cells with chromosomal aberrations and the frequency of cells with multiple chromosomal aberrations. The mice exposed to radiation in the accident zone demonstrated decreasing number of endotheliocytes in different parts of cerebrum and increased radiosensitivity of animals.
The above-described effects are observed in late cell generations and may be a product of low dose radiation-induced genomic instability.

---

* Main addressee: 4 Kosygin street, 119991 Moscow, Russia, Institute of Chemical Physics, RAS; Tel.: (7-495) 939-74-47; E-mail: pele@chph.ras.ru

Keywords: the Chernobyl accident, genomic instability , cytogenetic effects, cells in the tissue culture, human blood lymphocytes, survival of mice

The single Chernobyl accident had exposed millions of people to radiation: territories of about 150,000 sq. km with the population of 5,000,000 citizens of the former USSR and 45,000 sq. km in the Northern and Eastern Europe were contaminated.

Epidemiological investigations carried out all over the world on the groups of people exposed to high-dose radiation impacts (for example, examinations of survived in the nuclear disaster in Japan), radiobiological investigations of animals, and the study of radiobiological regularities at the cell level indicate that the main late effects are the effect on inheritance, occurrence of malignant tumors, cataract, reduction of the intelligence quotient for people exposed to radiation *in utera* [1 - 3].

Essentially, radiation effects on non-neoplastic diseases as, for example, cardiovascular and ischemic heart disease, pulmonary diseases, organic pathologies of cerebral vessels, encephalopathy, and disorders of the immune system [4 – 6].

It is significant that the effect of ionizing radiation induces such biological consequences as chromosomal aberrations, mutations, and other genetic apparatus disorders, which may be markers or precursors of a future disease [7 – 9].

Currently, 20 years passed since the Chernobyl accident, over 50 years after radioactive waste discharge from "Mayak" Plant to Techa river, and 60 years after atomic bombing of Japan.

It is not the time to judge about late consequences of the Chernobyl accident and occurrence of solid malignant tumors, but the official viewpoint of some Russian and international organizations is that this accident had no serious medical consequences (except for thyroid cancer) [10].

At the same time, 50 years after the accidents, the consequences of Japanese and Chelyabinsk populations are epidemiologically and radiobiologically studied still. The risks after low dose and low intensity radiation impacts are revised [11 – 15].

After the Chernobyl accident, a considerable cohort of people was exposed to low-dose radiation, so that Chernobyl initiated a new epoch in radiobiology development – the effect of low dose radiation has received much attention all over the world. A significant role in understanding the Chernobyl aftereffects was played by experiments on isolated cells, animals, lymphocytes of irradiated persons (liquidators and citizens of contaminated areas), which give an opportunity to understand the mechanisms of early and late consequences of low-dose radiation.

Therefore, the problem of development of the Chernobyl accident consequences generally comes to understanding of the low-dose ionizing radiation mechanisms.

There is national US program for investigating the action of low doses of radiation, designed for the ten–year period (USD 21 million per year). The study of biological effects at ionizing irradiation in doses about 0.1 Gy with low LPE is suggested. This program supports the fundamental investigations applying the methods of molecular biology, cellular biology and genetics for the purpose of studying the low-dose radiation action mechanisms [16].

At present, "non-target" effects at low-dose irradiation are described: genomic instability (GI), adaptive response (AR), bystander effect, hormesis, and the clastogenic effect – at the organism level [17 – 23].

Low dose and high dose radiation effects show different responses: the classical model of radiation effects suggests that a discrete cellular target has an individual response in accordance with the quantity nonrepaired or misrepaired DNA damages. According to modern ideas, the processes may operate at distances exceeding the cell size and are controlled by signal cell systems: to damage a cell direct hit is not certain, so it should be assumed that the dose-effect dependence is changed. Hence, the gene expression will be of fundamental importance, which may induce a malignant transformation without direct mutations. Unradiated cells demonstrate changes in gene expression, DNA repair, occurrence of chromosomal aberrations, mutations, and death; the radiation-induced genomic instability may be caused by the occurrence of malignant tumors. Also, in epidemiological investigations for determination risk study of the radiosensitive genes, which may be markers of genetic sensitivity of individuals and populations to be carried out [16, 24].

It is suggested that generally the problem of late consequences of the Chernobyl accident is reduced to understanding of the effects and mechanisms of low dose radiation. Hence, a new phenotype, another population with its specific properties appears. Of prime importance seems to be the genomic instability induced, which in turn induce other responses to external impacts, stress and radiation. Therefore, it may be suggested that genomic instability is the basic phenomenon determined in the Chernobyl accident consequences e.g. low dose radiation impact, for people living in contaminated areas and liquidators [25].

Genomic instability is a type of an alteration transferred via cell generations and inducing elevated frequency of mutations, chromosomal aberrations and death in descendants of irradiated cells. It is observed both *in vitro* and *in vivo*. GI may increase sensitivity to physical and chemical agents and the presence of it makes impossible extrapolation from high to low doses, because it may modify the biological effect and requires revision of the risk computation concept.

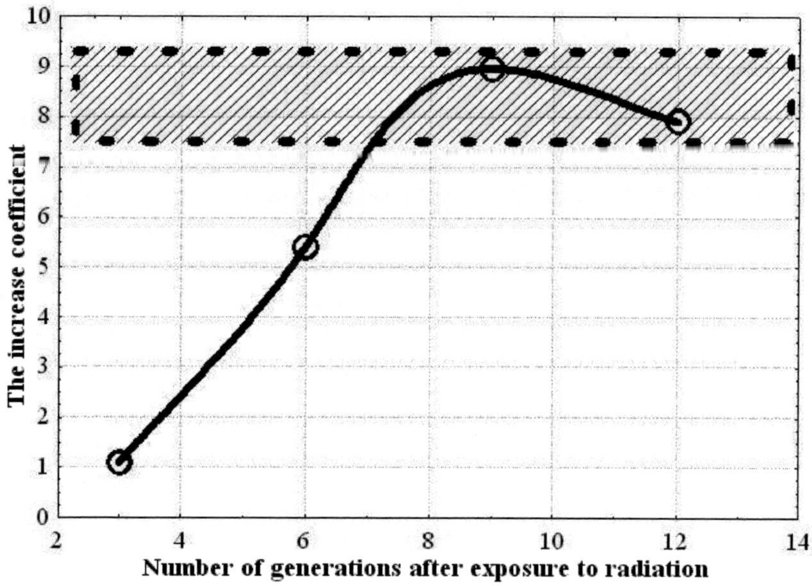

Figure 1. The population growth rate of cells exposed to the 10-km restriction zone of the Chernobyl accident (with 0.096 Gy total dose radiation)

This work sums up the results of the long-term investigations on GI cellular manifestations, performed in the areas contaminated by the Chernobyl accident (nearby the reactor at the border of the "Red forest"), in Bryansk Oblast (Klintsy, Vyshkov, Novozybkov), and among liquidators of the accident [25, 26]. The tests were performed on human and mammal cells in the tissue culture, children's and adults' lymphocytes, and small laboratory animals kept under industrial-induced contamination conditions.

GI manifestations determined *in cells in the tissue culture* (HeLa), exposed in the accident zone during 1, 4 and 6 days (with the radiation dose 0.083, 0.331 and 0.497 Gy, respectively, at the dose rate of γ-radiation within 100 – 300 mR/h) and then cultivated under common laboratory conditions during many generations, were the following (control cultures were exposed to the same environmental conditions on uncontaminated territories): the cell proliferation is suppressed during 6 – 7 days after the end of exposure in the Chernobyl zone (Figure 1); the control level is approached in the 8th generation only. On average, the number of cells (HeLa) in the control increases by ~7.5 times, whereas after exposure in the zone the elevation parameter is immediately 2-fold decreased. It may be suggested that a decrease of proliferation in descendants of irradiated cells is due to induced GI and cell death in late generations (Table 1). This assumption is also supported by the fact that within 24 generations a clonogenic ability of cells is decreased (Table 1). The same situation is observed for the giant cells, which quantity increases by 2 times or more and is not reduced during 20 generations (Table 2).

Table 1. The part of survived cells (clonogenic ability) in descendants of HeLa cells exposed to the 10-kilometer restriction zone of the Chernobyl accident (with total radiation dose 0.096 Gy)

Number of generations after exposure	Control	In the Chernobyl accident zone
6	0.21	0.14*
12	0.17	0.06*
18	0.31	0.25
21	0.36	0.15*
24	0.36	0.15*

* The reliable difference between control and exposed populations (χ^2 criterion)

Table 2. The number of giant cells (per 1000 examined cells) in HeLa cell descendants exposed to the 10-km restriction zone of the Chernobyl accident

Number of generations after exposure	Control	Exposure time – total dose	
		1 day - 0.024 Gy	6 days - 0.144 Gy
4	62	70	-
6	50	136*	-
9	62	80	-
14	60	85	-
16	49	62	127*
24	63	50	154*

* The reliable difference between control and exposed populations (χ^2 criterion)

One more urgent phenomenon can be belonged to GI manifestations in descendants of cells exposed to the accident zone is the increase of their radiosensitivity. An additional irradiation of the descendants of the exposed cells by 3.0 Gy dose decreases their survival (the clonogenic ability) below the control level (Table 3). The frequency of cell occurrence with micronuclei and giant cells was reliably exceeded during 9 – 12 generations (Table 4). The comparison of the dose-effect dependencies demonstrated the absence of a threshold for the test population (the data are not cited).

These results indicate also that long-term exposure to contaminated territories does not induce an adaptive response and elevation of radioresistance, because acute radiation impact on the cells exposed to the zone only increases their radiosensitivity (Tables 3 and 4).

Table 3. The part of survived HeLa cells (clonogenic ability) after exposure to the 10-km restriction zone of the Chernobyl accident and subsequent acute 3 Gy dose irradiation

Number of generations after radiation	Exposure time – total dose			
	2 days – 0.048 Gy		4 days – 0.096 Gy	
	3 Gy	0.048 Gy + 3 Gy	3 Gy	0.096 Gy + 3 Gy
6	0.16	-	0.21	0.14
12	0.10	0.03	0.17	0.06
18	0.26	0.17	0.31	-
21	0.30	0.16	0.36	0.15
24	0.27	0.16	0.36	0.15

Table 4. The number of cells with micronuclei and the number of giant cells in descendants of HeLa cells exposed to the 10-km zone of the Chernobyl accident (total radiation dose 0.096 Gy) and additionally irradiated by acute 3 Gy dose

Criterion	Number of generations after exposure	Control	In the accident zone
Number of cells with micronuclei per 1000 cells	3	109	185*
	9	106	127*
	12	101	123*
	15	94	105
	21	96	92
Number of giant cells per 1000 cells	3	107	167*
	9	95	109*
	12	94	99
	15	90	88
	21	84	86
	27	87	83

* The significant difference between control and exposed populations (χ^2 criterion)

On a culture of fibroblasts obtained from embryos of mice matched in the accident zone, a separate set of tests was performed. The cytogenetic analysis indicates a reliable increase of the number of cells with chromosome aberrations, which may reach 24.5% (Table 5).

Table 5. Cytogenetic disorders in the culture of embryo fibroblasts of mice after exposure to the 10-km restriction zone of the Chernobyl accident during 5 days (with 0.13 Gy total dose radiation)

Alternative	% of cells with damaged chromosomes	Damages in cells	Type of damage
Control	1 – 3	singular	deletions
In the Chernobyl accident zone	24.5	singular, multiple	deletions, fragments, translocations

Cells with multiple chromosome aberrations appear, reaching 8% of the total quantity of aberrant metaphases.

Thus, in the cells in the tissue culture exposure to the Chernobyl zone induces GI, which then in descendants of exposed cells is manifested as follows:

- proliferation rate decrease;
- cell death in generations;
- increasing quantity of cells with micronuclei;
- increasing quantity of giant cells;
- the absence of AR;
- the increased sensitivity to consequent radiation impact.

Naturally, a question is brought up if radiation separately may induce such effects. Therefore, model experiments on different cell lines irradiated by 10 – 40 cGy doses with the dose rate approaching the level of the Chernobyl tests were performed [27, 28]. As indicated, the late descendants of irradiated cells demonstrate reproductive death and increasing quantity of the cells with micronuclei. This allows a suggestion that currently described effects, obtained at the exposure in the 10-kilometer restriction zone of the Chernobyl accident, are generally explained by the long-term low dose radiation impact.

As considering the whole organism of mice exposed during different times in the accident zone (animals were kept in all-round cellular cages with a dosimeter beyond each cage), similar effects were observed. After the exposure to chronic radiation in 0.024 – 0.336 Gy doses (from 1 to 14 days) the animals were brought to Moscow and then after 2, 7 and 30 days repeatedly irradiated by 3, 5, 7 and 9 Gy doses. Control animals dwelt under the same environmental conditions, but in uncontaminated zone. Thereafter, survival of animals within 30 days was studied (Figure 2). It has been found that an additional irradiation by 9 Gy dose (2 days after the exposure to the accident zone) significantly increases mortality of the animals that reaches 100% on the 9th day. The effect of radiosensitivity increase significantly depends on the time gap between the exposure end and additional acute radiation impact. For example, if this period equals 2 days the death increase is observed in the period of the bone marrow syndrome, whereas for 30 days – gastrointestinal syndrome. These data show that for the whole organism the impact of long-term radiation (and other factors) in the accident zone causes an increase of radiosensitivity in mice [29].

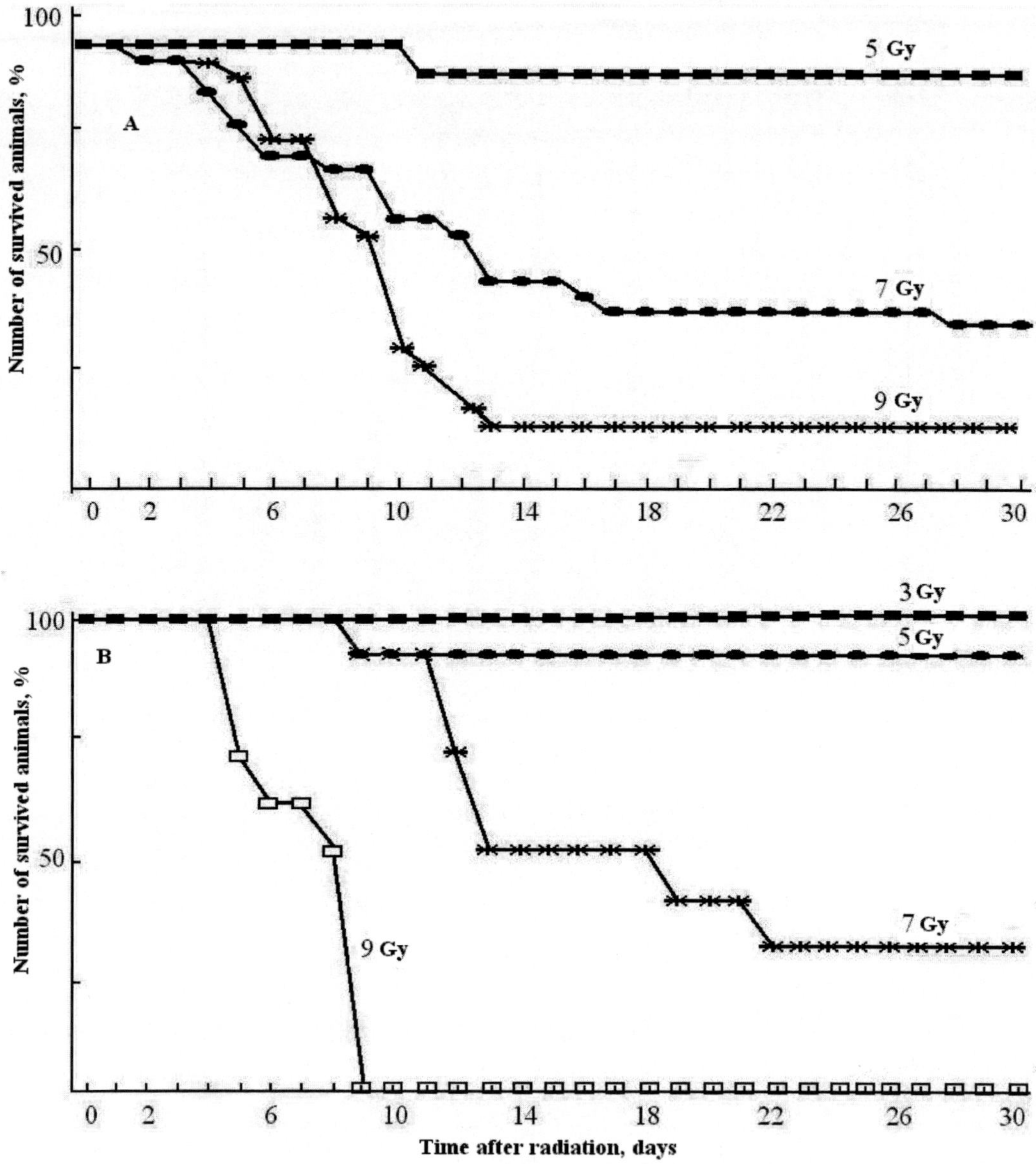

Figure 2. Survival of mice after exposure to different doses (A), to the Chernobyl accident zone (with 0.336 Gy total radiation dose) and additional acute radiation by different doses (B) 2 days after

In a separate set of experiments animals exposed to the accident zone were tested for the density of endotheliocytes in different parts of cerebrum using a fluorescent-histochemical method (Figure 3). The animals were also exposed to 2.0 Gy radiation (long-term e.g. 1 month exposure in the accident zone). A year after the exposure the density of endotheliocytes in different parts of cerebrum was decreased [29].

Thus, the tests on animals, as well as at the cellular level, demonstrate increasing radiosensitivity and late death of endotheliocytes in the cerebrum of mice. The effect of late death of endotheliocytes in the cerebrum may indicate cerebrovascular disorders, caused by exposure to the accident zone.

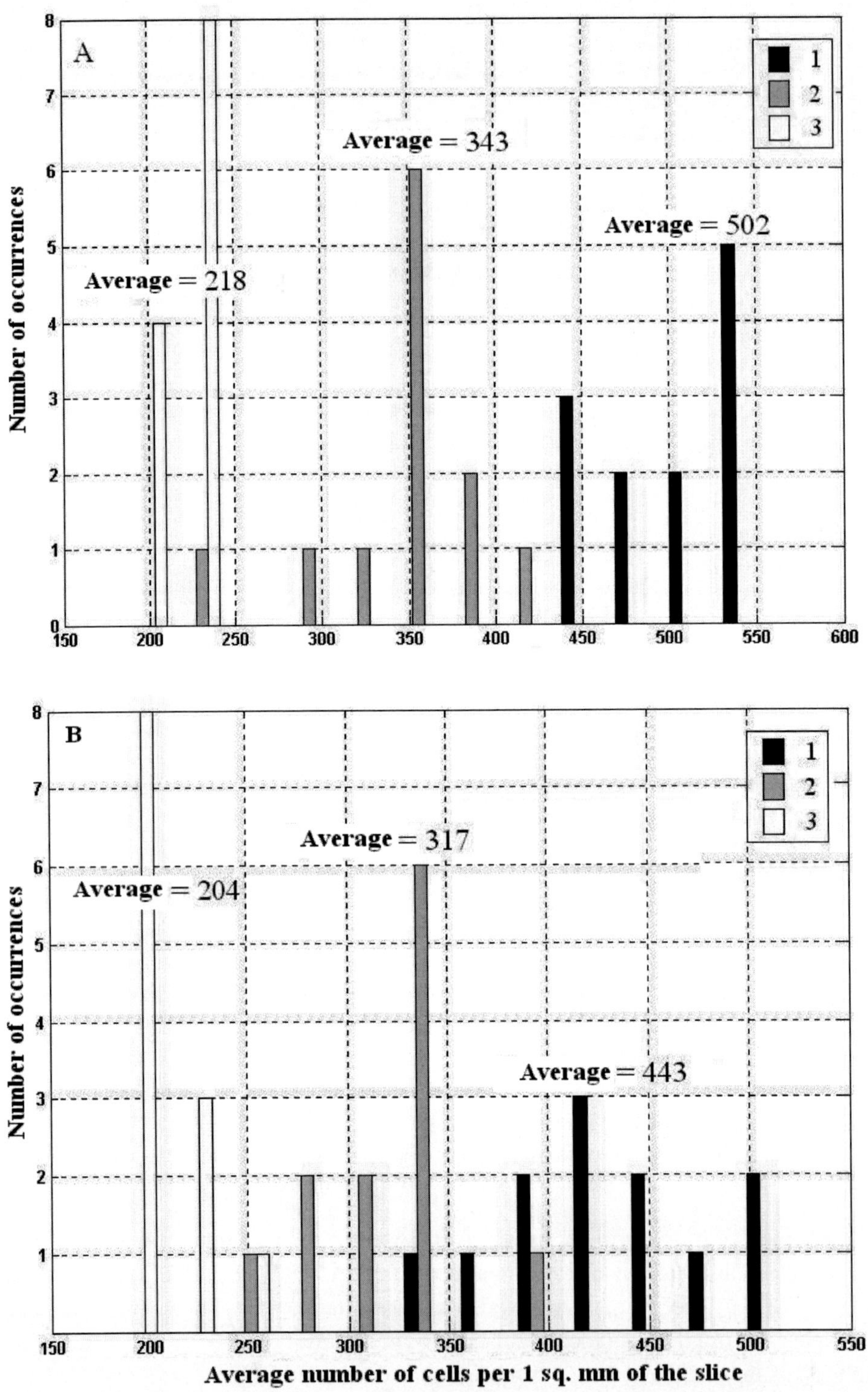

Figure 3. The distribution of endotheliocytes in the cerebrum of irradiated mice 14 months old (A) and 12 months after exposure to the accident zone (with 2 Gy total radiation dose) (B): 1 – cortex; 2 – septum; 3 – *corpus callosum*

For phytohemagglutinin (PHA) stimulated lymphocytes of children and adults living in the areas, contaminated by the Chernobyl accident (Vyshkov, Klintsy, Novozybkov – with the contamination density up to 40 Ci/km^2), and the liquidators, the spontaneous frequency of damaged lymphocytes, their radiosensitivity (after exposure to 1 Gy dose radiation), and the presence of adaptive response after exposure to 0.05 Gy (adapting) and 1 Gy (challenge) doses 5 hours after were studied. The study was implemented by cytochalasin B with the help of micronucleus test using the cytokinetic block. Of special attention is that children from Novozybkov city and adults and individuals from Vyshkov village demonstrated significantly decreased ability to PHA stimulation – the frequency of stimulated cells was decreased by 1.5 – 2 times. In the laboratory headed by A.A. Yarilin similar results for the same population were obtained (oral report).

It has been found that adults possessed spontaneous level of damages having no reliable difference from these from Moscow [25]. However, high individual variability is observed for several persons from contaminated regions. Among liquidators, several persons possessed rather high frequency of damaged lymphocytes. Among children, more than 2-fold increase of the spontaneous frequency of lymphocytes with micronuclei was observed (Figure 4).

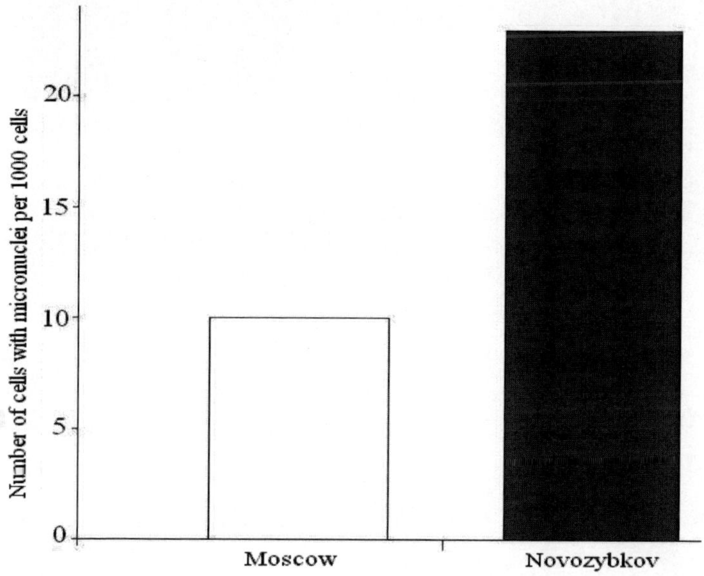

Figure 4. Spontaneous level of cytogenetic damages in lymphocytes of children living in contaminated area (Novozybkov city) and in Moscow

Neither in the liquidators, nor in the population of the contaminated areas, the exposure to 1 Gy dose radiation has induced elevation of radioresistance, i.e. in these regions AR is not induced by the long-term radiation impact [25]. On the other hand, examination of the citizens from the South-Ural Region dwelling near Techa river, where to radioactive wastes were discharged longer than 50 years ago, indicated a reliable elevation of radioresistance, i.e. chronic irradiation in these regions led to AR induction [30, 31]. Apparently, these differences in AR induction for dwellers of the contaminated territories may be stipulated by a broad spectrum of radionuclides (mostly strontium in Urals and cesium in Chernobyl), different periods of observation and radioactive contamination, differences in environmental conditions, human populations, different critical system of damaging, and other factors.

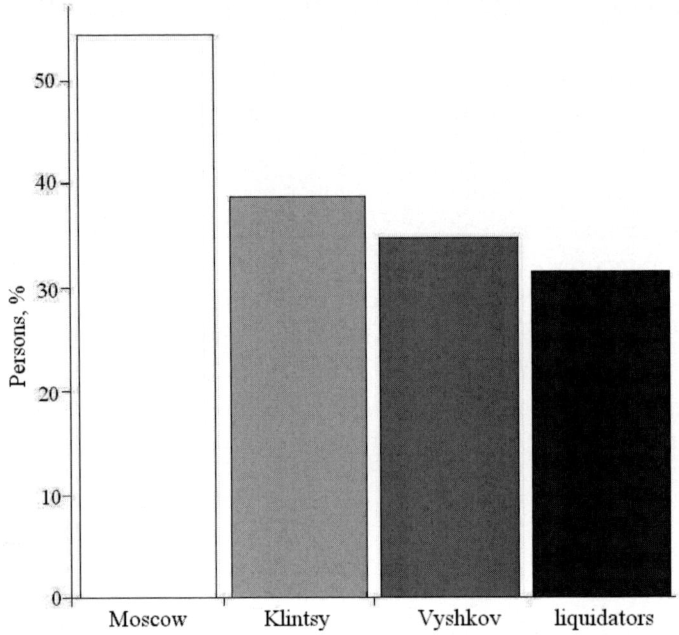

Figure 5. The part of persons (%) with adaptive response among Moscow citizens and dwellers of contaminated areas (Klintsy and Vyshkov villages), and liquidators

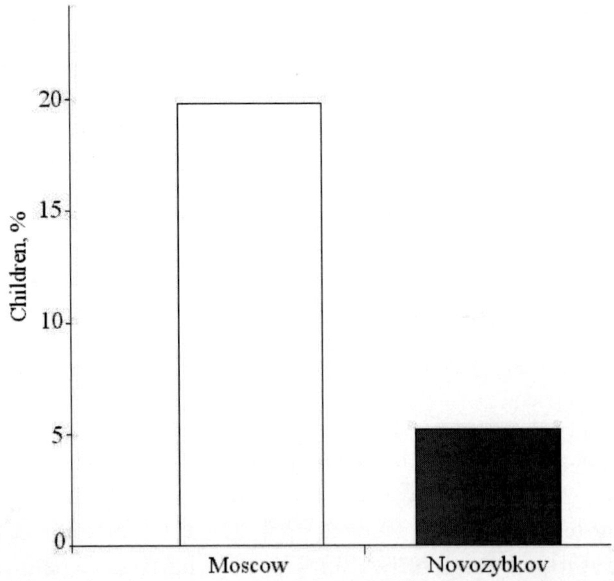

Figure 6. The part of persons (%) with adaptive response among children living in contaminated area (Novozybkov city) and in Moscow

As studying AR induced by additional irradiation by 0.05 Gy dose and registered after irradiation by 1 Gy dose, usually, liquidators and adults (Figure 5), and children (Figure 6)

dwelling on the territories contaminated by the Chernobyl accident demonstrated decreasing frequency of reliable AR occurrence. Similar situation is observed for the dwellers nearby Techa river [30].

Another effect is registered simultaneously: it is a phenomenon of increasing radiosensitivity among liquidators and all adult dwellers of radiation-contaminated territories (Figure 7), and children. This phenomenon appears important for the population and, apparently, depends on the initial concentration of damaged cells, environmental factors, defects in the repair systems, the factors of humoral and cellular immunity, individual features of the organism, and somatic diseases [32 – 34]. The role of the clastogenic factors in formation of increased radiosensitivity phenomenon may not also be excluded.

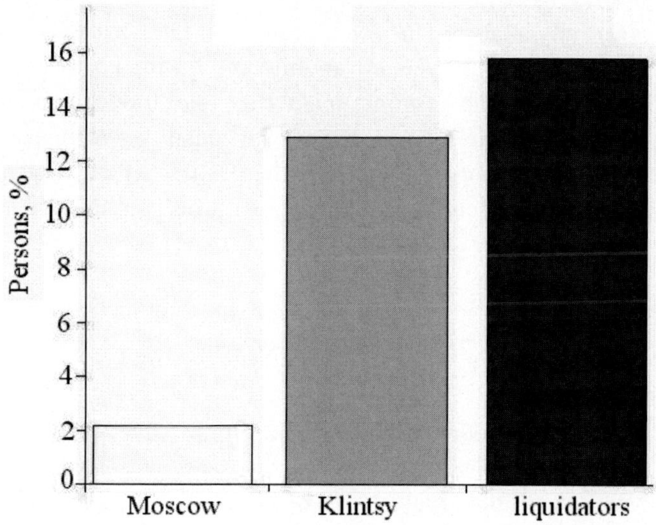

Figure 7. The part of persons (%) with reliable elevation of radiosensitivity after adaptive radiation impact among Moscow citizens, dwellers of contaminated Klintsy village and liquidators

Clastogenic factors are detected in animal and human blood serum, exposed to low dose radiation. They induce cell death, chromosome aberrations, mutations, etc. (for example, as serum of irradiated individuals is added to the tissue culture cells), initially described in [17]. However, high individual variability in the ability to form clastogenic factors is determined [35]. Apparently, one of the features of the existence of the man and other organisms on the territories, impacted by nuclear catastrophes, is the occurrence of individuals possessing high sensitivity to extreme factors after the effect of low dose chemical and physical agents.

Conclusion

The question is what are the consequences of the Chernobyl accident and what are the effects of low chronic radiation, which may be formulated basing on the analysis of the current tests?

They are the following:

- A population of cells, animals and, probably, people possessing specific features and phenotype occurs.
- This population is more sensitive to additional radiation impacts.
- This population demonstrates reduced ability to adaptive response.
- This population demonstrates high individual variability according to spontaneous level of damages, radiosensitivity, and adaptive response.
- This population demonstrates high frequency of occurrence of the phenomenon of elevated radiosensitivity after low dose radiation impact.

At present, we may only analyze late consequences of the Chernobyl disaster. The results obtained by the authors show that 20 years of monitoring for the post-Chernobyl effects is too short period. If we suggest that the genomic instability and induced damages of the genetic apparatus, radiosensitivity increase, the absence of adaptive response, damages of the cerebrum vessels are the risk factors and increase the probability of occurrence of cancer and many other nontumor diseases, then these pathological processes may develop in much more late periods.

REFERENCES

[1] Brenner D.J., Doll R., Goodhead D.T. et al., *Proc. Natl. Acad. Sci. USA*, 2003, vol. 100(24), pp. 13761 - 13766.

[2] Kiuru A., Auveren A., Luikkamaki M. *et al.*, *Radiat. Res.*, 2003, vol. 159(5), pp. 651 - 655.

[3] Preston D.L., Pierce D.A., Shimizu Y. et al., *Health Phys.*, 2003, vol. 85(1), pp. 43 - 46.

[4] Preston D.L., Shimuzi Y., Pierce D.A. *et al.*, *Radiat. Res.*, 2003, vol. 160(4), pp. 381 - 407.

[5] Yamada M., Wong F.L., Fijiwara S. et al., *Radiat. Res.*, 2004, vol. 161(6), pp. 622 - 632.

[6] Ivanov V.K., Maksyutov M.A., Chekin S.Yu. et al., *Radiats. Biol. Radioekol.*, 2005, vol. 45(3), pp. 261 – 270. (Rus)

[7] Holmberg M., *Leuk. Res.*, 1992, vol. 16(4), pp. 257 - 265.

[8] Snigiryova G, Braselmann H, Salassidis K. et al., *Int. J. Radiat. Res.*, 1997, vol. 71, pp. 119 - 127.

[9] Sevan'kaev A.V., Mikhailova G.F., Potetnya O.I. et al., *Radiats. Biol. Radioekol.*, 2005, vol. 45(1), pp. 5 – 16. (Rus)

[10] Parshkov E.M., Sokolov V.A., Tsyb A.F. et al., *Int. J. Low Radiation*, 2004, vol. 1(3), pp. 267 - 278.

[11] Kamada N., *J. Radiat. Res.*, 1991, vol. 32, pp. 172 - 179.

[12] Preston D.L., Kusumi S., Tomonaga M. *et al., Radiat. Res.*, 1994, vol. 137(2 suppl.), pp. 68 - 97.

[13] Little M.P., Weiss H.A., Boice J.D. et al., *Radiat. Res.*, 1999, vol. 152(3), pp. 280 - 292.

[14] *The Consequences of Industry-Induced Effect and the Problem of Rehabilitation of the Ural Region*, Ed. S.K. Shoigy, Moscow, Komtekhpprint, 2002, 287 p. (Rus)
[15] Little M.P. and Muirhead C.R., *Int. J. Low Radiation*, 2004, vol. 1(1), pp. 285 - 299.
[16] Brooks A.L., *Health Phys*., 2003, vol. 85(1), pp. 85 - 93.
[17] Parsons W.B., Watkins C.H., Pease G.L. *et al*., *Cancer*, 1954, vol. 7, pp. 179 - 189.
[18] Goh K. and Summer H., *Radiat. Res*., 1991, vol. 35, pp. 171 - 181.
[19] Little J.B., Azzam E.I., de Toledo S.M., and Nagasawa H., *Radiat. Protect. Dosimetry*, 2002, vol. 99, pp. 223 - 226.
[20] Prise K.M., Belyakov O.V., Newman H.C. et al., *Radiat. Protect. Dosimetry*, 2002, vol. 99, pp. 223 - 226.
[21] Kadhim M.A, Moore S.R, and Goodwin E.H., *Mutat. Res*., 2004, vol. 568, pp. 21 - 32.
[22] Mothersill C. and Seymour C., *Mutat. Res*., 2004, vol. 568, pp. 121 - 128.
[23] Streffer C., *Mutat. Res*., 2004, vol. 568, pp. 79 - 87.
[24] Amundson S.A., Lee R.A., Koch-Paiz C.A. et al., *Mol. Cancer*, 2003, vol. 1(6), pp. 445 - 452.
[25] Pelevina I.I., Gotlieb V.Ya., Kudryashova O.V. et al., *Radiats. Biol. Radioekol.*, 1996, vol. 4(36), pp. 546 – 560. (Rus)
[26] Pelevina I., Afanasiev G., Aleschenko A.V. et al., *Exerpta Medica Int. Congress*, Seria Elsevier Science, 1997, pp. 373 - 378.
[27] Gotlieb V.Ya., Pelevina I.I., Konoplya E.F. et al., *Radiobiologia*, 1991, vol. 31(3), pp. 318 – 325. (Rus)
[28] Antoshchina M.M., Ryabchenko N.I., Nosonova V.A. *et al.*, *Radiats. Biol. Radioekol.*, 2005, vol. 45(3), pp. 310 – 315. (Rus)
[29] Konradov A.A., Lyubimova N.V., and Pelevina I.I., *Radiats. Biol. Radioekol.*, 1993, vol. 33, Iss. 1(4), pp. 499 – 507. (Rus)
[30] Akleev A.V., Aleshchenko A.V., Gotlieb V.Ya. et al., *Radiats. Biol. Radioekol.*, 2004, vol. 44(4), pp. 426 – 431. (Rus)
[31] Akleev A.V, Aleschenko A.V., Gotlib V.Ya. *et al.*, *Jpn. J. Health Phys*., 2004, vol. 39, pp. 653 656.
[32] Pelevina I.I., Afanas'ev G.G., Aleshchenko A.V. et al., *Radiats. Biol. Radioekol.*, 1999, vol. 39(1), pp. 106 – 112. (Rus)
[33] Pelevina I.I., Aleshchenko A.V., Afanas'ev G.G. et al., *Radiats. Biol. Radioekol.*, 2000, vol. 40(5), pp. 544 – 549. (Rus)
[34] Pelevina I.I., Aleshchenko A.V., Gotlib V.Ya. et al., *Radiats. Biol. Radioekol.*, 2005, In print. (Rus)
[35] Gemignami F., Ballardin V., Maggiami P. et al., *Mutat. Res*., 1999, vol. 446(2), pp. 245 - 253.

INDEX

A

B

C

D

G

H

I

J

K

L

M

N

O

P

Q

R

S

T

U

V

W

Y

Z